D. R. Marshall

3-95

SCIENCE AND THE
RISE OF TECHNOLOGY
SINCE 1800

The Open University
Science and the Rise of
Technology since 1800

Course Team

Arts

C. A. Russell *Chairman*
N. G. Coley
D. C. Goodman
J. Ferguson
C. T. Harvie
A. J. B. Marwick
G. K. Roberts

Science

L. J. Haynes
S. P. R. Rose

Technology

G. H. Weaver

Institute of Educational Technology

C. J. Lawless

BBC

A. Clow
D. Nelson
P. Scroggs
N. Taylor

SCIENCE AND THE RISE OF TECHNOLOGY SINCE 1800

Edited by
C. A. Russell and
D. C. Goodman

at The Open University

John Wright and Sons Limited
in association with
The Open University Press

ISBN 0 7236 0336 7

Printed in Great Britain by
John Wright & Sons Ltd, at
The Stonebridge Press, Bristol BS4 5NU

Acknowledgements

For access to original material for the purposes of research and facsimile reproduction we are grateful to the following:

British Museum, Chemical Society, Royal Institution, Royal Society of Medicine, Institute of Civil Engineers, London Library, Bristol Public Libraries—Central Reference Library.

Permission to reproduce the extracts and illustrations shown in parentheses below has been given by the following:

Institution of Gas Engineers (4d); Science Museum, London (4k, 15l); Chemical Society (8b); Northwestern University Press (12a–d); Drake Well Museum (12a, d); Bettman Archive, New York (12b); *Petroleum Times* (12e, f); Sir Michael Perrin, C.B.E., F.R.I.C. (13d); ICI Ltd. (illustrations in 13d); *Nature* (13f); The Royal Institution of Naval Architects (15g); the Editor, *Science in War* (17a); *New Statesman* (17b); Her Majesty's Stationery Office (17d); British Society for Social Responsibility in Science (17e).

Finally we would like to acknowledge the assistance which we have received in the choice of documents from Dr. Frank Greenaway of the Science Museum, Dr. Ruth Hodgkinson of the Wellcome Institute of the History of Medicine, and Professor Steven Rose of the Open University, who suggested the material for the final chapter. We are grateful also to Miss Annabel Jones and Mr. Roger Day for much valuable assistance in preparing this book for the press.

Note

The punctuation, spelling and syntax of original material has been strictly reproduced in this anthology, even when its orthodoxy might be questionable.

C. A. Russell

D. C. Goodman

Abbreviations

Ann. Phil.
Annals of Philosophy

B.P.P.
British Parliamentary Papers

Chem. News
Chemical News

J. Chem. Soc.
Journal of the Chemical Society

J. Soc. Chem. Ind.
Journal of the Society of Chemical Industry

Phil. Mag.
Philosophical Magazine

Phil. Trans.
Philosophical Transactions of the Royal Society

Rep. Brit. Assoc. Adv. Sci.
Reports of the British Association for the Advancement of Science

Contents

1 INTRODUCTION

This anthology has been produced primarily to meet the needs of students of the Open University taking the course entitled 'Science and the Rise of Technology since 1800'. As a course reader it is closely integrated with the students' tuition received through correspondence and broadcasting. However, the anthology will be of general interest and can be read by anyone interested in the science and technology of the period.

The subject area is of course an enormous one, and no anthology of this size can make claims to be comprehensive. The essential task of making primary source materials available is only just beginning for the nineteenth-century history of science and technology. A few anthologies have appeared, and their contents have been taken into account in selecting the material for this book. We have tried to avoid overlaps as far as possible.

The principal aim of the University course is to explore the interaction between science and technology in the nineteenth and early twentieth centuries, and the passages in the Reader reflect this. While science and technology were perhaps never completely separated in the past, their intimate connection, now familiar, was not fully established until quite recently. Science, or natural philosophy as it was still commonly called in the early nineteenth century, was from the earliest times primarily concerned with the investigation of natural phenomena—an intellectual activity dedicated to the understanding of the natural world. These inquiries frequently had practical implications, for example in navigation and medicine. Moreover, individual natural philosophers (Bacon, Descartes, etc.) entertained considerable optimism as to the usefulness of their research for human affairs—alleviating drudgery, prolonging life, increasing wealth and generally increasing man's happiness on earth. But before the nineteenth century the sciences were insufficiently developed to provide the detailed information required to realize even the more modest of these ambitions.

Technology, or the practical activities associated with the traditional arts and crafts, therefore proceeded in a predominantly empirical fashion, largely uninformed by scientific knowledge. Craft experience and manual skill were responsible for the production of new materials as well as the construction of buildings and ships. Even during the early stages of the Industrial Revolution it is difficult to find any evidence that the tremendously important developments in iron metallurgy were caused by preliminary scientific inquiries. Similarly, most historians now agree that the invention of the first steam engines, including those of James Watt, was the result of expertise in mechanical crafts, not the results of applied scientific knowledge.

As the nineteenth century progressed, advances in scientific knowledge, especially in chemical science, provided the key to a greater understanding, and therefore eventual control, of a host of practical processes involved in manufactures and industry, in agriculture and engineering. Practices in the trades, adopted because they worked, without an understanding of why they worked, began to receive a scientific explanation. The scientist was increasingly called in as an expert to deal with technological problems: to investigate the causes of the collapse of bridges, to halt the spread of continental epidemics, to protect the public from food adulteration, to prevent explosions in the pits, to design ships for war and peace. The unprecedented population-growth which occurred in the nineteenth century (and which is still not fully explained) stimulated scientific inquiries into the supply and purification of water and the disposal of waste.

The social consequences of the growing application of scientific knowledge were profound and, in some cases, totally unexpected. A good example of this is electrical science, which was initially an area of pure scientific interest and without applications (apart from the lightning conductor and optimistic medical cures). Not even Faraday suspected the social transformations that would come from electrical lighting and power.

The importance of science in the life of the nation was appreciated to varying extents by nineteenth-century governments, though individual enterprise was more effective in England in promoting the cultivation of the sciences. The Royal Institution and numerous Mechanics' Institutes had ambitious programmes for diffusing scientific knowledge amongst the working classes, partly in the hope of improving the quality of manufactures and thereby leading to a greater distribution of wealth. These economic interests were accompanied by the determination not to allow man's intellectual capabilities to be dormant.

The State's interests in science derived from its importance in manufactures and in war; in each case superiority over foreign rivals was demanded. Nevertheless, in the case of nineteenth-century Britain, constant complaints were heard of the State's failure to promote science in the most obvious way open to it, by means of a suitable education. Various Royal Commissions were appointed in the second half of the century to inquire into the teaching at schools and universities. The evidence brought before these revealed how small a part science played in curricula at all levels. There were fears that this would soon lead to that loss in industrial leadership which Britain had enjoyed in the world since the eighteenth century; there was also anxiety over the military implications. Germany in particular appeared as the chief rival. The excellence of its elementary education, its numerous technical schools and the prominence of science in its universities were praised; there were demands for their imitation in Britain. But calls for a greater national promotion of scientific and technical knowledge were still heard in the early years of the twentieth century. The First World War finally presented Britain and other nations with a crisis which demanded urgent action in the creation of neglected scientific industries. The adversary was Germany, whose military power was based on a chemical industry whose scale far exceeded any other.

It would be unfair to place all the blame at the door of the nineteenth-century British governments, accusing them of inertia or ignorance. The difficulties of State intervention in a society which values private enterprise and individual liberty are well known. Opposition to State intervention in universities and schools in nineteenth-century Britain can be linked with the opposition experienced by those (e.g. Chadwick) advocating a national body to watch over public health. In both areas local interests were jealously maintained.

So far as English education was concerned, social attitudes were hardly favourable to an alteration of the syllabus to introduce a high content of scientific material. In the public schools and universities, classical studies were the principal subjects taught. The cultivated gentleman was the ideal product of these institutions; the competitive industrialist aware of the latest technical advances was not. The prejudice against manual activity, voiced in some passages of the ancient Greeks, seems to have persisted in nineteenth- (and twentieth-) century England. A pioneer of synthetic dyes, William Perkin (1838–1907), made the revealing remark that he had been accused of perverting students, by directing their interests away from purely intellectual inquiries to applied chemistry. He also felt the disapproval of pursuing knowledge for reasons of financial gain. Such social factors must be taken into account in attempting any explanation of the slow development of scientific and technical education in nineteenth-century England.

The application of scientific knowledge brought social consequences which would have shocked the naïve utopianism of Francis Bacon and later would-be philanthropists. Already in the nineteenth century complaints were made against pollution. The manufacture of Leblanc alkali at Liverpool in the 1820s was stopped by court order, following protests by landowners that plants and property were being destroyed by acidic fumes escaping from chimneys on the site. Eventually an 'Alkali Inspector' was appointed as a government official to check the waste products of alkali works, and to decide on acceptable levels of pollution. In the final chapter of the anthology we have thought it appropriate to include twentieth-century material on issues relating to the social responsibility of the scientist.

In selecting the material for the anthology, primary sources alone were considered, since these documents offer the most direct entry into the period and its problems. Only English texts have been used, partly because the book would have increased in bulk with the inclusion of the foreign original sources and their accompanying translations. However, in view of the vast quantity of English sources available, this did not represent a severe restriction in choice. Moreover, Britain's industrial dominance at the time calls for this emphasis. Several of the texts were selected, not only for their intrinsic importance, but because they are not readily available in other collections of this type. While most of the documents belong to the nineteenth century, the period covered is not precisely defined. A few eighteenth-century sources have been included to prepare the reader for the later developments. Some twentieth-century papers illustrate the continuity with nineteenth-century movements.

The relationship between science and technology in the nineteenth century is the main theme which has dictated the choice of material. This explains what may at first seem to be the surprising absence of any texts on Darwin. The theory of evolution was clearly not an area of scientific thought which led to applications in technology. On the other hand, chemistry has perhaps more connections with technology than any other science. The analysis of the material world, the study of the transformation of substances, and the creation of new materials are all chemical problems with profound implications for engineering, agriculture and medicine. For this reason many of the texts chosen are connected with chemistry. They should be comprehensible to readers who have no specialized chemical training.

The texts in chapter 3 give some indication of the relationship between science and technology in the eighteenth century. Two extracts describe the first working steam-engine of Thomas Savery. No evidence exists to suggest its invention was due to the application of pure science. The extracts contain no scientific theories. The success of the engine seems to be due to mechanical skills—in other words an example of technology without science. The same appears to be true of the subsequent development of steam-engines by Thomas Newcomen and James Watt. (It was left to the later nineteenth-century science of thermodynamics to explain the operations of the engines.)

Close links between eighteenth-century natural philosophy and technology did exist however in the field of scientific instrument-making. The growing demands for greater accuracy in carrying out measurements in scientific experiments (especially in electricity and chemistry) and for higher quality optical instruments could only be satisfied by highly skilled craftsmen. The artisan himself might also be very interested in natural philosophy. This was true of Benjamin Martin, a prolific writer, as well as one of the most famous instrument-makers of the time. He set out to explain the scientific principles behind the microscopes, astronomical instruments, etc. that he had made, gave directions for their use, and suggested various experiments with the apparatus, which he sold in his London shop. The passage in the anthology can therefore also be regarded as an illustration of a type of informal scientific education that helped to popularize science at a time when scientific and technical education did not otherwise exist.

William Herschel combined profound intellectual gifts with manual dexterity. Both natural philosopher and skilled craftsman, he constructed his own reflecting telescopes, carried out a vast number of observations, and proposed ingenious astronomical theories. The quality of his telescopes still receives high praise. The extract in the anthology describes some of his labours on his giant telescope.

In the nineteenth century striking industrial developments occurred as a result of scientific research into coal, the source of much of Britain's wealth. The texts in chapter 4 include pioneering chemical studies which resulted in the discovery of important new chemicals, such as benzene and naphthalene, and also caused the rise of a new industry in synthetic dyes.

The next chapter contains sources which illustrate the different opinions on the value of atomic theories in chemistry. While the more startling technological consequences of atomism belong to the nuclear power of the twentieth century, John Dalton's theory was recommended by nineteenth-century chemists as useful in industry as a quantitative check on yields. But much of the atomic discussions were confined to the speculative level: whether atoms really existed, or whether their consideration was necessary in practical chemistry. At the very end of the nineteenth century, the discovery of sub-atomic particles and of radioactivity showed that atoms were not indivisible as Dalton had claimed. The first suspicions of the enormous power contained in the atom soon began to be voiced.

Undoubtedly one of the greatest advances in scientific knowledge in the nineteenth century was in electricity. Previously this branch of science had been confined to electrostatics (chiefly the study of spark discharges). The electric current was not available in laboratories until the invention of the first electric battery, which Volta described in 1800. Almost at once the decomposition of water by the electric current was discovered. The text can be found in chapter 6. Soon Davy found potassium with the help of this new power. The moment of discovery is recorded in Davy's laboratory notebook and is reproduced here. At the theoretical level, electricity and chemistry had become intimately associated: Davy identified chemical affinity with electrical forces. In practice the way had been opened to the preparation of materials by electrochemical decomposition. The Castner–Kellner electrolytic process for the manufacture of caustic soda from salt was in operation at a works in Cheshire at the end of the nineteenth century.

The study of current electricity had other far-reaching technological consequences. In 1820 Oersted reported the surprising magnetic properties associated with electric currents (see chapter 7). This was the beginning of

Faraday's intensive researches, which eventually provided the fundamentals for an electrical industry: he invented the first electric motor and the first dynamo. The social consequences were to be immense, though they did not come until late in the century. First public buildings, then private houses began to be lit by electricity distributed from central generating stations. Electric power was slower to come to the factories, and for a while old steam-engines lingered on into the twentieth century. According to one story, Faraday was visited by Robert Peel, then Prime Minister, and questioned as to the uses of the dynamo. Faraday had been unable to foresee any. He had been conducting pure scientific inquiries, led on by his own passionate curiosity; the applications were not yet in his mind. The example is salutary—dramatic technological advances can result from pure scientific research, which at the time may seem devoid of applications. (See also chapter 14 for the developments in electrical communications.)

The educational issues are covered by the sources in chapter 9. Mrs. Marcet's charming *Conversations on Chemistry* can be compared with Benjamin Martin's dialogues (chapter 3). Her book went through many editions and helped to satisfy a growing popular interest in the sciences. Faraday read her book at an early age and always remembered it with affection. Some of the enthusiasm which set up Mechanics' Institutes for disseminating science among the working classes can be detected in Alfred Smith's address to an audience at Leeds in 1831. William Whewell, a leading philosopher, deeply interested in the sciences, also wrote on the importance of scientific education in universities. He also had a great admiration for the classics, and was not advocating the complete replacement of one by the other. A liberal education in the sense of an interest in knowledge for its own sake, apart from applications, was supported by him, and typically by Cardinal Newman. This has been a characteristic theme in English education, which has persisted alongside commitments to technical training for specific crafts and applied science.

The extracts from parliamentary papers reflect contemporary concern over the shortage of scientific and technical education in England; the frequent unfavourable comparisons with the situation abroad are conspicuous in the evidence given before the various Royal Commissions. The chapter also contains an early public examination in natural science. Closely connected with the educational debates were fears for Britain's industry and prestige (see chapter 16).

Further applications of scientific research are to be seen in public health. Chemical tests and microscopical observations revealed the alarming state of the nation's drinking water as well as the poisonous additions used in the adulteration of common food. The striking illustrations from the *Lancet* were part of that journal's campaign against malpractices in foodshops.

Sir John Simon is the most remarkable figure in the Victorian public health movement. A medical man, he became medical officer of health to the City of London, and then served on the Privy Council, providing information on matters relating to public health in yearly reports. Two extracts have been selected from these; they show Simon's efforts to fight disease through scientific research. The solution to the country's terrifying health problems, Simon believed, could only be found in science. During epidemics he therefore invited chemists and physicians to collaborate; the results were reported to the Privy Council.

Simon, like Chadwick before him, emphasized the importance of an adequate metropolitan water supply. The filth and stench were common causes of complaint at the time, especially in the river Thames, the source of London's drinking water. The connection with disastrous cholera epidemics was already suspected, before bacteriology had developed. The metropolitan water supply was greatly improved by the splendid system of sewers built by Bazalgette. They are still in use.

The chapter also includes the fundamental contributions of Lister to antiseptic surgery, and passages to illustrate the germ theory of disease.

Chapters 11–13 deal with the applications of chemical studies on various materials: sodium carbonate, petroleum by-products, new steels and, above all, the plastics. The latter, first made in the middle of the nineteenth century by the English chemist Parkes, have revolutionized twentieth-century technology, providing a vast range of materials with remarkable properties and widespread uses. The texts include those relating to the discovery of 'Parkesine' (a cellulose plastic), bakelite and polythene.

The nineteenth century is also notable for the increasing intrusion of science in those areas of technology that come under the collective heading of engineering. Demands to unite practice with sound scientific theory can be traced back to the engineers of the Renaissance (the best known example here is Leonardo da Vinci). However, at that time the sciences were not sufficiently developed to be able to provide this. This was the situation until the nineteenth century.

Two revealing texts in chapter 15 were written by men who were passionately convinced of the need to introduce more science into existing engineering, both to provide a rational understanding of current practices as well as to engender improvements. One of the authors, Rankine, was one of the first professors of engineering at a British university. The other passage is by Fairbairn, who was concerned with the fundamental problems of strengths of materials.

Other extracts illustrate the growing use of scientific and mathematical principles in branches of engineering which had previously relied heavily on empiricism: the construction of stable bridges and the design of ships. The increasing scientific content of engineering is also clearly shown by Rudolph Diesel's account of his engine. No longer simply an ingenious mechanic or inventor, common in the past, Diesel was a trained scientist, explicitly applying thermodynamics to design an efficient engine.

2 NATURAL PHILOSOPHER TO SCIENTIST

2a Two Centuries of *Philosophical Transactions*

By courtesy of the Chemical Society

[835] *Numb.* 167.

Beginning the fiveteenth Volume.

PHILOSOPHICAL TRANSACTIONS.

Jan. 28*th.* 168$\frac{4}{5}$.

The CONTENTS.

1. *Some Experiments about* Freezing, *and the Difference betwixt common* fresh-water Ice, *and that of* Sea-water; *also a probable Conjecture about the Original of the* Nitre *of* Ægypt: *by* Dr. Lister. 2. *A Letter from* Dr. Turbervile *of* Salisbury, *containing some considerable Observations in the Practise of Physic.* 3. *Observations on the* Cicindela Volans; *by R. Waller Esq. F. of the R. S.* 4. *A Letter from Mr.* J. Davis *concerning the* Wurtemburg Siphon. 5. *The Description of a* Siphon, *made to perform all the Effects exprest to be done by the* Wurtemburg Siphon: *by Dr.* Papin; *F. of the R. S.* 6. *A Catalogue of Experiments drawn up by Sr.* W. Petty, *præs. of the* Phil. Soc. *of* Dublin; *and by him presented to that* Society. 7. *A Letter from Mr. J.* Beaumont *jun, concerning a new way of* Cleaving Rocks. 8. *A Discourse proving from Experiments, that the larger the wheels of a Coach, &c. are, the more easily they are drawn over Stones, &c. lying in the way.* 9. *An account of a large præternaturall Glandulose Substance, found between the* Pericardium, *and* Heart *of an* Ox. 10. *Accounts of three Books;* 1. De Origine Fontium *per* R. Plott LL. D. *Musæi Ashmoleani* Oxoniæ *Præpositum, & R. S. S.* 2. *Medicina septentrionalis Collatitia, Opera* Theoph. Boneti. 3 Pechlini *Dialogus de* Potu Theæ.

Phil. Trans., 1684–5, **15**, 835.

By courtesy of the Chemical Society

CONTENTS

OF

VOL. LXXV. Part I.

Phil. Trans., 1785, **75**, 2.

By courtesy of the Chemical Society

CONTENTS

OF VOL. 176.

PART I.

PART II.

LIST OF ILLUSTRATIONS.

2b Swift's Satire on the Royal Society

This academy is not an entire single building, but a continuation of several houses on both sides of a street; which, growing waste, was purchased, and applied to that use. I was received very kindly by the warden, and went for many days to the academy. Every room hath in it one or more projectors; and, I believe, I could not be in fewer than five hundred rooms.

The first man I saw was of a meagre aspect, with sooty hands and face, his hair and beard long, ragged and singed in several places. His clothes, shirt, and skin were all of the same colour. He had been eight years upon a project for extracting sun-beams out of cucumbers, which were to be put into vials hermetically sealed, and let out to warm the air in raw inclement summers. He told me, he did not doubt, in eight years more, he should be able to supply the governor's gardens with sunshine at a reasonable rate; but he complained that his stock was low, and entreated me to give him something as an encouragement to ingenuity, especially since this had been a very dear season for cucumbers. I made him a small present, for my lord had furnished me with money on purpose, because he knew their practice of begging from all who go to see them.

I saw another at work to calcine ice into gunpowder, who likewise shewed me a treatise he had written concerning the malleability of fire, which he intended to publish.

There was a most ingenious architect, who had contrived a new method for building houses, by beginning at the roof, and working downwards to the foundation, which he justified to me, by the like practice of those two prudent insects, the bee and the spider.

There was a man born blind, who had several apprentices in his own condition: their employment was to mix colours for painters, which their master taught them to distinguish by feeling and smelling. It was, indeed, my misfortune to find them, at that time, not very perfect in their lessons, and the professor himself happened to be generally mistaken; this artist is much encouraged and esteemed by the whole fraternity.

In another apartment, I was highly pleased with a projector who had found a device of plowing the ground with hogs, to save the charges of ploughs, cattle, and labour. The method is this: in an acre of ground you bury, at six inches distance, and eight deep, a quantity of acorns, dates, chestnuts, and other mast, or vegetables, whereof these animals are fondest; then you drive six hundred, or more of them, into the field, where, in few days, they will root up the whole ground in search of their food, and make it fit for sowing; it is true, upon experiment, they found the charge and trouble very great, and they had little or no crop. However, it is not doubted that this invention may be capable of great improvement.

I went into another room, where the walls and ceiling were all hung round with cobwebs, except a narrow passage for the artist to go in and out. At my entrance, he called aloud to me not to disturb his webs. He lamented the fatal mistake the world had been so long in of using silkworms, while we had such plenty of domestic insects, who infinitely excelled the former, because they understood how to weave, as well as spin. And he proposed farther, that, by employing spiders, the charge of dying silks would be wholly saved; whereof I was fully convinced, when he shewed me a vast number of flies most beautifully coloured, wherewith he fed his spiders, assuring us that the webs would take a tincture from them; and, as he had them of all hues, he hoped to fit everybody's fancy, as soon as he could find proper food for the flies, of certain gums, oils, and other glutinous matter, to give a strength and consistence to the threads.

There was an astronomer, who had undertaken to place a sundial upon the great weather-cock on the town house, by adjusting the annual and diurnal motions of the earth and sun, so as to answer and coincide with all accidental turnings of the wind.

I had visited many other apartments, but shall not trouble my reader with all the curiosities I observed, being studious of brevity.

I had hitherto seen only one side of the academy, the other being appropriated to the advancers of speculative learning, of whom I shall say something, when I have mentioned one illustrious person more, who is called among them the universal artist. He told us he had been thirty years employing his thoughts for the improvement of human life. He had two large rooms full of wonderful curiosities, and fifty men at work. Some were condensing air into a dry tangible substance, by extracting the nitre, and letting the aqueous or fluid particles percolate; others softening marble for pillows and pin-cushions; others petrifying the hoofs of a living horse, to preserve them from foundering. The artist himself was at that time busy upon two great designs; the first to sow land with chaff, wherein he affirmed the true seminal virtue to be contained, as he demonstrated by several experiments which I was not skilful enough to comprehend. The other was, by a certain composition of gums, minerals, and vegetables, outwardly applied, to prevent the growth of wool upon two young lambs; and he hoped, in a reasonable time, to propagate the breed of naked sheep all over the kingdom.

JONATHAN SWIFT, *Gulliver's Travels*, Part III, London, 1726, chapter 5.

2c 'Winter'

Lo! from the livid East, or piercing North,
Thick Clouds ascend, in whose capacious Womb,
A vapoury Deluge lies, to Snow congeal'd;
Heavy, they roll their fleecy World along;
And the Sky saddens with th' impending Storm.
Thro' the hush'd Air, the whitening Shower descends,
At first, thin-wavering; till at last, the Flakes
Fall broad, and wide, and fast, dimming the Day,
With a continual Flow. See! sudden, hoar'd,
The Woods beneath the stainless Burden bow,
Blackning, along the mazy Stream it melts;
Earth's universal Face, deep-hid, and chill,
Is all one, dazzling, Waste. The Labourer-Ox
Stands cover'd o'er with Snow, and then demands
The Fruit of all his Toil. The Fowls of Heaven,
Tam'd by the cruel Season, croud around
The winnowing Store, and claim the little Boon,
That *Providence* allows. The foodless Wilds
Pour forth their brown *Inhabitants*; the Hare,
Tho' timorous of Heart, and hard beset
By Death, in various Forms, dark Snares, and Dogs,
And more unpitying Men, the Garden seeks,
Urg'd on by *fearless* Want. The bleating Kind
Eye the bleak Heavens, and next, the glistening Earth
With Looks of dumb Despair; then sad, dispers'd,
Dig, for the wither'd Herb, thro' Heaps of Snow.
Clear Frost succeeds, and thro' the blew Serene
For Sight too fine, th' Ætherial Nitre flies,
To bake the Glebe, and bind the slip'ry Flood.
This of the wintry Season is the Prime;
Pure are the Days, and lustrous are the Nights,
Brighten'd with starry Worlds, till then unseen.
Mean while, the Orient, darkly red, breathes forth
An Icy Gale, that, in its mid Career,
Arrests the bickering Stream. The nightly Sky,
And all her glowing Constellations pour
Their rigid Influence down: It freezes on
Till Morn, late-rising, o'er the drooping World,
Lifts her pale Eye, unjoyous: then appears
The various Labour of the silent Night,
The pendant Isicle, the Frost-Work fair,
Where thousand Figures rise, the crusted Snow,
Tho' white, made whiter, by the fining North.
On blithesome Frolics bent, the youthful Swains,

While every Work of Man is laid at Rest,
Rush o'er the watry Plains, and, shuddering, view
The fearful Deeps below: or with the Gun,
And faithful Spaniel, range the ravag'd Fields,
And, adding to the Ruins of the Year,
Distress the Feathery, or the Footed *Game.*

JAMES THOMSON, *The Seasons,* revised edition, London, 1744.

2d Natural History in Prose

LETTER I

THE parish of Selborne lies in the extreme eastern corner of the county of Hampshire, bordering on the county of Sussex, and not far from the county of Surrey; is about fifty miles south-west of London, in latitude fifty-one, and near mid-way between the towns of Alton and Petersfield. Being very large and extensive, it abuts on twelve parishes, two of which are in Sussex—viz., Trotton and Rogate. If you begin from the south and proceed westward, the adjacent parishes are Emshot, Newton Valence, Faringdon, Harteley Mauduit, Great Wardleham, Kingsley, Hedleigh, Bramshot, Trotton, Rogate, Lysse, and Greatham. The soils of this district are almost as various and diversified as the views and aspects. The high part of the south-west consists of a vast hill of chalk, rising three hundred feet above the village, and is divided into a sheep-down, the high wood and a long hanging wood, called The Hanger. The covert of this eminence is altogether beech, the most lovely of all forest trees, whether we consider its smooth rind or bark, its glossy foliage, or graceful pendulous boughs. The down, or sheepwalk, is a pleasing park-like spot, of about one mile by half that space, jutting out on the verge of the hill-country, where it begins to break down into the plains, and commanding a very engaging view, being an assemblage of hill, dale, woodlands, heath, and water. The prospect is bounded to the south-east and east by the vast range of mountains called the Sussex Downs, by Guild-down near Guildford, and by the Downs round Dorking, and Ryegate in Surrey, to the north-east, which altogether, with the country beyond Alton and Farnham, form a noble and extensive outline.

At the foot of this hill, one stage or step from the upland, lies the village, which consists of one single straggling street, three-quarters of a mile in length, in a sheltered vale, and running parallel with The Hanger. The houses are divided from the hill by a vein of stiff clay (good wheat-land), yet stand on a rock of white stone, little in appearance removed from chalk; but so far from being calcareous, that it endures extreme heat. Yet that the freestone still preserves somewhat that is analogous to chalk, is plain from the beeches which descend as low as those rocks extend, and no farther, and thrive as well on them, where the ground is steep, as on the chalks.

The cart-way of the village divides, in a remarkable manner, two very incongruous soils. To the south-west is a rank clay, that requires the labour of years to render it mellow; while the gardens to the north-east, and small enclosures behind, consist of a warm, forward, crumbling mould, called black malm, which seems highly saturated with vegetable and animal manure; and these may perhaps have been the original site of the town; while the woods and coverts might extend down to the opposite bank.

At each end of the village, which runs from south-east to north-west, arises a small rivulet: that at the north-west end frequently fails; but the other is a fine perennial spring, little influenced by drought or wet seasons, called Well-head.* This breaks out of some high grounds joining to Nore Hill, a noble chalk promontory, remarkable for sending forth two streams into two different seas. The one to the south becomes a branch of the Arun, running to Arundel, and so sailing into the British Channel; the other to the north, the Selborne stream, makes one branch of the Wey; and, meeting the Blackdown stream at Hedleigh, and the Alton and Farnham stream at Tilford-bridge, swells into a considerable river, navigable at Godalming; from whence it passes to Guildford, and so into the Thames at Weybridge; and thus at the Nore into the German Ocean.

Our wells, at an average, run to about sixty-three feet, and when sunk to that depth seldom fail; but produce a fine limpid water, soft to the taste and much commended by those who drink the pure element, but which does not lather well with soap.

To the north-west, north and east of the village, is a range of fair enclosures, consisting of what is called a white malm, a sort of rotten or rubble stone, which, when turned up to the frost and rain, moulders to pieces, and becomes manure to itself. This soil produces good wheat and clover.

Still on to the north-east, and a step lower, is a kind of white land, neither chalk nor clay, neither fit for pasture nor for the plough, yet kindly for hops, which root deep in the freestone, and have their poles and wood for charcoal growing just at hand. The white soil produces the brightest hops.

As the parish still inclines down towards Wolmer Forest, at the juncture of the clays and sand the soil becomes a wet, sandy loam, remarkable for timber, and infamous for roads. The oaks of Temple and Blackmoor stand high in the estimation of purveyors, and have furnished much naval timber; while the trees on the freestone grow large, but are what workmen call shaky, and so brittle as often to fall to pieces in sawing. Beyond the sandy loam the soil becomes a hungry lean sand, till it mingles with the forest; and will produce little without the assistance of lime and turnips.

LETTER II

In the court of Norton farmhouse, a manor-farm to the north-west of the village, on the white malms, stood within these twenty years a broad-leaved elm, or wych hazel, *Ulmus folio latissimo scabro* of Ray, which, though it had lost a considerable leading bough in the great storm in the year 1703, equal to a moderate tree, yet, when felled, contained eight loads of timber; and, being too bulky for a carriage, was sawn off at seven feet above the butt, where it measured near eight feet in the diameter. This elm I mention to show to what a bulk planted elms may attain; as this tree must certainly have been such from its situation.

In the centre of the village, and near the church, is a square piece of ground surrounded by houses, and vulgarly called 'The Plestor.' In the midst of this spot stood, in old times, a vast oak, with a short squat body, and huge horizontal arms extending almost to the extremity of the area. This venerable tree, surrounded with stone steps, and seats above them, was the delight of old and young, and a place of much resort in summer evenings; where the former sat in grave debate, while the latter frolicked and danced before them. Long might it have stood, had not the amazing tempest in 1703 overturned it at once, to the

* This spring produced, September 10th, 1781, after a severe hot summer, and a preceding dry spring and winter, nine gallons of water in a minute, which is 540 in an hour, and 12,960, or 216 hogsheads, in twenty-four hours, or one natural day. At this time many of the wells failed, and all the ponds in the vale were dry.

infinite regret of the inhabitants, and the vicar, who bestowed several pounds in setting it in its place again: but all his care could not avail; the tree sprouted for a time, then withered and died. This oak I mention to show to what a bulk planted oaks also may arrive; and planted this tree must certainly have been, as will appear from what will be said farther concerning this area, when we enter on the antiquities of Selborne.

On the Blackmoor estate there is a small wood called Losel's, of a few acres, that was lately furnished with a set of oaks of a peculiar growth and great value; they were tall and taper like firs, but standing near together had very small heads, only a little brush without any large limbs. About twenty years ago the bridge at the Toy, near Hampton Court, being much decayed, some trees were wanted for the repairs that were fifty feet long without bough, and would measure twelve inches diameter at the little end. Twenty such trees did a purveyor find in this little wood, with this advantage, that many of them answered the description at sixty feet. These trees were sold for twenty pounds apiece.

In the centre of this grove there stood an oak, which, though shapely and tall on the whole, bulged out into a large excrescence about the middle of the stem. On this a pair of ravens had fixed their residence for such a series of years, that the oak was distinguished by the title of the Raven-tree. Many were the attempts of the neighbouring youths to get at this eyry: the difficulty whetted their inclinations, and each was ambitious of surmounting the arduous task. But when they arrived at the swelling, it jutted out so in their way, and was so far beyond their grasp, that the most daring lads were awed, and acknowledged the undertaking to be too hazardous: so the ravens built on, nest upon nest, in perfect security, till the fatal day arrived in which the wood was to be levelled. It was the month of February, when those birds usually sit. The saw was applied to the butt,—the wedges were inserted into the opening,—the woods echoed to the heavy blows of the beetle or mall or mallet,—the tree nodded to its fall; but still the dam sat on. At last, when it gave way, the bird was flung from her nest; and, though her parental affection deserved a better fate, was whipped down by the twigs, which brought her dead to the ground.

GILBERT WHITE, *The Natural History and Antiquities of Selborne*, London, 1789, letters to Thomas Pennant, I and II.

2e Davy on Nitrous Oxide

A SHORT time after I began the study of Chemistry, in March 1798, my attention was directed to the dephlogisticated nitrous gas of Priestley, by Dr. Mitchill's Theory of Contagion.*

The fallacy of this theory was soon demonstrated, by a few coarse experiments made on small quantities of the gas procured from zinc and diluted nitrous acid. Wounds were exposed to its action, the bodies of animals were immersed in it without injury; and I breathed it mingled in small quantities

* Dr. Mitchill attempted to prove from some phenomenon connected with contagious diseases, that dephlogisticated nitrous gas which he called oxide of septon, was the principle of contagion, and capable of producing the most terrible effects when respired by animals in the minutest quantities, or even when applied to the skin or muscular fibre.

with common air, without remarkable effects. An inability to procure it in sufficient quantities, prevented me at this time from pursuing the experiments to any great extent. I communicated an account of them to Dr. Beddoes.

In 1799, my situation in the Medical Pneumatic Institution, made it my duty to investigate the physiological effects of the aëriform fluids, the properties of which presented a chance of useful agency. At this period I recommenced the investigation.

A considerable time elapsed before I was able to procure the gas in a state of purity, and my first experiments were made on the mixtures of nitrous oxide, nitrogen and nitrous gas, which are produced during metallic solutions.

In the beginning of March, I prepared a large quantity of impure nitrous oxide from the nitrous solution of zinc. Of this I often breathed the quantities of a quart and two quarts generally mingled with more than equal parts of oxygen or common air. In the most decisive of those trials, its effects appeared to be depressing, and I imagined that it produced a tendency to fainting: the pulse was certainly rendered slower under its operation.

At this time, Mr. Southey respired it in an highly diluted state; it occasioned a slight degree of giddiness, and considerably diminished the quickness of his pulse.

Mr. C. Coates likewise respired it highly diluted, with similar effects.

In April, I obtained nitrous oxide in a state of purity, and ascertained many of its chemical properties. Reflections upon these properties and upon the former trials, made me resolve to endeavour to inspire it in its pure form, for I saw no other way in which its respirability or powers could be determined.*

I was aware of the danger of this experiment. It certainly would never have been made if the hypothesis of Dr. Mitchill had in the least influenced my mind. I thought that the effects might be possibly depressing and painful, but there were many reasons which induced me to believe that a single inspiration of a gas apparently possessing no immediate action on the irritable fibre, could neither destroy nor immediately injure the powers of life.

On April 11th, I made the first inspiration of pure nitrous oxide; it passed into the bronchia without stimulating the glottis, and produced no uneasy feeling in the lungs.

The result of this experiment proved that the gas was respirable, and induced me to believe that a farther trial of its effects might be made without danger.

On April 16th, Dr. Kinglake being accidentally present, I breathed three quarts of nitrous oxide from and into a silk bag for more than half a minute, without previously closing my nose or exhausting my lungs.

The first inspirations occasioned a slight degree of giddiness. This was succeeded by an uncommon sense of fulness of the head, accompanied with loss of distinct sensation and voluntary power, a feeling analogous to that produced in the first stage of intoxication; but unattended by pleasurable sensation. Dr. Kinglake, who felt my pulse, informed me that it was rendered quicker and fuller.

This trial did not satisfy me with regard to its powers; comparing it with the former ones I was unable to determine whether the operation was stimulant or depressing.

I communicated the result to Dr. Beddoes; and on April the 17th, he was present, when the following experiment was made.

* I did not attempt to experiment upon animals, because they die nearly in equal times in non-respirable gases, and gases incapable of supporting life and possessed of no action on the venous blood.

Having previously closed my nostrils and exhausted my lungs, I breathed four quarts of nitrous oxide from and into a silk bag. The first feelings were similar to those produced in the last experiment; but in less than half a minute, the respiration being continued, they diminished gradually, and were succeeded by a sensation analogous to gentle pressure on all the muscles, attended by a highly pleasurable thrilling, particularly in the chest and the extremities. The objects around me became dazzling and my hearing more acute. Towards the last inspirations, the thrilling increased, the sense of muscular power became greater, and at last an irresistible propensity to action was indulged in; I recollect but indistinctly what followed; I know that my motions were various and violent.

These effects very soon ceased after respiration. In ten minutes, I had recovered my natural state of mind. The thrilling in the extremities, continued longer than the other sensations.*

This experiment was made in the morning; no languor or exhaustion was consequent, my feelings throughout the day were as usual, and I passed the night in undisturbed repose.

The next morning the recollections of the effects of the gas were very indistinct, and had not remarks written immediately after the experiment recalled them to my mind, I should have even doubted of their reality. I was willing indeed to attribute some of the strong emotion to the enthusiasm, which I supposed must have been necessarily connected with the perception of agreeable feelings, when I was prepared to experience painful sensations. Two experiments, however, made in the course of this day, with scepticism, convinced me that the effects were solely owing to the specific operation of the gas.

In each of them I breathed five quarts of nitrous oxide for rather a longer time than before. The sensations produced were similar, perhaps not quite so pleasurable; the muscular motions were much less violent.

Having thus ascertained the powers of the gas, I made many experiments to ascertain the length of time for which it might be breathed with safety, its effects on the pulse, and its general effects on the health when often respired.

I found that I could breathe nine quarts of nitrous oxide for three minutes, and twelve quarts for rather more than four. I could never breathe it in any quantity, so long as five minutes. Whenever its operation was carried to the highest extent, the pleasurable thrilling at its height about the middle of the experiment, gradually diminished; the sense of pressure on the muscles was lost; impressions ceased to be perceived; vivid ideas passed rapidly through the mind, and voluntary power was altogether destroyed, so that the mouth-piece generally dropped from my unclosed lips.

Whenever the gas was in a high state of purity, it tasted distinctly sweet to the tongue and palate, and had an agreeable odour. I often thought that it produced a feeling somewhat analogous to taste, in its application to my lungs. In one or two experiments, I perceived a distinct sense of warmth in my chest.

I never felt from it any thing like oppressive respiration: my inspirations became deep in proportion as I breathed it longer; but this phenomenon arose from increased energy of the muscles of respiration, and from a desire of increasing the pleasurable feelings.

Generally when I breathed from six to seven quarts, muscular motions were produced to a certain extent; sometimes I manifested my pleasure by stamping

* Dr. Beddoes has given some account of this experiment, in his 'Notice of some Observations made at the Medical Pneumatic Institution.' It was noticed in Mr. Nicholson's Phil. Journal for May 1799.

or laughing only; at other times, by dancing round the room and vociferating.

After the respiration of small doses, the exhilaration generally lasted for five or six minutes only. In one or two experiments when ten quarts had been breathed for near four minutes, an exhilaration and a sense of slight intoxication lasted for two or three hours.

On May 3d, to ascertain whether the gas would accelerate or retard the progress of sleep, I breathed at about 8 o'clock in the evening, 25 quarts of nitrous oxide, in quantities of six at a time, allowing but short intervals between each dose. The feelings were much less pleasurable than usual, and during the consumption of the two last doses, almost indifferent; indeed the gas was breathed rather too soon after its production, and contained some suspended acid vapour which stimulated the lungs so as to induce coughing.

After the experiments, for the first time I was somewhat depressed and debilitated; my propensity to sleep, however, came on at the usual hour, and as usual was indulged in, my repose was sound and unbroken.

Between May and July, I habitually breathed the gas, occasionally three or four times a day for a week together; at other periods, four or five times a week only.

The doses were generally from six to nine quarts; their effects appeared undiminished by habit, and were hardly ever exactly similar. Sometimes I had the feelings of intense intoxication, attended with but little pleasure; at other times, sublime emotions connected with highly vivid ideas; my pulse was generally increased in fulness, but rarely in velocity.

The general effects of its operation upon my health and state of mind, are extremely difficult of description; nor can I well discriminate between its agency and that of other physical and moral causes.

I slept much less than usual, and previous to sleep, my mind was long occupied by visible imagery. I had a constant desire of action, a restlessness, and an uneasy feeling about the præcordia analogous to the sickness of hope.

But perhaps these phenomena in some measure depended on the interest and labour connected with the experimental investigation relating to the production of nitrous oxide, by which I was at this time incessantly occupied.

My appetite was as usual, and my pulse not materially altered. Sometimes for an hour after the inspiration of the gas, I experienced a species of mental indolence* pleasing rather than otherwise, and never ending in listlessness.

During the last week in which I breathed it uniformly, I imagined that I had increased my sensibility of touch: my fingers were pained by any thing rough, and the tooth edge produced from slighter causes than usual. I was certainly more irritable, and felt more acutely from trifling circumstances. My bodily strength was rather diminished than increased.

At the end of July, I left off my habitual course of respiration; but I continued occasionally to breathe the gas, either for the sake of enjoyment, or with a view of ascertaining its operation under particular circumstances.

In one instance, when I had head-ache from indigestion, it was immediately removed by the effects of a large dose of gas; though it afterwards returned, but with much less violence. In a second instance, a slighter degree of head-ache was wholly removed by two doses of gas.

The power of the immediate operation of the gas in removing intense physical pain, I had a very good opportunity of ascertaining.

* Mild physical pleasure is perhaps always destructive to action. Almost all our powerful voluntary actions, arise either from hope, fear, or desire; and the most powerful from desire, which is an emotion produced by the coalescence of hope or ideal pleasure with physical pain.

In cutting one of the unlucky teeth called dentes sapientiae, I experienced an extensive inflammation of the gum, accompanied with great pain, which equally destroyed the power of repose, and of consistent action.

On the day when the inflammation was most troublesome, I breathed three large doses of nitrous oxide. The pain always diminished after the first four or five inspirations; the thrilling came on as usual, and uneasiness was for a few minutes swallowed up in pleasure. As the former state of mind however returned, the state of organ returned with it; and I once imagined that the pain was more severe after the experiment than before.

In August, I made many experiments with a view of ascertaining whether any analogy existed between the sensible effects of the different gases which are sooner or later fatal to life when respired, and those of nitrous oxide.

I respired four quarts of hydrogen* nearly pure, produced from zinc and muriatic acid, for near a minute, my lungs being previously exhausted and my nostrils carefully closed. The first six or seven inspirations produced no sensations whatever: in half a minute I perceived a disagreeable oppression of the chest, which obliged me to respire very quickly; this oppression gradually increased, till at last the pain of suffocation compelled me to leave off breathing. I felt no giddiness during or after the experiment; my pulse was rendered feebler and quicker; and a by-stander informed me that towards the last, my cheeks became purple.

In a second experiment, when the hydrogen was produced from iron and diluted sulphuric acid, I was unable to respire it for so long as three quarters of a minute; a transient giddiness and muscular debility were produced, the pulse was rendered very feeble, and the pain of suffocation was greater than before.

I breathed three quarts of nitrogen mingled with a very small portion of carbonic acid, for near a minute. It produced no alteration in my sensations for the first twenty seconds; then the painful sense of suffocation gradually came on, and increased rapidly in the last quarter of the minute, so as to oblige me to desist from the experiment. My pulse was rendered feebler and quicker. I felt no affection whatever in the head.

Mr. Watt's observations on the respiration of diluted hydrocarbonate by men, and Dr. Beddoes's experiments on the destruction of animals by pure hydrocarbonate, proved that its effects were highly deleterious.

As it destroyed life apparently by rendering the muscular fibre inirritable without producing any previous excitement, I was anxious to compare its sensible effects with those of nitrous oxide, which at this time I believed to destroy life by producing the highest possible excitement, ending in lesion of organization.

In the first experiment, I breathed for nearly a minute three quarts of hydrocarbonate mingled with nearly two quarts of atmospheric air.† It produced a slight giddiness and pain in the head, and a momentary loss of voluntary power: my pulse was rendered much quicker and feebler. These effects, however, went off in five minutes, and I had no return of giddiness.

Emboldened by this trial, in which the feelings were not unlike those I experienced in the first experiments on nitrous oxide, I resolved to breathe pure hydrocarbonate.

* Pure hydrogen has been often respired by different philosophers, particularly by Scheele, Fontana, and the adventurous and unfortunate Rosier.

† I believe it had never been breathed before by any individual, in a state so little diluted.

For this purpose, I introduced into a silk bag four quarts of gas nearly pure, which was carefully produced from the decomposition of water by charcoal an hour before, and which had a very strong and disagreeable smell.

My friend, Mr. James Tobin, Jun. being present, after a forced exhaustion of my lungs, the nose being accurately closed, I made three inspirations and expirations of the hydrocarbonate. The first inspiration produced a sort of numbness and loss of feeling in the chest and about the pectoral muscles. After the second inspiration, I lost all power of perceiving external things, and had no distinct sensation except a terrible oppression on the chest. During the third expiration, this feeling disappeared, I seemed sinking into annihilation, and had just power enough to drop the mouth-piece from my unclosed lips. A short interval must have passed during which I respired common air, before the objects about me were distinguishable. On recollecting myself, I faintly articulated, '*I do not think I shall die.*' Putting my finger on the wrist, I found my pulse thread-like and beating with excessive quickness.

In less than a minute, I was able to walk, and the painful oppression on the chest directed me to the open air.

After making a few steps which carried me to the garden, my head became giddy, my knees trembled, and I had just sufficient voluntary power to throw myself on the grass. Here the painful feeling of the chest increased with such violence as to threaten suffocation. At this moment, I asked for some nitrous oxide. Mr. Dwyer brought me a mixture of oxygen and nitrous oxide. I breathed this for a minute, and *believed* myself relieved. In five minutes, the painful feelings began gradually to diminish. In an hour they had nearly disappeared, and I felt only excessive weakness and a slight swimming of the head. My voice was very feeble and indistinct. This was at two o'clock in the afternoon.

I afterwards walked slowly for about half an hour, with Mr. Tobin, Jun. and on my return, was so much stronger and better, as to believe that the effects of the gas had disappeared; though my pulse was 120, and very feeble. I continued without pain for near three quarters of an hour, when the giddiness returned with such violence as to oblige me to lie on the bed; it was accompanied with nausea, loss of memory, and deficient sensation. In about an hour and half, the giddiness went off, and was succeeded by an excruciating pain in the forehead and between the eyes, with transient pains in the chest and extremities.

Towards night these affections gradually diminished. At ten,* no disagreeable feeling except weakness remained. I slept sound, and awoke in the morning very feeble and very hungry. No recurrence of the symptoms took place, and I had nearly recovered my strength by the evening.

I have been minute in the account of this experiment, because it proves, that hydrocarbonate acts as a sedative, i.e. that it produces diminution of vital action, and debility, without previously exciting. There is every reason to believe, that if I had taken four or five inspirations instead of three, they would have destroyed life immediately without producing any painful sensation. Perhaps most of the uneasy feelings after the experiment, were connected with the return of the healthy condition of organs.†

* I ought to observe, that between eight and ten, I took, by the advice of Dr. Beddoes, two or three doses of diluted nitric acid.

† By whatever cause the exhaustion of organs is produced, pain is almost uniformly connected with their returning health. Pain is rarely ever perceived in limbs debilitated by fatigue till after they have been for some hours at rest. Pain is uniformly connected with the recovery from the debility induced by typhus, often with the recovery from that produced by the stimulation of opium and alcohol.

About a week after this experiment, I attempted to respire carbonic acid, not being at the time acquainted with the experiments of Rosier.

I introduced into a silk bag four quarts of well washed carbonic acid produced from carbonate of ammonia* by heat, and after a complete voluntary exhaustion of my lungs, attempted to inspire it. I tasted strongly acid in the mouth and fauces, and produced a sense of burning at the top of the uvula. In vain I made powerful voluntary efforts to draw it into the windpipe; at the moment that the epiglottis was raised a little, a painful stimulation was induced, so as to close it spasmodically on the glottis; and thus in repeated trials I was prevented from taking a single particle of carbonic acid into my lungs.

I tried to breathe a mixture of two quarts of common air and three of carbonic acid, without success; it stimulated the epiglottis nearly in the same manner as pure carbonic acid, and was perfectly non-respirable.

I found that a mixture of three quarts of carbonic acid with seven of common air was respirable, I breathed it for near a minute. At the time, it produced a slight degree of giddiness, and an inclination to sleep. These effects, however, very rapidly disappeared after I had ceased to breathe,† and no other affections followed.

During the course of experiments on nitrous oxide, I several times breathed oxygen procured from manganese by heat, for from three to five minutes.

In respiring eight or ten quarts; for the first two or three minutes I could perceive no effects. Towards the end, even when I breathed very slowly, my respiration became oppressed, and I felt a sensation analogous to that produced by the want of fresh air; though but little of the oxygen had been consumed.

In one experiment when I breathed from and into a bag containing 20 quarts of oxygen for near six minutes, Dr. Kinglake felt my pulse, and found it not altered in volocity, but rather harder than before. I perceived no effects but those of oppression on the chest.‡

Having observed in my experiments upon venous blood, that nitrous gas rendered that fluid of a purple tinge, very like the colour generated in it by nitrous oxide; and finding no painful effects produced by the application of nitrous gas to the bare muscular fibre, I began to imagine that this gas might be breathed with impunity, provided it were possible in any way to free the lungs of common air before inspiration, so as to prevent the formation of nitrous acid.

* Carbonic acid is produced in this way in a high state of purity, and with great readiness.

† Carbonic acid possesses no action on arterial blood. Hence perhaps its slight effects when breathed mingled with large quantities of common air. Its effects are very marked upon venous blood! If it were thrown forcibly into the lungs of animals, the momentary application of it to the pulmonary venous blood would probably destroy life.

‡ In a conversation with Mr. Watt, relating to the powers of gases, that excellent philosopher told me he had for some time entertained a suspicion, that the effects attributed to oxygen produced from manganese by heat, in some measure depended upon nitrous acid suspended in the gas, formed during ignition by the union of some of the oxygen of the manganese with nitrogen likewise condensed in it.

In the course of experiments on nitrous acid, detailed in Research I, made in September, October, and December, 1799, I several times experienced a severe oppression on the chest and difficulty of respiration, not unanalogous to that produced by oxygen, but much more violent, from breathing an atmosphere loaded with nitrous acid vapour. This fact seemed to confirm Mr. Watt's suspicion. I confess, however, that I have never been able to detect any smell of nitrous acid, either by means of my own organs or those of others, during the production of oxygen; when the gas is suffered to pass into the atmosphere. The oxygen breathed in the experiments detailed in the text, had been for some days in contact with water.

On this supposition, during a fit of enthusiasm produced by the respiration of nitrous oxide, I resolved to endeavour to breathe nitrous gas.

114 cubic inches of nitrous gas were introduced into the large mercurial air-holder; two small silk bags of the capacity of seven quarts were filled with nitrous oxide.

After a forced exhaustion of my lungs, my nose being accurately closed, I made three inspirations and expirations of nitrous oxide in one of the bags, to free my lungs as much as possible from atmospheric oxygen; then, after a full expiration of the nitrous oxide, I transferred my mouth from the mouth-piece of the bag to that of the air-holder, and turning the stop-cock, attempted to inspire the nitrous gas. In passing through my mouth and fauces, it tasted astringent and highly disagreeable; it occasioned a sense of burning in the throat, and produced a spasm of the epiglottis so painful as to oblige me to desist instantly from attempts to inspire it. After moving my lips from the mouth-piece, when I opened them to inspire common air, aëriform nitrous acid was instantly formed in my mouth, which burnt the tongue and palate, injured the teeth, and produced an inflammation of the mucous membrane which lasted some hours.

As after the respiration of nitrous oxide in the experiments in the last Research, a small portion of the residual atmospheric air remained in the lungs, mingled with the gas, after forced expiration; it is most probable that a minute portion of nitrous acid was formed in this experiment, when the nitrous gas was taken into the mouth and fauces, which might produce its stimulating properties. If so, perhaps I owe my life to the circumstance; for supposing I had taken an inspiration of nitrous gas, and even that it had produced no positive effects, it is highly improbable, that by breathing nitrous oxide, I should have freed my lungs from it, so as to have prevented the formation of nitrous acid when I again inspired common air. I never design again to attempt so rash an experiment.

In the beginning of September I often respired nitrous oxide mingled with different proportions of common air or oxygen. The effects produced by the diluted gas were much less violent than those produced by pure nitrous oxide. They were generally pleasant: the thrilling was not often perceived, but a sense of exhilaration was almost constant.

Between September and the end of October, I made but few experiments on respiration, almost the whole of my time being devoted to chemical experiments on the production and analysis of nitrous oxide.

At this period my health being somewhat injured by the constant labour of experimenting, and the perpetual inhalation of the acid vapours of the laboratory, I went into Cornwall; where new associations of ideas and feelings, common exercise, a pure atmosphere, luxurious diet and moderate indulgence in wine, in a month restored me to health and vigour.

Nov. 27th; immediately after my return, being fatigued by a long journey, I respired nine quarts of nitrous oxide, having been precisely thirty-three days without breathing any. The feelings were different from those I had experienced in former experiments. After the first six or seven inspirations, I gradually began to lose the perception of external things, and a vivid and intense recollection of some former experiments passed through my mind, so that I called out '*what an amazing concatenation of ideas!*' I had no pleasurable feeling whatever, I used no muscular motion, nor did I feel any disposition to it; after a minute, when I made the note of the experiment, all the uncommon sensations had vanished; they were succeeded by a slight soreness in one of the arms and in the leg: in three minutes these affections likewise disappeared.

From this experiment I was inclined to suppose that my newly acquired health had diminished my susceptibility to the effects of the gas. About ten days after, however, I had an opportunity of proving the fallacy of this supposition.

Immediately after a journey of 126 miles, in which I had no sleep the preceding night, being much exhausted, I respired seven quarts of gas for near three minutes. It produced the usual pleasurable effects, and slight muscular motion. I continued exhilarated for some minutes afterwards; but in half an hour found myself neither more nor less exhausted than before the experiment. I had a great propensity to sleep.

I repeated the experiment four or five times in the following week, with similar effects. My susceptibility was certainly not diminished. I even thought that I was more affected than formerly by equal doses.

Though, except in one instance, when indeed the gas was impure, I had experienced no decisive exhaustion after excitement from nitrous oxide, yet still I was far from being satisfied that it was unanalogous to stimulants in general. No experiment had been made in which the excitement from nitrous oxide had been kept up for so great a length of time and carried to so great an extent as that in which it is uniformly succeeded by excessive debility under the agency of other powers.

It occurred to me, that supposing nitrous oxide to be a stimulant of the common class, it would follow that the debility produced in consequence of excessive stimulation by a known agent, ought to be *increased* after excitement from nitrous oxide.*

To ascertain whether this was the case, I made, on December 23d, at four P.M. the following experiment. I drank a bottle of wine in large draughts in less than eight minutes. Whilst I was drinking, I perceived a sense of fulness in the head, and throbbing of the arteries, not unanalogous to that produced in the first stage of nitrous oxide excitement. After I had finished the bottle, this fulness increased, the objects around me became dazzling, the power of distinct articulation was lost, and I was unable to walk steadily. At this moment the sensations were rather pleasurable than otherwise, the sense of fulness in the head soon however increased so as to become painful, and in less than an hour I sunk into a state of insensibility.†

In this situation I must have remained for two hours or two hours and a half.

I was awakened by head-ache and painful nausea. The nausea continued even after the contents of the stomach had been ejected. The pain in the head every minute increased; I was neither feverish nor thirsty; my bodily and mental debility were excessive, and the pulse feeble and quick.

In this state I breathed for near a minute and half five quarts of gas, which was brought to me by the operator for nitrous oxide; but as it produced no sensations whatever, and apparently rather increased my debility, I am almost convinced that it was from some accident, either common air, or very impure nitrous oxide.

Immediately after this trial, I respired 12 quarts of oxygen for near four minutes. It produced no alteration in my sensations at the time; but immediately after I imagined that I was a little exhilarated.

* In the same manner as the debility from intoxication by two bottles of wine is increased by a third.

† I ought to observe that my usual drink is water, that I had been little accustomed to take wine or spirits, and had never been completely intoxicated but once before in the course of my life. This will account for the powerful effects of a single bottle of wine.

The head-ache and debility still however continuing with violence, I examined some nitrous oxide which had been prepared in the morning, and finding it very pure, respired seven quarts of it for two minutes and half.

I was unconscious of head-ache after the third inspiration; the usual pleasurable thrilling was produced, voluntary power was destroyed, and vivid ideas rapidly passed through my mind; I made strides across the room, and continued for some minutes much exhilarated. Immediately after the exhilaration had disappeared, I felt a slight return of the head-ache; it was connected with transient nausea. After two minutes, when a small quantity of acidified wine had been thrown from the stomach, both the nausea and head-ache disappeared; but langour and depression not very different in degree from those existing before the experiment, succeeded. They, however, gradually went off before bed time. I slept sound the whole of the night except for a few minutes, during which I was kept awake by a trifling head-ache. In the morning, I had no longer any debility. No head-ache or giddiness came on after I had arisen, and my appetite was very great.

This experiment proved, that debility from intoxication was not increased by excitement from nitrous oxide. The head-ache and depression, it is probable would have continued longer if it had not been administered. Is it not likely that the slight nausea following the effects of the gas was produced by new excitability given to the stomach?

To ascertain with certainty, whether the most extensive action of nitrous oxide compatible with life, was capable of producing debility, I resolved to breathe the gas for such a time and in such quantities, as to produce excitement equal in duration and superior in intensity to that occasioned by high intoxication from opium or alcohol.

To habituate myself to the excitement, and to carry it on gradually, on December 26th, I was inclosed in an air-tight breathing-box,* of the capacity of about 9 cubic feet and half, in the presence of Dr. Kinglake.

After I had taken a situation in which I could by means of a curved thermometer inserted under the arm, and a stop-watch, ascertain the alterations in my pulse and animal heat, 20 quarts of nitrous oxide were thrown into the box.

For three minutes I experienced no alteration in my sensations, though immediately after the introduction of the nitrous oxide the smell and taste of it were very evident.†

In four minutes I began to feel a slight glow in the cheeks, and a generally diffused warmth over the chest, though the temperature of the box was not quite 50°. I had neglected to feel my pulse before I went in; at this time it was 104 and hard, the animal heat was 98°. In ten minutes the animal heat was near 99°, in a quarter of an hour 99·5°, when the pulse was 102, and fuller than before.

At this period 20 quarts more of nitrous oxide were thrown into the box, and well mingled with the mass of air by agitation.

In 25 minutes the animal heat was 100°, pulse 124. In 30 minutes, 20 quarts more of gas were introduced.

My sensations were now pleasant; I had a generally diffused warmth without the slightest moisture of the skin, a sense of exhilaration similar to that produced by a small dose of wine, and a disposition to muscular motion and to merriment.

* The plan of this box was communicated by Mr. Watt. An account of it will be detailed in the *Researches*.

† The nitrous oxide was too diluted to act much; it was mingled with near 22 times its bulk of atmospheric air.

In three quarters of an hour the pulse was 104, and animal heat not quite 99·5°, the temperature of the chamber was 64°. The pleasurable feelings continued to increase, the pulse became fuller and slower, till in about an hour it was 88°, when the animal heat was 99°.

20 quarts more of air were admitted. I had now a great disposition to laugh; luminous points seemed frequently to pass before my eyes, my hearing was certainly more acute, and I felt a pleasant lightness and power of exertion in my muscles. In a short time the symptoms became stationary; breathing was rather oppressed, and on account of the great desire of action, rest was painful.

I now came out of the box, having been in precisely an hour and quarter.

The moment after, I began to respire 20 quarts of unmingled nitrous oxide. A thrilling, extending from the chest to the extremities, was almost immediately produced. I felt a sense of tangible extension highly pleasurable in every limb; my visible impressions were dazzling, and apparently magnified, I heard distinctly every sound in the room, and was perfectly aware of my situation.* By degrees, as the pleasurable sensations increased, I lost all connection with external things; trains of vivid visible images rapidly passed through my mind, and were connected with words in such a manner, as to produce perceptions perfectly novel. I existed in a world of newly connected and newly modified ideas. I theorised—I imagined that I made discoveries. When I was awakened from this semi-delirious trance by Dr. Kinglake, who took the bag from my mouth, indignation and pride were the first feelings produced by the sight of the persons about me. My emotions were enthusiastic and sublime; and for a minute I walked round the room, perfectly regardless of what was said to me. As I recovered my former state of mind, I felt an inclination to communicate the discoveries I had made during the experiment. I endeavoured to recall the ideas, they were feeble and indistinct; one collection of terms, however, presented itself: and with the most intense belief and prophetic manner, I exclaimed to Dr. Kinglake, '*Nothing exists but thoughts!—the universe is composed of impressions, ideas, pleasures and pains!*'

About three minutes and half only had elapsed during this experiment, though the time as measured by the relative vividness of the recollected ideas, appeared to me much longer.

Not more than half of the nitrous oxide was consumed. After a minute, before the thrilling of the extremities had disappeared, I breathed the remainder. Similar sensations were again produced; I was quickly thrown into the pleasurable trance, and continued in it longer than before. For many minutes after the experiment, I experienced the thrilling in the extremities, the exhilaration continued nearly two hours. For a much longer time I experienced the mild enjoyment before described connected with indolence; no depression or feebleness followed. I ate my dinner with great appetite and found myself lively and disposed to action immediately after. I passed the evening in executing experiments. At night I found myself unusually cheerful and active; and the hours between eleven and two, were spent in copying the foregoing detail from the common-place book, and in arranging the experiments. In bed I enjoyed profound repose. When I awoke in the morning, it was with consciousness of pleasurable existence, and this consciousness more or less continued through the day.

Since December, I have very often breathed nitrous oxide. My susceptibility to its power is rather increased than diminished. I find six quarts a full dose,

* In all these experiments, after the first minute, my cheeks became purple.

and I am rarely able to respire it in any quantity for more than two minutes and half.

The mode of its operation is somewhat altered. It is indeed very different at different times.

I am scarcely ever excited into violent muscular action; the emotions are generally much less intense and sublime than in the former experiments, and not often connected with thrilling in the extremities.

When troubled with indigestion, I have been two or three times unpleasantly affected after the excitement of the gas. Cardialgia, eructations, and unpleasant fulness of the head were produced.

I have often felt very great pleasure when breathing it alone, in darkness and silence, occupied only by ideal existence. In two or three instances when I have breathed it amidst noise, the sense of hearing has been painfully affected even by moderate intensity of sound. The light of the sun has sometimes been disagreeably dazzling. I have once or twice felt an uneasy sense of tension in the cheeks and transient pains in the teeth.

Whenever I have breathed the gas after excitement from moral or physical causes, the delight has been often intense and sublime.

On May 5th, at night, after walking for an hour amidst the scenery of the Avon, at this period rendered exquisitely beautiful by bright moonshine; my mind being in a state of agreeable feeling, I respired six quarts of newly prepared nitrous oxide.

The thrilling was very rapidly produced. The objects around me were perfectly distinct, and the light of the candle not as usual dazzling. The pleasurable sensation was at first local, and perceived in the lips and about the cheeks. It gradually, however, diffused itself over the whole body, and in the middle of the experiment was for a moment so intense and pure as to absorb existence. At this moment, and not before, I lost consciousness; it was, however, quickly restored, and I endeavoured to make a by-stander acquainted with the pleasure I experienced by laughing and stamping. I had no vivid ideas. The thrilling and the pleasurable feeling continued for many minutes; I felt two hours afterwards, a slight recurrence of them, in the intermediate state between sleeping and waking; and I had during the whole of the night, vivid and agreeable dreams. I awoke in the morning with the feeling of restless energy, or that desire of action connected with no definite object, which I had often experienced in the course of experiments in 1799.

I have two or three times since respired nitrous oxide under similar circumstances; but never with equal pleasure.

During the last fortnight, I have breathed it very often; the effects have been powerful and the sensations uncommon; but pleasurable only in a slight degree.

I ought to have observed that a desire to breathe the gas is always awakened in me by the sight of a person breathing, or even by that of an air-bag or an air-holder.

I have this day, June 5th, respired four large doses of gas. The first two taken in the morning acted very powerfully; but produced no thrilling nor other pleasurable feelings. The effects of the third, breathed immediately after a hearty dinner, were pleasant, but neither intense nor intoxicating. The fourth was respired at night in darkness and silence after the occurrence of a circumstance which had produced some anxiety. This dose affected me powerfully and pleasantly; a slight thrilling in the extremities was produced; an exhilaration continued for some time, and I have had but little return of uneasiness; (11 P.M.)

From the nature of the language of feeling, the preceding detail contains many imperfections; I have endeavoured to give as accurate an account as possible of the strange effects of nitrous oxide, by making use of terms standing for the most similar common feelings.

We are incapable of recollecting pleasures and pains of sense.* It is impossible to reason concerning them, except by means of terms which have been associated with them at the moment of their existence, and which are afterwards called up amidst trains of concomitant ideas.

When pleasures or pains are new or connected with new ideas, they can never be intelligibly detailed unless associated during their existence with terms standing for analogous feelings.

I have sometimes experienced from nitrous oxide, sensations similar to no others, and they have consequently been indescribable. This has been likewise often the case with other persons. Of two paralytic patients who were asked what they felt after breathing nitrous oxide, the first answered, '*I do not know how, but very queer*.' The second said, '*I felt like the sound of a harp*.' Probably in the one case, no analogous feelings had ever occurred. In the other, the pleasurable thrillings were similar to the sensations produced by music; and hence, they were connected with terms formerly applied to music.

* Physical pleasure and pain generally occur connected with a compound impression, i.e. an organ and some object. When the idea left by the compound impression, is called up by being linked accidentally to some other idea or impression, no recurrence, or the slightest possible, of the pleasure or pain in any form will take place. But when the compound impression itself exists *without* the physical pleasure or pain, it will awaken ideal or intellectual pleasure or pain, i.e. hope or fear. So that physical pleasure and pain are to hope and fear, what impressions are to ideas. For instance, assuming no accidental association, the child does not fear the fire before he is burnt. When he puts his finger to the fire he feels the physical pain of burning, which is connected with a visible compound impression, the fire and his finger. Now when the compound idea of the fire and his finger, left by the compound impression are called up by his mother, saying, '*You have burnt your finger*,' nothing like fear or the pain of burning is connected with it. But when the finger is brought near the fire, i.e. when the compound impression again exists, the ideal pain of burning or the passion of fear is awakened, and it becomes connected with those very actions which removed the finger from the fire.

HUMPHRY DAVY, *Researches on Nitrous Oxide* (1800), in *Collected Works*, London, 1839, vol. iii.

3 SCIENCE AND TECHNOLOGY IN THE INDUSTRIAL REVOLUTION

3a 'Raising Water by the Impellent Force of Fire'

By courtesy of the British Museum

TO THE

Gentlemen Adventurers

IN THE

Mines of ENGLAND.

I Am very ſenſible a great many among you do as yet look on my *Invention* of raiſing *Water* by the impellent force of *Fire*, a uſeleſs ſort of a *Project*, that never can anſwer my *Deſigns* or *Pretenſions*; and that it is altogether impoſſible that ſuch an *Engine* as this can be wrought *under-ground*, and ſucceed in the *Raiſing* of *Water*, and *Dreining* your *Mines*, ſo as to deſerve any Incouragement from you. I am not very fond of lying under the *Scandal* of a bare *Projector*, and therefore preſent you here with a *Draught* of my *Machine*, and lay before you the Uſes of it, and leave it to your Conſideration, whether it be worth your while to make uſe of it or no. I can eaſily give grains of allowance for your *Suſpicions*, becauſe I know very well what *Miſcarriages* there have been by People *Ignorant* of what they *pretend* to. Theſe I know have been ſo *frequent*, ſo fair and *promiſing at firſt*, but ſo ſhort of *performing* what they pretended to, that your *Prudence* and *Diſcretion* will not now ſuffer you to *believe* any thing without a demonſtration, your Appetites to *new Inventions* of this nature, having been *baulkt* too often; yet after all, I muſt beg you not to *condemn* me, before you have *read* what I have to *ſay* for my *ſelf*, and let not the *failures* of *others* prejudice me, or be placed to my account. I have often lamented the want of underſtanding the true *Powers of Nature*, which misfortune, has of late put ſome on making ſuch vaſt *Engines* and *Machines*, both *troubleſome* and *expenſive*, yet of *no* manner of *uſe*, inaſmuch as the *old Engines*,uſed many *Ages* paſt, far exceeded them. And I fear, whoever by the *old Cauſes* of Motion pretends to *Improvements* within this *laſt Century*, does *betray* his *Knowledge* and Judgment. For more then an hundred years ſince,*Men* and *Horſes* would Raiſe by *Engines* then made, as much *Water* as they have ever ſince done,or I believe ever will, or according to the *Law of Nature* ever can do. And though my *Thoughts* have been long imployed about *Water-works*, I ſhould never have pretended to any *Invention* of that kind, had I not *happily* found out this *new*, but yet a much *Stronger* and *Cheaper Force* or Cauſe of Motion, than any before made uſe of. But finding this of *Rarefaction by Fire*, the conſideration of the Difficulties the *Miners* and *Colliers* labour under by the frequent *Diſorders*, Cumberſomneſs, and in general of *Water-Engines*; Incouraged me to *invent* Engines *to work* by this *new force*,that tho' I was obliged to *incounter* the *oddeſt* and almoſt *inſuperable* Difficulties; I ſpared neither *Time*, *Pains* nor *Money*, till I had abſolutely *Conquer'd them*. I hope this will at leaſt incourage you to *read* over this *ſmall Treatiſe* I now put into your hands, for the farther and more particular Information of the *Nature* and *Uſes* of this*Engine* for raiſing *Water* by the *Force of Fire*: After which I ſhall patiently *ſubmit* to any *Judgment* you ſhall pleaſe to *paſs* upon *me* or the *Invention*, and may have reaſon to believe you will not any longer ſuffer your *Judgments* to be impoſed on by *thoſe*, whoſe *Profit* and *Intereſt* it may ſeem to be, to *Condemn* both right or wrong; I mean *ſuch* who make your common Ginns, and their Friends and Acquaintance among you; tho' I am very ſure the promotion of the *Uſe* of this *Engine* is their true Intereſt, as I very plainly prove thus:

The cheaper *Water* is drawn, the more is the *Miner* incouraged to adventure; the more the *Miner* adventures, the more *Pits* or *Shafts* muſt be *ſunk*; the more *Shafts* or *Pits* are ſunk, the more *Wood-work* will be neceſſarily imployed in *Timbering them*, or ſupporting the ſides from *falling* in where the *Earth* is looſe; beſides *Windeleſſes* and all other *Utenſils* of Wood *uſed* in *Mines*, or the *Trades* depending *thereon* muſt be more, which by increaſing the *Carpenters Trade* in general, will make them ſufficient *amends* for the *loſs* of a ſmall part of that Branch of their *Trade*, called *Ginn-making*. As for *Pump-making*, that part of the *Trade* will be much *Improved* by my *Engine*. For I muſt uſe Board and Timber for *Pipes*; and have conſiderable Imployment for *Pump-makers* and *Carpenters* for Timber uſed about my *Engine*. But ſhall never *Imploy* any other *Perſon* in making *Pipes*, or any other Carpenters work I ſhall have to do, but the *Perſon* who was before *imployed* in the *Work*, or ſuch as ſhall be recommended, as a *Perſon* imployed in the *Mines* of the *Country* whereſoever I ſhall fix *Engines*; provided they will work as cheap, and fairly, and obſerve the Orders and Directions given them. For my *Deſign* is not in the leaſt to *prejudice* the *Artificers*, or indeed any *other* ſort of *People*, by this *Invention*; but on the contrary, is intended for the *Benefit* and *Advantage* of *Mankind* in general, eſpecially the *People* of my own *Nation*; and wherein, you *Gentlemen* concern'd in *Mines*, may, if you pleaſe, *Reap* the greateſt *Profit*.

And altho' I do not queſtion but the *Plan*, and *Draught* of my *Engine*, will be very well and readily underſtood with *many* Gentlemen, by the *Deſcription* here given; yet it will require a longer Time in *Others*, to imploy their Minds and Thoughts more Intenſely about it; eſpecially ſuch as have not been *familiar* and acquainted with *things* of this *kind*. But ſhould the *Engine*, to the Apprehenſion of ſome, ſeem *intricate* and difficult to be worked, after all the *Deſcription* I have given of it in this *Book*; yet I can, and do aſſure them, That the *attending* and working the *Engine* is ſo far from being ſo, that it is familiar and *eaſie* to be learned by thoſe of the *meaneſt Capacity*, in a very little time; insomuch, that I have Boys of 13 or 14 years of *Age*, who now *attend* and *work it* to perfection, and were *taught* to do it in a few days; and I have known some *learn* to work the Engine in *half an hour*. We have a Proverb, That *Interest* never lyes: And I am assured that you *Gentlemen* of the *Mines* and *Collyeries*, when you have once made this *Engine* familiar in your Works, and to your Selves and Servants; not only the *Profit*, but abundance of other *Advantages* and Conveniencies, which you will find to attend your Works in the *Use* thereof, will create in you a favourable Opinion of the Labours of

Your real Friend

London, Sept. 22. 1701.

and humble Servant

Tho. Savery.

T. Savery, *The Miner's Friend*, London, 1702, p. 1.

3b Savery's Engine

By courtesy of the British Museum

Chap. II.

Of the Uses that this Engine may be applied unto.

IT may be supposed that there are few *People* among us so *ignorant*, but must necessarily know of what *value* the *Falls* of *Water* are in most Places, as being *applicable* to *Mills*; which are made after *various* Kinds and *Forms*, according to the *different* Genius and *Abilities* of the *Mill-right*; for *Mill-work* being in a manner infinitely *diversified*; and had I leisure to Comment thereon, and give you an Account, not only of the vast *variety* that I have seen and *heard* of; but (when incouraged) what may yet be brought to *work* by a steady *Stream*, and the *Rotation* or Circular Motion of a *Water-wheel*, it would *swell* these Papers to a much larger *Volumn* than was at first designed, and *frustrate* my intended *Brevity*. I only just hint this to shew what

Uſe this *Engine* may be put to in *working* of *Mills*, eſpecially where *Coals* are cheap.

I have only this to urge, That *Water* in its *Fall* from any determinate *Height*, has ſimply a *Force* anſwerable and *equal* to the *Force* that *Raiſes* it. So that an *Engine* which will *Raiſe* as much *Water* as *two Horſes*, working together at one time in ſuch a *work*, can do, and for which there muſt be conſtantly kept *ten* or *twelve* Horſes for doing the ſame. Then I ſay, ſuch an *Engine* will do the *work* or *labour* of *ten* or *twelve* Horſes; and whereas this *Engine* may be made *large* enough to do the *work* required in imploying *eight*, *ten*, *fifteen*, or *twenty* Horſes to be conſtantly maintained and kept for doiog *ſuch a work*; it will be improper to ſtint or *confine* its Uſes and Operation in reſpect of *Water-Mills*.

2. It may be of great Uſe for *Palaces*, for the Nobilities or Gentlemens *Houſes*: For by a *Ciſtern* on the *top* of a *Houſe*, you may with a great deal of *Eaſe*, and little Charge, *throw* what Quantity of Water you have occaſion for to the *top* of any *Houſe*, which Water in its fall, makes you what *ſorts* of *Fountains* you pleaſe, and ſupplys any *Room* in the *Houſe*. And it is of excellent *Uſe* in caſe of *Fire*, of which more hereafter.

3. Nothing can be more *fit* for ſerving Cities, and Towns *with water*, except a *Crank-work* by the *force* of a *River*. In the Compoſing ſuch ſort of *Engines*, I think no Perſon hath excelled the Ingenious Mr. *George Sorocold*. But where *they* are forced to Uſe *Horſes*, or any *other* ſtrength, I believe no Ingenious Perſon will deny this *Engine* to have the *Preference* in all *Reſpects*, being of more *Univerſal* Uſe than any yet *Diſcovered* or Invented.

4. As for *Draining* Fens and *Marſhes*, &c. I ſuppoſe I need ſay no more than this, That that *Force* which will *Raiſe* great Quantities of *Water* a height of above 80 Foot, muſt neceſſarily *deliver* a much *greater* Quantity at a *leſſer* height. And that it is much *cheaper*, and every way *eaſier*, eſpecially where Coals are Water-borne, to *continue* the Diſcharge of any *Quantities* of Water by our *Engine*, than it can

be done by any *Horse-Engines* whatsoever.

5. I believe it may be made very *useful* to Ships, but I dare not meddle with that matter; and leave it to the Judgment of those who are the best Judges of Maritain Affairs.

6. For *Draining* of *Mines* and *Coal-Pits*, the Use of the *Engine* will sufficiently *recommend* it self, in raising *Water* so *easie* and *cheap*; and I do not doubt, but that in a few *years*, it will be a *means* of making our *Mining Trade*, which is no *small part* of the *Wealth* of this *Kingdom*, *double*, if not *treble* to what it now is. And if such vast Quantities of *Lead*, *Tin*, and *Coals* are now *yearly* exported, under the *Difficulties* of such an *immense Charge* and *Pains* as the *Miners*, &c. are now at to Discharge *their water*, how *much more* may be hereafter *Exported*, when the *Charge* will be very much *lessen'd* by the Use of this *Engine*, every way fitted for the Use of Mines? For the far *greater* part of our *richest Mines* and *Coal-Pits*, are *liable* to *two* grand *Inconveniencies*, and thereby rendred *useless*; *viz.* The Eruption and Excess of *Subterranious* Waters, as not being worth the *Expence* of *Draining them* by the great *Charge* of *Horses*, or *hand* Labour. Or secondly, Fatal *Damps*, by which *many* are struck *blind*, *lame*, or *dead* in these *Subterraneous Cavities*, if the *Mine* is wanting of a due *Circulation* of Air. Now *both* these Inconveniencies *are* naturally *remedied* by the *work* of this *Engine* of raising *water* by the impellant Force of FIRE.

For the *Water*. Be the *Mine* never so *deep*, each *Engine* working it 60, 70, or 80 foot high, by applying or *setting* the *Engines one* over another, as shall be shew'd at large hereafter in the following Pages, you may by a sufficient number of *Engines* keep the *bottom* of any *Mine dry*; and when once you know how large your *Feeder* or *Spring* is, it is very *easie* to know what siz'd *Engine*, or what number of *Engines* will do your Business.

The *Coals* used in this *Engine* is of as little value, as the *Coals* commonly burned on the *Mouths* of the *Coal-Pits* are: For an *Engine* of a three Inch-bore, or thereabout, *working* the Water up 60 *foot* high, requires a *Fire-place* of not above *twenty* Inches deep, and about *fourteen* or *fifteen* Inches wide, which will occaſion ſo ſmall a Conſumption, that in a Coal-Pit it is of no account, as we have Experienced. And in all parts of *England* where there are *Mines*; *Coals* are ſo *cheap*, that the Charge of them is not to be mentioned, when we conſider the vaſt quantity of *Water* raiſed, by the inconſiderable *value* of the *Coals* uſed and *burnt* in ſo ſmall a *Furnace.* What the quantity of Coals uſed for one Engine in a year is, cannot eaſily be aſcertained, becauſe of the different Nature of the ſeveral ſorts of Coals.

As for the Cure of *Damps* by this *Engine*, the Air perpetually crowding into the *Aſh-hole* and *Fire-place*, as it is natural for it to do, and with a moſt impetuous Force diſcharged with the *Smoak* at the top of the Chimney, the *contigious* Air is ſucceſſively following it; ſo that not only all *Steams* or Vapours whatſoever, that may or can *ariſe*, muſt naturally *force* its way through the *Fire*, and ſo be *diſcharged* at the top with the *Smoak*. But this Motion of the *Fire* will occaſion the *freſh Air* to deſcend from above, down all the *Pits*, and every *where* elſe in the *Mine*, but down the *Chimney*; provided you have a heading Drift, or Paſſage from all the Shafts, or Pits in the Work, to that place where the *Engine* ſtands, whether the *Mouth* of the ſaid *Pit* and Chimney be *lower* or *higher* than the Mouths of any of the reſt of the *Pits* or *Shafts* in the ſame *Work*, it matters not; for here will be a perpetual *Circulation* of the *Air*, and with that *ſwiftneſs*, as is hardly to be *believed.* This I have tryed, and know to be true; ſo leave the Ingenious Miner to his own Judgment, Whether when all the Air is in a ſwift Motion, that any Stagnation of Air (which has always been adjudged the cauſe of *Damps*) can happen in any *Pit*.

T. SAVERY, *The Miner's Friend*, London, 1702, p. 28.

3c Teaching Instrumentation in the Eighteenth Century

DIALOGUE III. On the Nature, Construction, and Use of the Solar Microscope

Cleonicus

I HAVE hitherto entertained you in the best Manner I am able, with an Account of microscopic Constructions of different Forms and Kinds; nor have I finished that Subject yet, as there remains one which far exceeds all the rest in its magnifying Power, or the astonishing Appearance it gives of the smallest Objects.

Euphros. This, I presume, is the Solar Microscope, which I have heard you speak of very often, on that Account, —— is not this the Instrument now before me on the Table?

Cleon. That is the Solar Microscope, whose Nature and Construction I shall now explain to you; and then, afterwards, entertain you with some Experiments of its Use.

Euphros. It is called the Solar Microscope, I suppose, from its being used with the Sun, or rather, that it is intirely useless without the Sun-beams; for I have heard you say, that it is applied to Use in a Room that it is darkened, and then only when the Sun shines. But the Method of its Application in the *Camera Obscura*, as you call it, is a Thing as yet I have little or no Idea of, and is what, I presume, you intend to instruct me in.

Cleon. This Morning is devoted to that Purpose. —— The Room is darkened and ready for Use, and the Sun will probably continue till I have dispatched, in a few Words, the Description and Use of the several Parts of the Microscope. —— You will therefore observe, this Instrument consists of the following Parts; (1.) A long plain Glass in a Brass Frame, which goes on the Outside of the Window to receive the Sun-beams which fall obliquely thereon. (2.) The square Plate, or Frame of Brass, is, by a Couple of Screws, fastened to a Window-shutter, containing a Hole for the Looking-glasses to pass through. (3.) In the Middle of this Plate lies a Wheel with Teeth, concealed, with a circular Rim of Brass placed over it, and is moveable by a Pinion which you observe in one Angle of the Plate. (4.) In the Central Hole of this Circular Wheel, or Piece of Brass on the Outside, towards the Looking-glass, is fixed a Convex Lens, whose focal Distance is about 8 or 10 inches. (5.) In the Hole of the same Plate, on the Face of the Instrument, is screwed the Tube or Body of the Microscope with its Drawer or sliding Part, as you there see. (6.) On the End of this Slider is screwed a single Microscope of the first or second Form, which I have before described to you, and thus you see all the component Parts of this curious Machine as they are put together.

Euphros. Well; but, methinks, I should be glad to see the particular Use of each of these Parts, before we enter the dark Room, where I cannot so well see the Manner of their Application.

Cleon. This I can shew you just as well in a light, as in a dark Room, the Light is no Obstruction to any Thing but forming a Picture of the Object; I have provided a Board with a Hole of the same Size with that in the Window-shutter, to which I shall apply the Machine in its several Parts successively. And,

First, I put the Looking-glass through the Hole of the Board, and then screw the Plate firmly on to the frame. This done, then,

Secondly, I turn the Pinion which moves the Wheel with the Looking-glass on it, towards the Sun, by which Means the Rays of the Sun are reflected from the Looking-glass, through the Convex Lens into the Room; as here you see by the Experiment.

Thirdly, It is necessary in the Use of this Instrument, that the Sun-beams should be made to go strait across the Room, or parallel to the Floor, and to fall Perpendicular on the opposite Wall or Side of the Room. Now there is but one Position of the Looking-glass that will reflect the Rays in that Manner, and therefore, you see here is a small Wheel, and endless Screw fixed on the other Part of the Plate, for moving the Looking-glass up and down, till it comes into the necessary Position, and then you observe.

Fourthly, The Beams of Light are refracted through the Convex Lens to its Focus, which, you plainly see, is at the Distance of about 10 Inches from the Glass.

Fifthly, From the focal Point you observe the Rays proceed diverging, and form a large Circle of Light on the opposite Side of the Room.

Sixthly, I screw on the Tube with the small Microscope, and place the Slider therein with the Objects intended to be magnified.

Seventhly, This done, I move the Drawer forward and backward, till I bring the Object in the Slider within a small Distance of the Focus of the Convex-glass, where you may observe it is very much illuminated in the denser Part of the Cone of Rays: Then,

Eighthly, and lastly, I apply the small Magnifiers to the single Microscope, and by the adjusting Screw, move them to such a Distance from the Object, as shall shew their Image very plain, and large, upon a Sheet hung on the opposite Side of the Room: And thus having shewn you how every Part is applied to use, I shall repair, with you, to the Room made dark, there fix the Instrument, and shew you its surprizing Effects.

Euphros. I shall, with great Pleasure, attend you on this Occasion, to see how Darkness itself conduces to perfect these Experiments, which depend on the strongest Light.

Cleon. Here, my *Euphrosyne,* enter the Room, and place yourself in the Chair I have there set for you, you will there be near me, to see the several Processes in the Application of the Instrument; and at a proper Distance to view the Image to the best Advantage.

Euphros. I am seated, as you direct. —— You may proceed to your Operation, as soon as you please.

Cleon. The first Object I shall put in the Microscope is, a transparent Scale of a Sole-fish —— observe its Image upon the Sheet.

Euphros. Its Image! —— I see something prodigiously large, indeed; and you amaze me, when you say it is only the Scale of a Sole: —— It appears at least 12 or 15 Feet long; and 6 or 8 Feet wide; —— it fills almost the Side of the Room. —— How thick set with long sharp Points at one End! its Substance transparent, and variegated with beautiful Undulations. —— Were the whole Fish magnified in the same Proportion, it must surely be as big as a Whale.

Cleon. You might have said, my *Euphrosyne,* as big as a hundred Whales; since the Fish, magnified as much as its Scale, would appear very near half a Mile long. —— By this Instance, you see how Nature has provided for the Defence of the Sole-fish, since the external Part of every Scale is armed with those sharp pointed Spines, or Thorns.

Euphros. I have often observed, that the Surface of those Fish were very rough to the Hand; but never knew the Reason of it before.* —— What is the next Object, *Cleonicus*?

* See a microscopic View of the Skin and Scales of a Soal-fish, in our Miscellany, Page 681, for *December* 1757.

Cleon. I have shewn you the magnified Scale of a Fish, and I will next shew the Scale of an Eel, magnified in the same Proportion.
Euphros. The Scale of an Eel, *Cleonicus*, that Expression seems to have a strange Sound: I am so great a *Virtuosa*, that I never before heard of, or knew there was any such Thing.
Cleon. That may be, and numberless others besides yourself are not only Ignorant of it, but will not believe it when it is told them: Nay, the *Jews* insist that it is contrary to the *Mosaic* Law, by which they are forbidden to eat of that delicious Dish, as they apprehend that Eels come under the Denomination of that Sort of Fish which have no Scales: —— But observe, on the Sheet, the Picture of one in its full Dimensions and Perfection.
Euphros. A prodigious Spectacle this, —— it is nearly as large as the Scale of a Soal, but of a very different Form; —— its Substance seems to be one continued Piece of Net-work, proceeding in long oval Rings, as it were, from a common Center: —— I observe no Bristles, or Points, in this, nor can I tell on what Part it grows on the Body.
Cleon. These Scales, my *Euphrosyne*, differ from the Scales of common Fish, in the following Particulars: *First*, they lye not on the external Surface of the Eel, but are contained under one common Covering, or Skin. *Secondly*, they do not lye one over another, as the Scales of a common Fish; but are placed once by the other, as thick as they can lye, through the whole Skin of the Eel. *Thirdly*, their reticular Form, or Net-like Texture, is such as we observe in no other Sort of common Fish-Scales; beside many other Peculiarities which I shall observe to you at another Season. —— You will next behold the wonderful Appearance of a certain Species of Eels themselves, real living Animals.
Euphros. Living Animals, *Cleonicus*, —— they have really so much Life and Motion, and monstrous Size withal, that I have scarce Courage enough to keep my Seat. Who could have thought to have seen, in a Moment's Time, that Sheet transformed into a Lake of Conger Eels? —— They are vastly numerous, indeed, —— and larger in Size than I have ever seen any Thing of the Kind; 5, 6, or 7 Feet long, and 3 or 4 Inches thick: —— Their quick tortuous and serpentine Motion are really amazing and stupendious. —— Pray, what are these Creatures in themselves, and how came you by them?
Cleon. These are really a Species of small Eels, or *Anguiculæ*; they are found very common in sour Paste, or stale Vinegar, and some other Fluids, of which I shall have the Pleasure to entertain you, one Day or other, with the common Microscope, when you will find that, notwithstanding the Appearance they now make by the solar Microscope, they are scarcely visible to the naked Eye; such Objects of Minuature are the mighty Serpents you now see.*
Euphros. They are every Way wonderful, I find. —— What do you intend to surprize me with next, *Cleonicus*?
Cleon. You will justly admire the wonderful Mechanism in the Eye of a Fly.
Euphros. A glorious Object, indeed! —— the finest Piece of Lattice-work my Eyes have ever yet beheld: —— How grand and beautiful are the Works of Nature, and what great Variety in the Structure of the same Organ! —— In the Surface of this small Eye, I observe Myriads of those beautiful six-sided Figures. —— Pray, *Cleonicus*, what Purpose do they serve, in this visual Organ, in the Fly?
Cleon. We are not yet acquainted with the Design of Nature in every Part of her Operation: But I shall recommend you to read the Conjectures of some of

* See Figure 8, in the Plate of Microscopic Animalculæ. Page 161 in the Miscellany, *October* 1755.

the best Naturalists upon this particular Subject, rather than give you my own Opinion concerning it.* —— Every Part of a Fly is equally wonderful, as you will e're long be satisfied by the common Microscope; but they are not all equally proper Subjects for the solar Microscope: But the Wings of most Flies may be shewn to great Advantage in this Instrument, particularly those that are membranous and transparent; nay, even the hard Wings of the Scarab Kind are some of them, in the smaller Species, most excellent Objects for this Purpose, —— see one of them fill the whole Side of the Room.

Euphros. Is this, *Cleonicus*, the Wing of a small, contemptible Scarab? —— This is a sufficient Instance, to prove how necessary it is to apply the Instruments of Art for giving us right Notions, and just Ideas of the Works of Nature. —— I think this is the richest, and most glorious Spectacle I ever saw, and can scarce help looking upon it as the Height of Nature's Fancy, in Point of Grandeur; —— it seems all an embroidered Ground of Gold, embossed and studded with Diamonds, in such beautiful and regular Squares of Cheques and Compartments, as nothing but an actual View of such a high finished Piece could ever raise an Idea of. —— Pray, how large may the Animal be, that is thus so richly bedeck'd and adorn'd?

Cleon. This small Beetle you may have often observed crawling about upon rotten Wood, old Walls, *&c.* of so diminutive a Size, as not to exceed one Tenth, or Two, of an Inch, and therefore escaped your Notice: But were these Minute Creatures to become magnified at once, *two thousand* Times in Length and Breadth, as is the Case of the Wing you now see, what stupendous Admiration would they excite in all Mankind! —— The highest Wisdom hath directed you to contemplate the Gaiety of the Lilly, and other fine Flowers of the Garden, with a Remark, at the same Time, that *Solomon*, in all his Glory, was not arrayed like one of these. But if, as I said, those Animals were to appear at large, we should find Nature much more profuse in the Richness of her Ornaments and Embellishments to them; and that the Robes they are invested with, do far surpass not only the Glory of *Solomon*, but all the Pomp and Grandeur that all the Courts in the World can afford in Works of Art.

Euphros. I should never be tired with feasting my Eyes with these delightful Scenes; but I must not detain you too long in Particulars. What is the next Object you intend shall gratify my Curiosity?

Cleon. It is what they call the *Exuviæ*, or cast off Skin of a Spider. —— See the wonderous Appearance on the Sheet.

Euphros. Prodigious! nay, even shocking: —— What a hideous Apparition is this! a Head three Feet long, —— with 8 Eyes as big as Saucers; —— 8 monstrous hairy Legs, as big as Mill-posts: —— What Appearance must such a Spider as this make, with every Part magnified two thousand Times in Length and Breadth, as I now see it, alive and in Motion.

Cleon. Your Sex are generally intimidated enough with this Animal in its natural Size and Appearance; but were a Spider, as large as you now see this, to drop from the Clouds in the Middle of *Smithfield*, on a Market-day, it would scare not only the Ladies, but the Butchers themselves, and find Oxen to feed on instead of Flies.

Euphros. I don't know but that might be the Case, as I have known a Spider to have conquer'd an Ant, and devour it, tho' nearly as big as itself: But enough of this tremendous Creature. What does your Microscope next display upon the Sheet?

* See the magnified Head and Eyes of a Bee, in a large Copperplate Print, with some Account of it, in the Miscellany, Page 303, for *January* 1760.

Cleon. A Spectacle that will fill you with equal Amazement and Delight. —— See Rivulets of fluid Particles running over all the Side of the Room.

Euphros. What can this mean? —— I guess, by what I have heard you say, that this is the Circulation of the Blood in animal Vessels. —— But what an incredible Number of Streams there are, and with what Velocity do they flow in every Direction through the Scene! —— Pray, *Cleonicus*, what Part of the Animal is it that appears thus large and transparent, and in such a wonderful Degree of Motion?

Cleon. It is the transparent Tail of a Water-Newt, which I judge to be the best Subject of all for this Purpose, and are found in every Pond of Water. The Tail of a Tad-pole will do nearly as well; as also the transparent Web and Toes in the Foot of a very small Frog; the Tail of that Fish we call the Loach; and many other Subjects may be innocently applied for this Purpose, without those cruel and inhuman Methods, practised by some who appear to have more the Form, than the Rationality of Men.*

Euphros. I abhor the Thoughts of Cruelty, in every Shape, and should rather be thought to have no Curiosity, or Taste at all, than a vicious One. —— I am thoroughly sensible, by this Experiment, how the Blood flows through the Arteries to the extreme Parts, and from thence returns again by the Veins; I see the very Form and Size of the Particles of Blood: And therefore keep the Creature no longer in such a Condition, which, tho' it may give me much Pleasure, may be very uneasy to itself. —— I now wait your following Exhibition.

Cleon. Keep your Eye stedfast upon the Sheet, and you will soon see a common Animal appear in quite a new Light.

Euphros. This is a terrible Appearance, indeed: —— What hideous Monster do I see? —— How large his Head, his jointed Horns, its bulbous Eyes, and monstrous Legs and Claws, the Motions of which are very terrible in Appearance: —— I suspect, by the Form, that this frightful Creature is the LOUSE. —— But what is that red Spot I see in the Forepart of its Body, attended with an odd Kind of Motion?

Cleon. It is a living Louse that you see, and the Spot which you mention shews the Place and Pulsation of the Heart; from which, you observe a Motion of a vital Fluid through the middle Part of the *Abdomen*, or large, hinder Part of the Body: But you see nothing like the Circulation of the Blood in other larger Animals, by distinct Arteries and Veins. How the Circulation of the Fluids are carried on in this, does not, even by this Instrument, appear.†

Euphros. Well; an uglier Thing I scarce ever saw, they are Objects of our Aversion in their natural Minute State; but what an horrible Aspect they have now! —— I have viewed this Monster long enough. Pray, what Appearance does a Flea make?

Cleon. I have one at Hand to shew you. —— There, you see it.

Euphros. I do, sure enough: —— It is as large as an Ox, —— it seems to be a Creature covered with a Coat of Mail or Tortoishell, —— its Appearance is rather uncouth, than formidable, and a good deal in a Shape of a Lobster, as you formerly told me: —— What an amazing Head it has, with a large black Eye. —— But what is most surprizing in the Structure of this Animal is, the peculiar and wonderful Form of its Legs, by which one may easily see how

* The barbarous Practice of *dissecting a Frog alive*, to shew the Circulation of the Blood in the Mesentery, is here refer'd to.

† See a Print and Description of the Louse, Page 893, of the Miscellaneous, for *October* 1758.

Nature has prepared it for moving in the Manner it does. —— I can likewise perceive, in the Body of this Animal, a Motion of the Fluids somewhat like that in the Louse.* —— But now for another Object, *Cleonicus*, if you please.

Cleon. I shall now entertain you with a few extraordinary Views of the vegetable Kind. —— Look at the Sheet, and tell me what you think you see there, my *Euphrosyne*.

Euphros. You have suddenly transformed the Scene to something like a wonderful Sort of Stars, —— some of which are Opake, others Transparent; —— they are perfectly round, and large as cannon Balls: —— They are vastly numerous, and make a delightful Appearance, truly. —— Pray, what can it be, of a vegetable Nature, that I see?

Cleon. It is what the Botanists call the *Farina*, or that dusty Matter which grows on the Apices in the Middle of the Flowers, of which you see great Quantities in the Tulip, the Lilly, and many others: But that which I have in the Microscope, is the Farina of a Holyhock, and is found in all of the Mallow Kind; for in every Plant of a different Species, those Particles have a peculiar Size and Form, as you may observe at your Leisure in the common Microscope. —— Now cast your Eye again on the Sheet, and you will see an Object that will give you Pleasure to view it.

Euphros. This is a curious Phænomenon, indeed: —— This has, in Truth, the Appearance of Vegetation itself: —— How beautifully coloured, and regularly branched and ramified in every Part! —— Pray, what do you call this, *Cleonicus*?

Cleon. It is a Species of a very fine Sea Plant, or Coralline, of which there are innumerable Kinds, and all make the same beautiful Appearance you now see; for not only the Earth, but the Sea is productive of Vegetables, Plants, and Trees of various Kinds and Sizes; all which afford the most curious Objects for the Microscope. —— Again, you see the Scene is changed, and another Part of a Vegetable exposed to your View.

Euphros. I think the Appearance of it very fine and delicate; but what it is, I cannot conjecture, as I never remember to have seen any Thing in a Plant of this curious Texture.

Cleon. It is a very thin transverse Slice of the Pith of Elder, and tho' we know so little of the Use of that Part of the Plant, yet you may be assured that something very considerable depends upon so elegant, and quite a geometrical Structure. —— The next Object has a similar Appearance, but the Pores much smaller.

Euphros. Be this what it will, it is vastly beautiful; —— the Pores, it is true, are much smaller, and more numerous, and therefore a more compact Substance than the Pith; but what Part of the Plant it can be, I am as much at a Loss to guess as I was before.

Cleon. This is a very thin Piece of Cork, which, you know, is a vegetable Substance, as it is the outer Part of the Bark of a Cork-Tree; and tho' we cannot so well tell the Use of these Pores in Vegetation, yet, as they render Cork a soft and yielding Substance, they make it a very fit Medium for many Uses in Life, particularly for corking of Bottles, and sundry other Uses for Fishermen, *&c.* —— Another Object well deserves your Notice: Look upon the Sheet, and you will see it.

Euphros. I see something in Appearance very regular, full of Holes of a larger and lesser Size, and disposed in a very beautiful Order: —— But this, like all other Objects in this new View, becomes quite unknown to me, however common it may be to the vulgar Sight.

* See a Print and Description of the Flea, in the Miscellany for *April* 1759, Page 74.

Cleon. This, my *Euphrosyne*, is the essential Part of a Plant, it is a very thin Slice of Oak, cut transversely, so as to render it quite Transparent; for, by this, it appears, that the Body of a Plant consists almost entirely of long, hollow Tubes, or Vessels, the Orifices of which, when cut asunder, are what makes that variegated Appearance of Ringlets of different Sizes upon the Sheet. —— Those of the largest Sort are the Orifices of Airpipes which is that System of Vessels which circulate the Air through the Body of a Plant; and though some of them appear now near a Foot in Diameter, yet, in themselves; they are scarce visible to the naked Eye. —— The other Order of smaller Orifices are those of the Sap-Vessels, which circulate the nutricious Juices of the Earth to all the Parts of the Plant; and it is these that make the very Substance of the Plant itself, and constitute those cylindric Shells of Wood which appear in cutting the Tree asunder, each of which is the Product of one Year's Growth: All which I shall more particularly discourse of, to you, at another Time. —— At present, these Specimens must suffice for the Use of the solar Microscope, and for giving you a just Idea and Esteem for such an excellent Invention, and which exceed all others of the optical Kind.
Euphros. I am much obliged to you for the Trouble I have given you, and for your Confinement so long in a dark Room; and I shall always look upon the Experiments of the Microscope, and especially of this Sort, as the most sublime and rational Amusements of my Life.

B. MARTIN, *The Young Gentleman and Lady's Philosophy*, London, 1759–63, vol. 2, p. 182.

3d Herschel's Reflecting Telescope

XVIII. *Description of a Forty-feet Reflecting Telescope. By* William Herschel, *LL.D. F.R.S.*

Read June 11, 1795.

THE uncommon size of my forty-feet reflecting telescope will render a description of it not unacceptable to lovers of astronomy. I shall therefore endeavour to give as complete an idea of its construction as the limited compass of this paper will permit, and hope that, with the assistance of the annexed drawings, the mechanism of it will be sufficiently intelligible to such as have been in the habit of viewing machines and mechanical works.

It will be necessary to mention a few circumstances that led the way to the construction of this large instrument, in the execution of which two very material requisites were necessary: namely, the support of a very considerable expence, and a competent experience and practice in mechanical and optical operations.

When I resided at Bath I had long been acquainted with the theory of optics and mechanics, and wanted only that experience which is so necessary in the practical part of these sciences. This I acquired by degrees at that place, where in my leisure hours, by way of amusement, I made for myself several 2-feet, 5-feet, 7-feet, 10-feet, and 20-feet NEWTONIAN telescopes; besides others of the GREGORIAN form, of 8 inches, 12 inches, 18 inches, 2 feet, 3 feet, 5 feet, and 10 feet focal length. My way of doing these instruments at that time, when the direct method of giving the figure of any of the conic sections to specula was still unknown to me, was, to have many mirrors of each sort cast, and to finish them all as well as I could; then to select by trial the best of them, which I preserved; the rest were put by to be repolished. In this manner I made not less than 200, 7-feet; 150, 10-feet; and about 80, 20-feet mirrors; not to mention those of the GREGORIAN form, or of the construction of Dr. SMITH's reflecting microscope, of which I also made a great number.

My mechanical amusements went hand in hand with the optical ones. The number of stands I invented for these telescopes it would not be easy to assign. I contrived and delineated them of different forms, and executed the most promising of the designs. To these labours we owe my 7-feet NEWTONIAN telescope-stand, which was brought to its present convenient construction about 17 years ago; a description and engraving of which I intend to take some future opportunity of presenting to the Royal Society. In the year 1781 I began also to construct a 30-feet aërial reflector; and after having invented and executed a stand for it, I cast the mirror, which was moulded up so as to come out 36 inches in diameter. The composition of my metal being a little too brittle, it cracked in the cooling. I cast it a second time, but here the furnace, which I had built in my house for the purpose, gave way, and the metal ran into the fire.

These accidents put a temporary stop to my design, and as the discovery of the GEORGIAN planet soon after introduced me to the patronage of our most gracious King, the great work I had in view was for a while postponed.

In the year 1783 I finished a very good 20-feet reflector with a large aperture, and mounted it upon the plan of my present telescope. After two years observation with it, the great advantage of such apertures appeared so clearly to me, that I recurred to my former intention of increasing them still farther; and being now sufficiently provided with experience in the work I wished to undertake, the President of our Royal Society, who is always ready to promote useful undertakings, had the goodness to lay my design before the King. His Majesty was graciously pleased to approve of it, and with his usual liberality to support it with his royal bounty.

In consequence of this arrangement I began to construct the 40-feet telescope, which is the subject of this paper, about the latter end of the year 1785. The wood-work of the stand, and machines for giving the required motions to the instrument, were immediately put in hand, and forwarded with all convenient expedition. In the whole of the apparatus none but common workmen were employed, for I made drawings of every part of it, by which it was easy to execute the work, as I constantly inspected and directed every person's labour; though sometimes there were not less than 40 different workmen employed at the same time.

While the stand of the telescope was preparing I also began the construction of the great mirror, of which I inspected the casting, grinding, and polishing; and the work was in this manner carried on with no other interruption than what was occasioned by the removal of all the apparatus and materials from Clay-hall, where I then lived, to my present situation at Slough.

Here soon after my arrival, I began to lay the foundation, upon which by degrees the whole structure was raised as it now stands; and the speculum being highly polished and put into the tube, I had the first view through it on Feb. 19, 1787. I do not however date the completing of the instrument till much later; for the first speculum, by a mismanagement of the person who cast it, came out thinner on the centre of the back than was intended, and on account of its weakness would not permit a good figure to be given to it. A second mirror was cast Jan. 26, 1788; but it cracked in cooling. Feb. 16, we recast it with particular attention to the shape of the back, and it proved to be of a proper degree of strength. Oct. 24, it was brought to a pretty good figure and polish, and I observed the planet Saturn with it. But not being satisfied, I continued to work upon it till Aug. 27, 1789, when it was tried upon the fixed stars, and I found it to give a pretty sharp image. Large stars were a little affected with scattered light, owing to many remaining scratches in the mirror.

Aug. the 28th, 1789. Having brought the telescope to the parallel of Saturn, I discovered a sixth satellite of that planet; and also saw the spots upon Saturn, better than I had ever seen them before, so that I may date the finishing of the 40-feet telescope from that time.

WILLIAM HERSCHEL, 'Description of a Forty-feet Reflecting Telescope', *Phil. Trans.*, 1795, **85**, 347.

4 COAL
The Basis of Nineteenth-century Technology

4a Old Mining Engine in Staffordshire

By courtesy of the British Museum

B.P.P., 1844, (13) plate 3.

4b Clayton's Experiments in Lancashire (1684)

V. An Experiment concerning the Spirit of Coals, *being part of a Letter to the* Hon. Rob. Boyle, *Esq; from the late Rev.* John Clayton, *D.D. communicated by the Right Rev. Father in God* Robert *Lord Bishop of* Corke *to the Right Hon.* John *Earl of* Egmont, *F.R.S.*

HAVING seen a Ditch within two Miles from *Wigan* in *Lancashire*, wherein the Water would seemingly burn like Brandy, the Flame of which was so fierce, that several Strangers have boiled Eggs over it; the People thereabouts indeed affirm, that about 30 Years ago it would have boiled a Piece of Beef; and that whereas much Rain formerly made it burn much fiercer, now after Rain it would scarce burn at all. It was after a long-continued Season of Rain that I came to see the Place, and make some Experiments, and found accordingly, that a lighted Paper, though it were waved all over the Ditch, the Water would not take Fire. I then hired a Person to make a Dam in the Ditch, and fling out the Water, in order to try whether the Steam which arose from the Ditch would then take Fire, but found it would not. I still, however, pursued my Experiment, and made him dig deeper; and when he had dug about the Depth of half a Yard, we found a shelly Coal, and the Candle being then put down into the Hole, the Air catched Fire, and continued burning.

I observed that there had formerly been Coal-pits in the same Close of Ground; and I then got some Coal from one of the Pits nearest thereunto, which I distilled in a Retort in an open Fire. At first there came over only *Phlegm*, afterwards a black *Oil*, and then likewise a *Spirit* arose, which I could noways condense, but it forced my Lute, or broke my Glasses. Once, when it had forced the Lute, coming close thereto, in order to try to repair it, I observed that the Spirit which issued out caught Fire at the Flame of the Candle, and continued burning with Violence as it issued out, in a Stream, which I blew out, and lighted again, alternately, for several times. I then had a Mind to try if I could save any of this Spirit, in order to do which I took a turbinated Receiver, and putting a Candle to the Pipe of the Receiver whilst the Spirit arose, I observed that it catched Flame, and continued burning at the End of the Pipe, though you could not discern what fed the Flame: I then blew it out, and lighted it again several times; after which I fixed a Bladder, squeezed and void of Air, to the Pipe of the Receiver. The *Oil* and *Phlegm* descended into the Receiver, but the Spirit, still ascending, blew up the Bladder. I then filled a good many Bladders therewith, and might have filled an inconceivable Number more; for the Spirit continued to rise for several Hours, and filled the Bladders almost as fast as a Man could have blown them with his Mouth; and yet the Quantity of Coals I distilled were inconsiderable.

I kept this Spirit in the Bladders a considerable time, and endeavour'd several ways to condense it, but in vain. And when I had a Mind to divert Strangers or Friends, I have frequently taken one of these Bladders, and pricking a Hole therein with a Pin, and compressing gently the Bladder near the Flame of a Candle till it once took Fire, it would then continue flaming till all the Spirit was compressed out of the Bladder; which was the more surprising, because no one could discern any Difference in the Appearance between these Bladders and those which are filled with common Air.

But then I found, that this Spirit must be kept in good thick Bladders, as in those of an Ox, or the like; for if I filled Calves Bladders therewith, it would lose its Inflammability in 24 Hours, though the Bladder became not relax at all.

J. Clayton, 'An Experiment concerning the Spirit of Coals', *Phil. Trans.*, 1739, **41**, 59.

4c Murdoch's Experiments on Coal-gas

X. An Account of the Application of the Gas from Coal to economical Purposes. By Mr. William Murdoch. *Communicated by* the Right Hon. Sir Joseph Banks, *Bart. K.B.P.R.S.*

Read February 25, 1808.

The facts and results intended to be communicated in this Paper, are founded upon observations made, during the present winter, at the cotton manufactory of Messrs. PHILIPS and LEE at Manchester, where the light obtained by the combustion of the gas from coal is used upon a very large scale; the apparatus for its production and application having been prepared by me at the works of Messrs. BOULTON, WATT, and Co. at Soho.

The whole of the rooms of this cotton mill, which is, I believe, the most extensive in the United Kingdom, as well as its counting-houses and store-rooms, and the adjacent dwelling-house of Mr. LEE, are lighted with the gas from coal. The total quantity of light used during the hours of burning, has been ascertained, by a comparison of shadows, to be about equal to the light which 2500 mould candles of six in the pound would give; each of the candles, with which the comparison was made consuming at the rate of 4-10ths of an ounce (175 grains) of tallow per hour.

The quantity of light is necessarily liable to some variation, from the difficulty of adjusting all the flames, so as to be perfectly equal at all times; but the admirable precision and exactness with which the business of this mill is conducted, afforded as excellent an opportunity of making the comparative trials I had in view, as is perhaps likely to be ever obtained in general practice. And the experiments being made upon so large a scale, and for a considerable period of time, may, I think, be assumed as a sufficiently accurate standard for determining the advantages to be expected from the use of the gas lights under favourable circumstances.

It is not my intention, in the present Paper, to enter into a particular description of the apparatus employed for producing the gas; but I may observe generally, that the coal is distilled in large iron retorts, which during the winter season are kept constantly at work, except during the intervals of charging; and that the gas, as it rises from them, is conveyed by iron pipes into large reservoirs, or gazometers, where it is washed and purified, previous to its being conveyed through other pipes, called mains, to the mill. These mains branch off into a variety of ramifications (forming a total length of several miles), and diminish in size, as the quantity of gas required to be passed through them becomes less. The burners, where the gas is consumed, are connected with the above mains, by short tubes, each of which is furnished with a cock to regulate the admission of the gas to each burner, and to shut it totally off when requisite. This latter operation may likewise be instantaneously performed, throughout the whole of the burners in each room, by turning a cock, with which each main is provided, near its entrance into the room. . . .

The introduction of this species of light into the establishment of Messrs. PHILIPS and LEE, has been gradual; beginning in the year 1805, with two rooms of the mill, the counting-houses, and Mr. LEE's dwelling-house. After which, it was extended through the whole manufactory, as expeditiously as the apparatus could be prepared.

At first, some inconvenience was experienced from the smell of the unconsumed, or imperfectly purified gas, which may in a great measure be attributed to the introduction of successive improvements in the construction of the apparatus, as the work proceeded. But since its completion, and since the persons to whose care it is confided, have become familiar with its management, this inconvenience has been obviated, not only in the mill, but also in Mr. LEE's house, which is most brilliantly illuminated with it, to the exclusion of every other species of artificial light.

The peculiar softness and clearness of this light, with its almost unvarying intensity, have brought it into great favour with the work people. And its being free from the inconvenience and danger, resulting from the sparks and frequent snuffing of candles, is a circumstance of material importance, as tending to diminish the hazard of fire, to which cotton mills are known to be much exposed.

The above particulars, it is conceived, contain such information, as may tend to illustrate the general advantages attending the use of the gas light; but nevertheless the Royal Society may perhaps not deem it uninteresting to be apprized of the circumstances which originally gave rise in my mind to its application, as an economical substitute for oils and tallow.

It is now nearly sixteen years, since, in a course of experiments I was making at Redruth in Cornwall, upon the quantities and qualities of the gases produced by distillation from different mineral and vegetable substances, I was induced by some observations I had previously made upon the burning of coal, to try the combustible property of the gases produced from it, as well as from peat, wood, and other inflammable substances. And being struck with the great quantities of gas which they afforded, as well as with the brilliancy of the light, and the facility of its production, I instituted several experiments with a view of ascertaining the cost at which it might be obtained, compared with that of equal quantities of light yielded by oils and tallow.

My apparatus consisted of an iron retort, with tinned copper and iron tubes through which the gas was conducted to a considerable distance; and there, as well as at intermediate points, was burned through apertures of varied forms and dimensions. The experiments were made upon coal of different qualities, which I procured from distant parts of the kingdom, for the purpose of ascertaining which would give the most economical results. The gas was also washed with water, and other means were employed to purify it.

In the year 1798, I removed from Cornwall to Messrs. BOULTON, WATT, and Co's. works for the manufactory of steam engines at the Soho Foundry, and there I constructed an apparatus upon a larger scale, which during many successive nights was applied to the lighting of their principal building, and various new methods were practised, of washing and purifying the gas.

These experiments were continued with some interruptions, until the peace of 1802, when a public display of this light was made by me in the illumination of Mr. BOULTON's manufactory at Soho, upon that occasion.

Since that period, I have, under the sanction of Messrs. BOULTON, WATT, and Co. extended the apparatus at Soho Foundry, so as to give light to all the principal shops, where it is in regular use, to the exclusion of other artificial light; but I have preferred giving the results from Messrs. PHILIPS' and LEE's apparatus, both on account of its greater extent, and the greater uniformity of the lights, which rendered the comparison with candles less difficult.

At the time I commenced my experiments, I was certainly unacquainted with the circumstance of the gas from coal having been observed by others to be capable of combustion; but I am since informed, that the current of gas escaping from Lord DUNDONALD's tar ovens had been frequently fired; and I find that Dr. CLAYTON, in a Paper in Volume XLI of the Transactions of the Royal Society, so long ago as the year 1739, gave an account of some observations and experiments made by him, which clearly manifest his knowledge of the inflammable property of the gas, which he denominates 'the spirit of coals;' but the idea of applying it as an economical substitute for oils and tallow, does not appear to have occurred to this gentleman, and I believe I may, without presuming too much, claim both the first idea of applying, and the first actual application of this gas to economical purposes.

W. MURDOCH, 'An Account of the Application of the Gas from Coal to Economical Purposes', *Phil. Trans.*, 1808, **98**, 124.

4d Winsor's Demonstration Apparatus

By courtesy of the Institution of Gas Engineers

D. E. Chandler, *Lighting by Gas*, London, 1936, p. 16.

4e Committee on the Gas-Light and Coke Company's Bill

Minutes of Evidence taken before the Committee, to whom the Bill to incorporate certain Persons for procuring Coke, Oil, Tar, Pitch, Ammoniacal Liquor, Essential Oil, and Inflammable Air, from Coal; and for other purposes; was committed

COMMITTEE ON THE GAS-LIGHT AND COKE COMPANY'S BILL. VENERIS, 5° DIE MAIJ 1809. SIR JAMES HALL IN THE CHAIR

The Petition of Mr. Murdock, was read.

Agent for the Petitioner - - -	Mr. James Weston.
Counsel for the Bill - - - -	Messrs. Warren & Harrison.
Counsel for the Petition - - -	Mr. Brougham.

Mr. Accum
Called, and examined by Mr. Harrison, Counsel for the Bill

Mr. Accum I Believe you are a chemist, and have been for some time engaged in assisting the Committee in experiments on the consumption of coal?—I have.

State to the Committee the process by which coal is decomposed, and the products that are derived from that decomposition.—The coal is submitted in close vessels to the action of the heat of ignition, and by that process these products are obtained.

Will you state the products you allude to?—The products are inflammable gas, tar, pitch, essential oil, and an ammoniacal liquor, and in the distillatory vessel remains behind the well-known substance coke.

What quantity of coke is produced in proportion to the coal used?—It may be made to increase to double the bulk of the coal.

What is the quantity of coke produced by the decomposition of a chaldron of coals, according to Mr. Winsor's process?—One chaldron of coal will produce one chaldron and three quarters of coke.

Mr. Accum

Are you speaking of relative quantities by measure?—I am.

Have you made experiments on the coke produced in one of Mr. Winsor's stoves?—I have made them repeatedly.

State to the Committee the result of those experiments.—With regard to the coke, I have compared it with other combustible matters usually employed; I have compared it with coal, and from those experiments I have reason to state that the quantity of heat given by a given quantity of coke, if you compare them weight for weight, is as three to one, provided the coke has been well prepared; but this must depend greatly upon the nature of the coke produced: if the coke have been well prepared, the proportion I found usually was at three to two.

Is the stove made by Mr. Winsor so constructed as to prepare coke of good quality?—It is.

Is it equal or superior to coke made by the ordinary process?—It is superior to that kind of coke which I have hitherto been able to procure in the market.

Now will you state its relative power of producing heat by bulk?—It must depend upon the shape of the coke. If large quantities are given in the form of a cylinder, it must make a material difference than if given in the form of a globe or other shapes; I could certainly say that coke is as three to two, by experiments I have frequently made.

Compare a bushel of coal and a bushel of coke together, and what will be the result?—It would require three bushels of coal to produce the same quantity of heat as is produced by two bushels of coke.

State any experiments you have made to ascertain that fact.—I have taken a measured quantity of water, and I have applied heat to that water by means of coke, until all the water was evaporated. In this way I learnt the quantity of one kind of fuel necessary to evaporate such a given quantity of water as a standard, for a certain quantity of heat is absolutely necessary to produce the effect. I have then made the same experiment with coal, and I found that to evaporate a like quantity of water it required two of coke to three of coal; that is, three bushels of coal to evaporate a given quantity of water, whereas two bushels of coke are only required.

Have you ascertained its quality, by using it in the smelting of ores?—I have.

What is the result of any experiment you have made in smelting, as to the quality of the coke?—I have smelted a given quantity of lead ore, and several other kinds of ore, as ore of copper. I have tried it in ore of iron likewise, and I found that when coke was made use of, instead of common coal, I could effect the process in half the time that I could when coal was employed, and little more than half the quantity of coke was necessary to accomplish my object.

Have you made any experiments with coke made in Mr. Winsor's stove, and coke you have procured in the course of your operations as a chemist, so as to ascertain the relative value of the one and the other?—The coke which I have been able to obtain in the market was of so indifferent a nature, that it hardly could be compared with the coke obtained from Mr. Winsor's stove.

You have no difficulty in saying that the coke obtained from Mr. Winsor's stove is of very superior quality?—It is far superior.

Is a coke fire easily lighted for the use of houses and families?—Equally easy as a coal fire, provided the coke has not been charred too much, or has not been exposed to too intense a heat.

It is a fire that can be used comfortably and pleasantly in private houses, both for purposes of cooking and for use in rooms?—I should have no objection to use it in my own family, and in my own laboratory, and should give the

Mr. Accum preference on account of the less trouble it requires, and its longer duration and intensity of heat.

I believe it produces no smoke?—There is no smoke at all, hardly.

And of course it does not make the same dirt?—Certainly.

Is there any unpleasant smell attending the use of it in private rooms?—None in the coke of Mr. Winsor, but in the coke of commerce there is a considerable sulphureous odour.

Does the coal which is called in the North 'Slack,' which is considered as refuse coal, make coke of good quality?—It answers the purpose exceedingly well, for it is the same substance as the coal, only comminuted, that is, in small particles.

Does the construction of Mr. Winsor's stove produce different kinds of coke?—Coke may be produced according to the will of the proprietor by means of that stove.

But I am asking now if two descriptions of coke, one nearer the fire, and the other further from the action of the fire, are not produced in the ordinary operation of decomposing a quantity of coal for producing gas light?—Certainly there are.

Will you state the difference; whether one is not better for the founder, and the other for private use?—The one kind which is produced, and which has been exposed to a less heat, possesses the property of burning in a pleasant flame, and will therefore answer the purpose of the kitchen for culinary purposes, and wherever a coke fire is required, which is easily lighted, and will yield a pleasant fire. The other, on the contrary, is superior for the purpose of the smelter, for the laboratory, for the furnace, and in all situations where intense heat is demanded, and particularly if the bellows be applied to assist the combustion of it.

Has the coke made in Mr. Winsor's stove been subjected to the action of the blast pipe, so as to ascertain if it be fit for all purposes of smelting where the most powerful heat is required?—The fusing of the metallic ores was effected in that manner by means of a blast.

B.P.P., 1809, (3).

4f An Early Nineteenth-century Gas-works

The principal parts of the apparatus are the retort, hydraulic main, cooler, purifier, and gasometer. The design on the following page shews the arrangement of the apparatus when looking at the front of the furnace, or the anterior ends of the retorts:—*a* is the furnace, with a set of five retorts, from each of which a tube proceeds upwards perpendicularly to a height of from ten to twelve feet; it then takes a curve downwards, and enters the hydraulic main *b*. The latter is an iron tube of from twelve to fourteen inches in diameter, placed horizontally: the extremities of the tubes from the retorts dip about two inches below the surface of the liquid in the hydraulic main. From one end of this tube, a little above the centre, proceeds the tube *c*, into a close cistern *d*, in which the tar and ammoniacal liquor are received and separated. From the top of the opposite end of the hydraulic main proceeds the tube *e*, by which the gas is conducted into the cooler *f*, and having deposited here a further quantity of condensable matter, the gas is passed on to the lime purifier *g*, and from thence, commonly, to the gasometer *h*, but sometimes through other apparatus, not represented in the annexed figure, but which will be described afterwards. The construction of each of the parts of a coal-gas apparatus admits of certain modifications, and requires to be considered minutely.

By courtesy of the Chemical Society

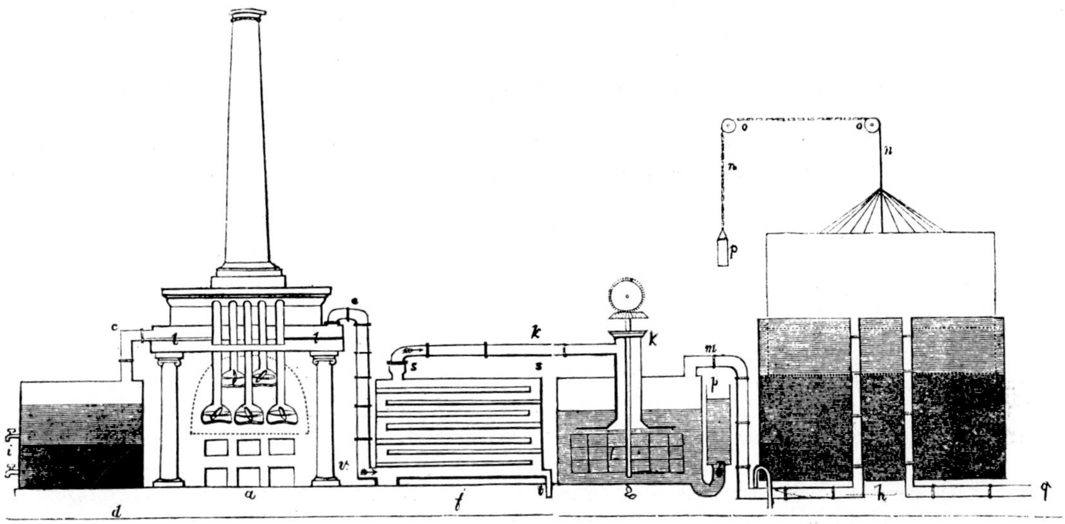

E. A. Parnell, *Applied Chemistry*, London, 1844, p. 63.

4g Inflation and Coal-tar

WHAT SHALL WE DO WITH OUR TAR?

BY LEWIS T. WRIGHT, C.E.

Among the various products of the distillation of coal in gas-works is a black, more or less viscid material known to us as tar. In some other manufactures tars are produced, but they are not discussed in the present paper.

This substance, that in the early history of gas undertakings was a waste product, became, through the discoveries of chemists and the enterprise of manufacturers, a storehouse of wealth and beauty. A few years ago, when the purchasing powers of the world for articles of necessity and luxury alike seemed to be almost without a limit, some of the products obtainable from coal-tar held a very considerable commercial value, and tar reached the price of 65s. per ton; yet, when the diminished demands of the purchasing populations first became evident to the great bulk of larger industries, tar products maintained their prices for a short time as though they were not to be affected as other commodities; but in a little while the channels through which so much had poured now began to evidence signs of choking, and increasing pressure became more and more necessary to get the products to flow away. Rapid then was the fall in prices—the irresistible force of which no efforts could stay.

The first individual to feel this enormous reduction of prices was the tar distiller; but the shadow cast over his shoulder by coming events fell across the threshold of the gas-maker, who suffered all the pangs of alarm before being actually smitten with the full blast of the storm, and then loud and indiscriminate, but quite unjust, complaints were hurled against tar distillers, dealers, and merchants, for what was thought to be their share in the fall of prices. When these were high and certain gas undertakings were receiving large sums for their tar and liquor, much injudicious jubilation was indulged in, and the inflated statements emanating from certain corners of the gas world became distorted and magnified images in the eyes of the public, who regarded gas-works as establishments wherein coal-gas was produced for nothing, and who firmly believed that the selling price of the gas was pure net profit. The enterprises called forth by the desire of capitalists to have a share in these streams of pure unalloyed profit did much to intensify, and even precipitate, the immense depression of gas products we are now so much deploring. . . .

We may fairly assume that the total quantity of tar going into distillers' hands is 97,175,000 gallons. It would surprise many people if they were told that tar only contains about 5 per cent. of substances having positive value as materials entering into chemical manipulation, and that the great bulk goes for common purposes, such as creosoting, lubrication, and fuel.

Surely the day will arrive when the great bulk, instead of a select portion of the coal-tar hydrocarbons, will have positive chemical uses, and who can doubt but there are immense triumphs yet to be gained by our chemists and manufacturers in this direction? . . .

It would be useless here to enter into the history of all the attempts that have been made to work up tar as a gas making material. Every gas maker has tried to make gas out of tar and failed. The obstructions with this material are even more obstinate than with the oils, and no one yet has been found who could deal with them. The results I have obtained with a pair of retorts charged with lumps of caustic lime average 10,700 cubic feet of 12½ candle gas. Tar as a gas-making material is, as every one is aware, dear at a gift. . . .

That there has been an over-production of tar, or that the price has been regulated by any other causes than the laws of supply and effective demand, I do not believe. There is no evidence of serious accumulation of stocks of tar or tar products. All seem to be absorbed in their proper markets. Should it not be a source of gratification to us that in this severe commercial crisis the industries, acting as intermediaries between tar makers and consumers of the finally manufactured forms of tar products, have been able to find purchasers for their articles? There are no signs of an extinction of the uses of these things, and when the present rage of competition passes away, there is every reason to believe that the manufacturers in question will again become profitable to all concerned. In order to arrive at a decision, whether or not to burn tar, a gas undertaking can only be influenced by plain business considerations of its economy or otherwise as a fuel. To burn it out of pique, because a certain price cannot be realised for it, or with the idea of forcing up the price by restricting the market, is unworthy of a public body.

LEWIS T. WRIGHT, 'What Shall We Do with Our Tar?' *J. Soc. Chem. Ind.*, 1886, **5**, 558.

4h Kidd's Examination of Naphthalene

XVI. Observations on Naphthaline, a peculiar substance resembling a concrete essential oil, which is apparently produced during the decomposition of coal tar, by exposure to a red heat. By J. Kidd, *M.D. Professor of Chemistry, Oxford. Communicated by* W. H. Wollaston, *M.D.F.R.S.*

Read March 8, 1821.

ALTHOUGH the existence, and many of the properties of the substance abovementioned, have been already noticed in two of the Philosophical Journals of this country,* there has not yet appeared, as far as I can discover, any systematic description of the mode by which it may be obtained, or of its relation to the substance from which it is produced; on which account I have been induced to offer to the Royal Society the following observations respecting these points of its history.

In the experiments which led, in the present instance, to the detection of the substance in question, it was proposed to effect the decomposition of coal tar by passing its vapour through an ignited iron tube; and, in order to increase to the utmost the extent of the ignited surface, that portion of the tube which was constantly kept up to a red heat, was filled, in the first instance, with a series of hollow iron cylinders open at both extremities, and successively decreasing in diameter, so as to be included one within another. In other instances these cylinders were removed, and their place supplied by sand, or by pieces of

* Thomson's Annals of Philosophy, January, 1820, page 74; and Mr. Brande's Quarterly Journal, January, 1820, page 287.

well burnt coke, or by pieces of brick; but it was found that the interstices between the cylinders, or between the particles of sand, &c. were so soon choaked up with carbon from the decomposition of the tar, as to be rendered absolutely impervious to the gas produced during the decomposition; so that it became necessary to pass the vapour of the tar simply through the tube itself.

Connected with the tube in which the tar was decomposed was a vessel, in which any undecomposed vapour of the tar, or any products resulting from its decomposition, might be condensed; and at the end of every experiment this condensing vessel was found to contain an aqueous fluid having an ammoniacal odour, and a dark coloured liquid, resembling tar in appearance. . . .

A pint of this dark coloured liquid was submitted to very slow distillation in a large glass retort connected with a large glass receiver, from the interior of which all communication with the external air was excluded by means of a common safety valve. The heat was supplied from the flame of an Argand gas burner, and was so slight as scarcely to inconvenience the naked hand, when held over it immediately under the bottom of the retort.

The same degree of heat was applied constantly during forty hours; at the end of which time there had distilled into the receiver rather more than half a pint of a liquid, which consisted of two perfectly distinct portions, which, however, had uniformly passed over together from the very commencement of the distillation.

The uppermost of these portions, in appearance, resembled pale olive oil, and amounted to not quite a quarter of a pint. The lowermost portion resembled water, but was not perfectly transparent, and amounted to rather more than a quarter of a pint: but there is ground for believing, from the results of subsequent distillations, that the proportion of the aqueous product is variable; and that it is greater when the distillation is carried on slowly, than when it is carried on rapidly.

After the above-mentioned products had passed over, a concrete substance as white as snow began to collect in dispersed crystalline flocculi, in the upper part of the body and neck of the retort, so as in a short time almost wholly to obstruct the passage; the oily fluid and the water continuing to pass over at the same time, but much more slowly than before. . . .

Properties of the white concrete substance

Taste, pungent and aromatic.

It is particularly characterized by its odour, which is faintly aromatic, and not unlike that of the narcissus and some other fragrant flowers. This odour is readily diffused through the surrounding atmosphere to the distance of several feet, and obstinately adheres for a long time to any substance to which it has been communicated.

When in its purest state, and reduced to powder, it is exceedingly smooth and slightly unctuous to the touch; is perfectly white, and of a silvery lustre.

Sp. gr. rather greater than that of water.

It does not very readily evaporate at the common atmospherical temperature: for, a comparison being made between this substance and camphor, in the quantity of half a grain of each in a very minute state of division, it was found that the camphor had entirely disappeared at the end of 18 hours, while the substance in question had not disappeared entirely at the end of 4 days.

A quantity of it being exposed to heat, in a glass vessel, soon melted; but did not begin to boil till the temperature had reached 410° of FAHRENHEIT: the heat being then withdrawn, it remained liquid till cooled down to 180; at which point the lowest portion was seen suddenly to congeal: the remaining portion congealed gradually; and when the whole had become solid, its temperature was 170°. The structure of the congealed mass was distinctly crystalline, and the crystalline laminæ were slightly flexible.

It is not very readily inflamed; but when inflamed it burns rapidly, and emits an unusually copious and dense smoke, which soon breaks into distinct particles that fall down in every direction. . . .

Of all the characters of the white concrete substance described in this section, its ready disposition to crystallize is perhaps the most remarkable. . . .

It remains for me to propose a name for the white concrete substance which has been described in this paper: and, unless a more appropriate term should be suggested by others, I would propose to call it naphthaline.

J. Kidd, 'Observations on Naphthaline . . . ', *Phil. Trans.*, 1825, **111**, 209.

4i Faraday's Discovery of Benzene

XX. On new compounds of carbon and hydrogen, and on certain other products obtained during the decomposition of oil by heat. By M. Faraday, *F.R.S. Cor. Mem. Royal Academy of Sciences of Paris, &c.*

Read June 16, 1825.

The object of the paper which I have the honour of submitting at this time to the attention of the Royal Society, is to describe particularly two new compounds of carbon and hydrogen, and generally, other products obtained during the decomposition of oil by heat. My attention was first called to the substances formed in oil at moderate and at high temperatures, in the year 1820; and since then I have endeavoured to lay hold of every opportunity for obtaining information on the subject. A particularly favourable one has been afforded me lately through the kindness of Mr. Gordon, who has furnished me with considerable quantities of a fluid obtained during the compression of oil gas, of which I had some years since possessed small portions, sufficient to excite great interest, but not to satisfy it.

It is now generally known, that in the operations of the Portable Gas Company, when the oil gas used is compressed in the vessels, a fluid is deposited, which may be drawn off and preserved in the liquid state. The pressure applied amounts to 30 atmospheres; and in the operation, the gas previously contained in a gasometer over water, first passes into a large strong receiver, and from it, by pipes, into the portable vessels. It is in the receiver that the condensation principally takes place; and it is from that vessel that the liquid I have worked with has been taken. The fluid is drawn off at the bottom by opening a conical valve: at first a portion of water generally comes out, and then the liquid. It effervesces as it issues forth; and by the difference of refractive power it may be seen, that a dense transparent vapour is descending through the air from the aperture. The effervescence immediately ceases; and the liquid may be readily retained in ordinary stoppered, or even corked bottles; a thin phial being sufficiently strong to confine it. I understand that 1000 cubical feet of good gas yield nearly one gallon of the fluid. . . .

A thermometer being introduced into another portion of the fluid, heat was applied, so as to keep the temperature just at the boiling point. When the vessel containing it was opened, it began to boil at 60° F. As the more volatile portions were dissipated, the temperature rose: before a tenth part had been thrown off, the temperature was above 100°. The heat continued gradually to rise, and before the substance was all volatized, it had attained 250°.

With the hope of separating some distinct substances from this evident mixture, a quantity of it was distilled, and the vapours condensed at a temperature of 0° into separate portions, the receiver being changed with each rise of 10° in the retort, and the liquid retained in a state of incipient ebullition. In this way a succession of products were obtained; but they were by no means constant; for the portions, for instance, which came over when the fluid was boiling from 160° to 170°, when redistilled, began to boil at 130°, and a part remained which did not rise under 200°. By repeatedly rectifying all these portions, and adding similar products together, I was able to diminish these differences of temperature, and at last bring them more nearly to resemble a series of substances of different volatility. During these operations I had occasion to remark, that the boiling point was more constant at, or between 176° and 190°, than at any other temperature; large quantities of fluid distilling over without any change in the degree; whilst in other parts of the series it was constantly rising. This induced me to search in the products obtained between these points for some definite substance, and I ultimately succeeded in separating a new compound of carbon and hydrogen, which I may by anticipation distinguish as bi-carburet of hydrogen.

BI-CARBURET OF HYDROGEN

This substance was obtained in the first instance in the following manner: tubes containing portions of the above rectified products were introduced into a freezing mixture at 0°; many of them became turbid, probably from the presence of water; one, received at 176°, (by which is meant that that was the boiling point of the contents of the retort when it came over) became partly solid, crystals forming round the side, and a fluid remaining in the centre; whilst two other portions, one received at 186°, and the other at 190°, became quite hard. A cold glass rod being introduced into one of these tubes, the mass within was found to resist considerable pressure; but by breaking it down, a solid part was thrust to the bottom of the tube, whilst a fluid remained above: the fluid was poured off, and in this way the solid portion partly purified. The contents of the tube were then allowed to fuse, were introduced into a larger and stronger tube, furnished with another which entered loosely within it, both being closed of course at the lower end; then again lowering the temperature of the whole to 0°, bibulous paper was introduced, and pressed on to the surface of the solid substance in the large tube by the end of the smaller one. In this way much fluid was removed by successive portions of paper, and a solid substance remained, which did not become fluid until raised to 28° or 29°. To complete the separation of the permanently fluid part, the substance was allowed to melt, then cast into a cake in a tin foil mould, and pressed between many folds of bibulous paper in a BRAMAH's press, care having been taken to cool the paper, tin foil, flannel, boards, and other things used, as near to 0° as possible, to prevent solution of the solid substance in the fluid part to be removed. It was ultimately distilled from off caustic lime, to separate any water it might contain. . . .

Bi-carburet of hydrogen appears in common circumstances as a colourless transparent liquid, having an odour resembling that of oil gas, and partaking also of that of almonds. Its specific gravity is nearly 0.85 at 60°. When cooled to about 32° it crystallizes, becoming solid; and the portions which are on the sides of the glass exhibit dendritical forms. By having tubes containing thin solid films of it in ice-cold water, and allowing the temperature to rise slowly, its fusing point was found to be very nearly 42° F.; but when liquid it may, like water and some saline solutions, be cooled much below that point before any part becomes solid. It contracts very much on congealing, 9 parts in bulk becoming 8 very nearly; hence its specific gravity in that state is about 0.956. At 0° it appears as a white or transparent substance, brittle, pulverulent, and of the hardness nearly of loaf sugar. . . .

With regard to the composition of this substance, my experiments tend to prove it a binary compound of carbon and hydrogen, two proportionals of the former element being united to one of the latter. The absence of oxygen is proved by the inaction of potassium, and the results obtained when passed through a red hot tube.

The following is a result obtained when it was passed in vapour over heated oxide of copper. 0.776 grains of the substance produced 5.6 cubic inches of carbonic acid gas, at a temperature of 60°, and pressure 29.98 inches; and 0.58 grains of water were collected. The 5.6 cubic inches of gas are equivalent to 0.711704 grains of carbon by calculation, and the 0.58 grains of water to 0.064444 of hydrogen.

Carbon . .	0.711704 or 11.44
Hydrogen .	0.064444 or 1.

These quantities nearly equal in weight the weight of the substance used; and making the hydrogen 1, the carbon is not far removed from 12, or two proportionals. . . .

The hydrogen present therefore in this compound is equivalent to 3 volumes, though condensed into one volume in union with the carbon; and of the latter elements there are present six proportionals, or 36 by weight. A volume therefore of the substance in vapour contains

Carbon	-	$6 \times 6 = 36$
Hydrogen	-	$1 \times 3 = 3$
		39

and its weight or specific gravity will be 39, hydrogen being 1. Other experiments of the same kind gave results according with these.

M. FARADAY, 'On New Compounds of Carbon and Hydrogen', *Phil. Trans.*, 1825, **115**, 440.

4j Mansfield's Researches on Coal-tar

XXI. Researches on Coal Tar. By Charles Blachford Mansfield, *B.A.*

PART I

Introductory remarks.—The products of the destructive distillation of matters of complex and indeterminate composition are deficient in the interest which attaches to products, the constitutional relations of which to the bodies from whose decomposition they result, may be traced in the phenomena attending the process, or in the nature of the original substances. The absence of retrospective interest is, however, compensated to some extent in the case of pyrogenous products, which are obtained in large quantities from established processes in the arts; firstly, when they are viewed as primitive materials demanding, with the rest of nature, to be interpreted and applied by human industry, (for the artificial is only a particular case of, and nowise distinct from, the natural); and, secondly, by the prospective interest connected with the distinct substances of much scientific importance, which may be isolated from them in abundance, but which could only be obtained in minute quantities from the sources which afforded them in the first instance. It is remarkable, however, that modern chemistry, with its rapid strides, has left on this extensive field but few foot-prints, whose impressions are distinct and likely to be durable. Although numerous researches, connected with very eminent names, show that the importance of a correct natural history of the substances known as tars, has not escaped the consideration of chemists, yet the meagre and indefinite details on these subjects, which are to be found in our scientific manuals, indicate that much yet remains to be accomplished. . . .

These considerations would render it unnecessary that Dr. A. W. Hofmann should back with much persuasion his suggestions to a student to undertake the further examination of coal tar. The researches, a portion of which it is the object of the present paper to communicate, were commenced at the request of Dr. Hofmann, and were conducted, thus far, chiefly in the laboratory of the College of Chemistry, under his valued and friendly advice. It has been, perhaps, the tedium of the methods necessary to effect a separation of mixed hydrocarbons from each other, which has deterred experienced chemists from devoting their time to disentangling the oils here treated of: and, perhaps, to have conducted the innumerable distillations necessary for this purpose in a laboratory imperfectly furnished with gas and other conveniences, would have been a task too laborious to have been persisted in. . . .

Of the examination of the light coal naphtha.—The method, which was actually adopted in the laboratory, for the separation of the substances in the *coal*-oil, will be first described, as it is, perhaps, the only method which promises much security for the detection of all the fluids present in such mixtures. It is in principle the same as that which Faraday applied to the dissection of the *oil*-oil. The details here given will render unnecessary many repetitions, which would otherwise be requisite in future communications, in which it is intended to relate the results of an investigation of the other tar products. Subsequently, more compendious methods will be pointed out, by which some of the bodies, once detected, may afterwards be practically prepared for experiment. The object of this investigation being the neutral oils, the first step was to obtain them free from basic and acid substances. For this purpose, two gallons of the crude light coal naphtha were agitated frequently with a gallon of a cold mixture, of about nine parts of water, with one of sulphuric acid, which was so far diluted to destroy any tendency it might possess to decompose, or to form compounds with, the hydrocarbons. After the separation of the fluids, the oil was drawn off from the acid, and washed with water. It was then similarly treated with a dilute solution of potash, and again washed. This treatment with acid and alkali having been repeated, the oil had now lost much of its bad smell; an improvement in this respect having followed the application of each of these agents.*

* Before treatment with acid, the light oil would tinge a piece of white deal yellow, or, if the wood had been previously dipped in hydrochloric acid, purplish red. The first of these reactions was probably due to some salts of aniline; the second, perhaps, to the substance indicated by Runge as pyrrol. It had also the power of destroying the texture of the human cuticle, which Reichenbach mentions as belonging to creosote (Annalen der Pharmacie, Bd. 4, s. 275), and which is possessed more strongly by the crude heavy coal naphtha; so that the skin of the hands, after being touched with it, becomes dry and scales off. These properties had disappeared from the purified oil, which had, however, but little diminished in bulk.

The light naphtha thus purified, was distilled in convenient quantities of about a quart at a time in glass retorts, a thermometer being kept immersed in the fluid, and regular ebullition being constantly maintained. The boiling commenced at about 100° (212° F.). The thermometer continuing to rise constantly as the distillation proceeded; when the temperature had reached 150° (302° F.), about two-thirds of the contents of the retorts had distilled over, the greater portion of the remainder coming over below 200° (392° F.). When the mercury stood at a little above 200° (392° F.), the distillate began to solidify on cooling, and was now heavier than water. Attention is for the present restricted to the fluids received previously to the appearance of solid matter in the condenser. From similar results obtained from other samples, it was evident that, what is usually called 'coal naphtha,' consists chiefly of a mixture of oils having boiling points below 200° (392° F.)* (and, consequently, more volatile than naphthaline, which boils at 212°) (414° F.); so that the subject of present inquiry is conveniently defined. The distillates were mostly colourless, or nearly so, a pale yellow appearing in the latter portions, the brown colouring matter remaining concentrated in the residue. During the distillation, the receivers were changed at regular intervals of five degrees (9° F.) of increasing temperature.

A series of products were thus obtained, consisting of mixtures gradually decreasing in volatility. These were then submitted to a second distillation, in turn, commencing with the first received; the vessels were again changed at every 5th degree of heat (9° F.) in the retorts. . . .

Of the preparation of benzole from coal naphtha.—The fifth series of the distillations, just spoken of, yielded about a pint of fluid, which had boiled between 80° and 90° (176°–194° F.). It was clear and colourless, of a powerful spirituous odour, in which a refined sense would recognise the fragrance of benzole, though a vulgar organ might be reminded of coal gas.

A small quantity of the fluid received on the fifth distillation between 85° and 90° (185° and 194° F.), being exposed to cold, at about five degrees below 0° (32° F.), in a test-tube, soon became in part solid and crystalline. The fraction received between 80° and 85° (176° and 185° F.) behaved similarly, while specimens corresponding to temperatures below 80° (176° F.) and above 90° (194° F.), shewed no signs of solidification, except such of the least volatile as contained naphthaline, which they partially deposited. On being submitted, in a freezing mixture, to a temperature of − 20° (4° F.) the sample representing the boiling *line* 75°–80° (167°–176° F.) yielded a portion of crystalline matter, and those from 80°–90° (176°–194° F.) became almost entirely solid, while those from 90°–95° (194°–203° F.), and upwards, still suffered not the slightest change of state.

It was now almost certain that benzole was secured, and that the solidification was due to that substance. It was evident, however, that the crystalline product thus obtained would be a mixture of the desired matter with substances having boiling points more or less near to that of the hydrocarbon, in which they were present as impurities; and, that, unless these fluids had the same freezing point as benzole, or would form compounds with it which are solid at the same temperature, they might be separated from it by pressure and filtration. The whole of those fractions of the fifth order of distillates, which had been received between 80° and 90° (176°–184° F.), were treated accordingly.

* The heavy coal naphtha consists chiefly of hydrocarbons, boiling between 200° and 300° (392°–572° F.), with more or less of paranaphthaline in solution.

A very simple apparatus—one of Beart's coffee-pots—effected the purpose. This machine consists of a hollow cylindrical vessel, into which a perforated piston fits tightly, supporting a piece of strong cloth, so as to form an apparatus, in which rapid filtration is effected by the atmospheric pressure, when the piston is forcibly raised through the fluid. The application is obvious. The filter-piston is placed at the bottom of the vessel, which is immersed in a mixture of snow and salt. The fluid to be purified is now poured into the pot, or, better still, is first exposed, in large test-tubes, to the freezing mixture, and from them, when solidified, is transferred to the filter. The crystalline mass may be expelled from the tubes very readily, by holding their mouths downwards, and slightly warming their sides and closed ends with the hand, when a minute quantity, becoming vapourized, will expand, and will force out the solid in the form of mould candles of benzole. The temperature is kept at about 10° or 12° below 0° (14°–11° F.) in the filtering vessel, while its contents are stirred and broken down, to prevent passages being left for the air through the mass when the piston is raised, it being essential that the pressure of the substance against the filter should be as great as can be ensured.

When the piston is drawn up, a crystalline mass, like snow in whiteness and texture, is separated from the fluid oil which has passed through the filter. This product is placed in a funnel, the bulb of a thermometer being buried in it, and the temperature is allowed to rise slowly while the thawing fluid is received below. The melting proceeds rapidly at first; more slowly as the temperature approaches 0° (32° F.), which point it will pass before the whole has fused. The receivers are changed at intervals of increasing temperature, and their contents, being separately subjected to another refrigeration, and again liquefied in like manner, yield further portions which remain solid above 0° (32° F.). These last, together with that which in the first instance was still congealed at 0° (32° F.), prove to be nearly pure benzole. It is usually found that the thermometer indicates 4° or 5° above 0° (39°–41° F.) before all the substance in contact with its bulb is fused. The portion that quits the solid state last, is found to be a fluid which crystallizes now at a temperature a little above 0° (32° F.), and boils *constantly* at 80° (175° F.) in glass vessels with pieces of platinum wire at the bottom, distilling over without leaving a trace of residue in the retort. It is, therefore, a pure substance.

It remains to confirm the identity of this substance with benzole, which its properties so strongly indicate. . . .

Of a practical mode of preparing benzole.—The boiling point of benzole is the same as that of alcohol of sp. gr. 0·825 (176°* F.=80° C.); it is evident, therefore, that any of the summary processes of rectification which are practised by distillers in the manufacture of alcoholic spirits, are applicable to the separation of benzole from the less volatile fluids of the naphtha. The method now to be described is, one which extracts nearly the whole of this spirituous hydrocarbon, by the very slight expenditure of time.

The light coal naphtha (of which it is convenient to take a sample which came over at the commencement of the distillation of the tar, which contains the benzole less diluted with substances having higher boiling points) is placed in a metal retort, which is surmounted by an open vessel filled with water, and containing a worm or chamber, into which the vapour of the naphtha passes directly from the retort, and so arranged that the less volatile fluids, which will be condensed in it, will flow back into the retort, or into a separate receiver,

* Ure, Dictionary of Chemistry, ed. 1835, art. Alcohol, p. 131.

while the fluids more volatile than water will pass on in vapour to another condenser, which is kept as cold as possible. The water surrounding the still head will gradually rise in temperature as the operation proceeds, and will at last boil; and when this takes place, (or if the heating of the water be checked at any period of the process, by addition of cold water,) distillation will cease, for no fluid remains in the retort whose vapour is not condensible at the temperature of the head, which can never rise above 100° (212° F.). The distillate being rectified a second time in such an apparatus, keeping now the temperature of the head at or a little below 80° (176° F.), and reserving the part which comes over before the temperature in the retort has passed 90° (194° F.), a large proportion of a very volatile oil will be obtained, which, when exposed to a temperature of −20° (4° F.), will become solid to at least half its bulk. Of course, if the apparatus be double, the same result may be obtained by a single operation. A comparatively small quantity of a similar product may be obtained by receiving separately the first two or three gallons that run off from the common stills on the rectification of some hundred gallons of the light naphtha.

C. B. Mansfield, 'Researches on Coal Tar', *J. Chem. Soc.*, 1849, **1**, 244.

4k Mansfield's Distillation Apparatus

Crown Copyright. Science Museum, London

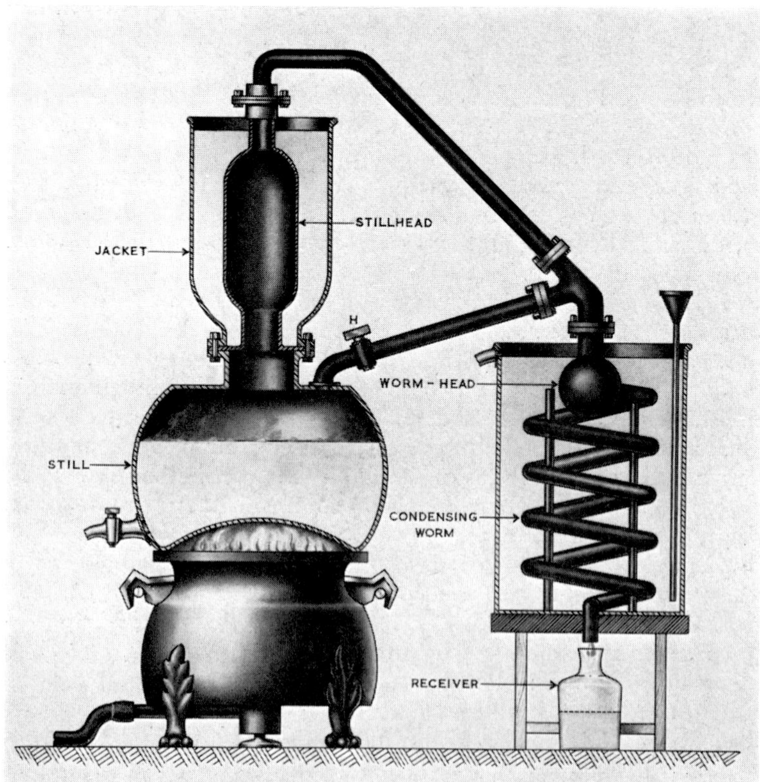

Print at the Science Museum.

41 The First Synthetic Dyestuffs

The Origin of the Coal-Tar Colour Industry, and the Contributions of Hofmann and his Pupils. By W. H. Perkin, *Ph.D., D.C.L., F.R.S.*

. . . This industry holds an unique position in the history of chemical industries, as it was entirely the outcome of scientific research. We have to go back to 1825, when Faraday discovered benzene, or, as he then termed it 'bicarburetted hydrogen,' for the first investigation which clearly bears upon the subject. Faraday separated the hydrocarbon from the liquid products condensed on compressing the gas obtained from oil. A year later (1826), Unverdorben obtained aniline by the mere distillation of indigo, and called it 'crystalline.' Runge afterwards obtained it from coal-tar oil, and having observed that it produced a violet-blue coloration with chloride of lime, called it 'kyanol.' It was subsequently obtained from indigo by Fritsche by distilling this colouring matter with caustic alkali. We then come to the important work of Mitscherlich, who obtained the hydrocarbon benzene from a new source, namely, benzoic acid,—whence the name, and produced from this nitrobenzene. Zinin subsequently found that *benzidam*, as he termed it, could be produced by the action of sulphuretted hydrogen in presence of ammonia on an alcoholic solution of nitrobenzene.

This brings us to the commencement of Hofmann's researches on aniline, a substance which he used sometimes to speak of as his 'first love.' In his first published paper he showed that Unverdorben's crystalline, Runge's kyanol, Fritsche's aniline, and Zinin's benzidam were all the same compound, for which he afterwards selected Fritsche's name, aniline. Later on, Hofmann and Muspratt prepared toluidine from toluene from tolu balsam.

The work on the separation of aniline from tar was done before the date of Hofmann's coming to this country, viz., in 1843. After his arrival here in 1845, he continued his researches, and, to realise something of his indomitable perseverance, it is necessary to remember that, until the coal-tar colour industry was established, practically all the aniline he used in his numerous enquiries was procured by the laborious and costly process of distilling indigo with potash.

In 1843, organic chemistry was still in its infancy, and coal-tar naphtha had not yet been investigated. Runge had isolated carbolic acid, pyrrol, kyanol or aniline, and leucol or quinoline. Naphthalene was well known to exist in tar, having been separated by Garden, as early as 1820. Dumas had discovered paranaphthalene or anthracene, and chrysene and pyrene had been referred to by Laurent, but these were very doubtful compounds. This was about all that was known of the composition of coal-tar at that time. Hofmann showed, in 1845, that benzene must exist in the naphtha, as he found that aniline could be produced from it, but he never separated this hydrocarbon; shortly afterwards, however, he induced his pupil, Charles Mansfield—of whom he always spoke in the highest terms—to undertake the investigation of the liquid hydrocarbons of coal-tar.

On reading over the account of Mansfield's investigation, and bearing in mind that in those days fractional distillation was conducted in old fashioned glass retorts with the thermometer in the liquid, it is impossible not to admire the patience and perseverance he exercised, as well as the systematic and skilful manner in which he worked.

All who have undertaken fractional distillations, even with all our present knowledge and improved apparatus, know how difficult it is to detect and isolate products in a mixture such as coal-tar naphtha. Yet Mansfield obtained benzene in a pure state, and toluene sufficiently so for Hofmann to prepare toluidine from it. He also obtained pseudocumene, and was led to believe in the existence of xylene. In describing his work, he modestly remarks:—

> It has been perhaps the tedium of the methods necessary to effect a separation of mixed hydrocarbons from each other, which has deterred experienced chemists from devoting their time to disentangling the oils here treated of: and perhaps to have conducted the innumerable distillations necessary for this purpose in a laboratory imperfectly furnished with gas and other conveniences, would have been a task too laborious to have been persisted in. (*J. Chem. Soc.*, 1849, **1**, 246.)

Amongst the enquiries carried on by Hofmann, in the early days of the Royal College of Chemistry, were those classical 'researches regarding the molecular constitution of the volatile organic bases,' in which he succeeded in displacing the hydrogen of the NH_2-group by different alcohol radicals, eventually obtaining also the ammonium compounds. In the first of these (*J. Chem. Soc.*, **3**, 1851) he describes ethylaniline (p. 284), and diethylaniline (p. 288), also methylaniline (p. 295). The method used in these researches, of substituting hydrogen in amines by means of the iodides and bromides of the alcohol radicles, and also the substituted anilines which were obtained, although not connected with the foundation of the coal-tar colour industry, have been of great value in its after development. These few references to observations on the early work carried on at the Royal College of Chemistry, for the sake of science *only*, show, in fact, what valuable material was produced for the coming new industry; indeed, without the research of Mansfield, it could never have become an industry. . . .

I entered the Royal College of Chemistry when I was in my fifteenth year, at the time when that institution became part of the School of Mines, but I only took up the study of chemistry. After seeing Dr. Hofmann with my father, the first person I encountered in the laboratory was the Assistant, Mr. W. Crookes, who set me to study the reactions of the metals.

There was no theatre at the Royal College then, and the students had to go to the Museum of Practical Geology in Jermyn Street to hear the lectures on chemistry, which involved a rather serious loss of time; but the lectures made up for this, as Hofmann spared no pains in making them as interesting, instructive, and perfect as he possibly could, illustrating, as far as practicable, everything by experiment, so that the facts were firmly impressed upon the mind. At that time he also had a very efficient lecture assistant, the late Mr. Witt. Hofmann was good enough to let me attend these lectures a second time.

When going through the ordinary course of qualitative and quantitative analysis, the students working at research appeared to me to be superior beings, something beyond ordinary persons; and being possessed with a desire to join their ranks, the ordinary course, and also gas analysis by Bunsen's method, was quickly gone through. Hofmann then set me to work at research, and very curiously gave me as a subject the hydrocarbon anthracene, or, as it was generally called in those days, paranaphthalene. To obtain this, pitch was taken as the starting point, but as it was found that this method of preparation was a very tedious one to carry out in the laboratory, Hofmann kindly obtained some of the crude product for me from Mr. Cliff, of Bethels Tar Works. As is well known, Hofmann—especially at that period—was much interested in the formation of organic bases from hydrocarbons, and the object of my investigation was to produce, if possible, a nitro-compound, and then convert this into a base by reduction. However, anthracene refused to give a nitro-compound, and consequently no base could be obtained, but, in the course of my work, I prepared the compound we now know as anthraquinone, and also the chlorine and bromine derivatives of anthracene. But these substances could not be got to yield intelligible results on analysis, and at that time it never occurred either to Hofmann or myself that there was any likelihood of Dumas and Laurent's formulæ for the hydrocarbon (*i.e.*, $C_{15}H_{12}$) being incorrect. The consequence was that this research was set aside, but I shall show further on that the experience I then gained was of great importance to me several years later, when I commenced to work at the production of alizarin.

Hofmann next set me to work to study the action of chloride of cyanogen on naphthylamine in the same way that he had examined the action of this gas on aniline. In those days there were no depôts where pure products for research could be obtained as there now are, and for this inquiry even the naphthalene had to be purified in the laboratory; this research was soon completed, but was not written out and published until nearly 12 months afterwards. It was brought before this Society when the meetings were held in Mr. Pepper's house in Cavendish Square.

Hofmann had a marvellous power of stimulating his students, and of imparting to them his own enthusiasm; he took the strongest personal interest in their work, visiting three or four times in the week even those who were going through the reactions, while those engaged in research work were seen daily by him, and if anything of special interest was going on, more than once in the day. His power of directing research was also most remarkable; with the aid of a few watch glasses, a glass rod, and a small gas flame he would make a number of experiments, and from the information thus gained tell his students how to proceed with their work. I well remember how one day, when the work was going on very satisfactorily with most of us and several new products had been obtained, he came up and commenced examining a product of the nitration of phenol one of the students had obtained by steam distillation; taking a little of the substance in a watch glass, he treated it with caustic alkali, and at once obtained a beautiful scarlet salt of what we now know to be orthonitrophenol. Several of us were standing by at the time, and, looking up at us in his characteristic and enthusiastic way, he at once exlaimed, 'Gentlemen, new bodies are *floating* in the air.' I mention this just as an example of the way in which he used to stimulate us by his own example. . . .

At this period much interest was taken in the artificial formation of natural organic substances; but at the time I was at the Royal College of Chemistry, although the theory of compound radicles, the doctrine of substitution, &c., were occupying much attention, very little was known of the internal structure of compounds and the conceptions as to the method by which one compound might be formed from another was necessarily very crude.

Thus, in the Report of the Royal College of Chemistry, published in 1849, Hofmann refers to the artificial formation of quinine as a great desideratum. . . .

As a young chemist I was ambitious enough to wish to work on this subject of the artificial formation of natural organic compounds. Probably from reading the above remarks on the importance of forming quinine, I began to think how it might be accomplished, and was led by the then popular additive and subtractive method to the idea that it might be formed from toluidine by first adding to its composition C_3H_4 by substituting allyl for hydrogen, thus forming allyltoluidine, and then removing two hydrogen atoms and adding 2 atoms of oxygen, thus

$$\underbrace{2(C_{10}H_{13}N)}_{\text{Allyltoluidine.}} + 3O = \underbrace{C_{20}H_{24}N_2O_2}_{\text{Quinine.}} + H_2O$$

The allyltoluidine having been prepared by the action of allyl iodide on toluidine, was converted into a salt and treated with potassium dichromate; no quinine was formed, but only a dirty reddish-brown precipitate. Unpromising though this result was, I was interested in the action, and thought it desirable to treat a more simple base in the same manner. Aniline was selected, and its sulphate was treated with potassium dichromate; in this instance a black precipitate was obtained, and, on examination, this precipitate was found to contain the colouring matter since so well known as *aniline purple* or *mauve*, and by a number of other names. All these experiments were made during the Easter vacation of 1856 in my rough laboratory at home. Very soon after the discovery of this colouring matter, I found that it had the properties of a dye, and that it resisted the action of light remarkably well.

After the vacation, experiments were continued in the evenings when I had returned from the Royal College of Chemistry, and combustions were made of the colouring matter. I showed it to my friend Church, with whom I had been working, on his visiting my laboratory, and who, from his artistic tastes, had a great interest in colouring matters, and he thought it might be valuable and encouraged me to continue to work upon it; but its evident costliness and the difficulties of preparing aniline on the large scale, made the probability of its proving of practical value appear very doubtful. Through a friend, I then got an introduction to Messrs. Pullar, of Perth, and sent them some specimens of dyed silk. On June 12, 1856, I received the following reply:—

> If your discovery does not make the goods too expensive, it is decidedly one of the most valuable that has come out for a very long time. This colour is one which has been very much wanted in all classes of goods, and could not be obtained fast on silks, and only at great expense on cotton yarns. I enclose you pattern of the *best* lilac we have on cotton—it is dyed only by one house in the United Kingdom, but even this is not quite fast, and does not stand the tests that yours does, and fades by exposure to air. On silk the colour has always been fugitive: it is done with cudbear or archil, and then blued to shade.

This somewhat lengthy extract is quoted because it gives a glimpse at the state of the dyeing trade in reference to this shade of colour at that period.

This first report was very satisfactory; the 'if' with which it commenced was, however, a doubtful point.

During the summer vacation, however, the preparation of the colouring matter on a very small, technical scale was undertaken, my brother (the late T. D. Perkin) assisting me in the operations, and, after preparing a few ounces of product, the results were thought sufficiently promising to make it desirable to patent the process for the preparation of this colouring matter. This was done on August 26, 1856 (Patent No. 1984). A visit was then made to Messrs. Pullar's and experiments on cotton dyeing were made, but, as no suitable mordants were known for this colouring matter, only the pale shades of colour, produced by the natural affinity of the dye for the vegetable fibre, were obtained; these, however, were admired. Experiments on calico printing were also made at some print works, but fears were entertained that it would be too dear, and, although it proved to be one of the most serviceable colours as regards fastness, yet the printers were not satisfied with it because it would not resist the action of chloride of lime like madder purple.

Although the results were not so encouraging as could be wished, I was persuaded of the importance of the colouring matter, and the result was that, in October, I sought an interview with my old master, Hofmann, and told him of the discovery of this dye, showing him patterns dyed with it, at the same time saying that as I was going to undertake its manufacture, I was sorry that

I should have to leave the Royal College of Chemistry. At this he appeared much annoyed, and spoke in a very discouraging manner, making me feel that perhaps I might be taking a false step which might ruin my future prospects. I have sometimes thought that, appreciating the difficulties of producing such compounds as aniline and this colouring matter on the large scale, Hofmann perhaps anticipated that the undertaking would be a failure, and was sorry to think that I should be so foolish as to leave my scientific work for such an object, especially as I was then but a lad of 18 years of age; and I must confess that one of my great fears on entering into technical work was that it might prevent my continuing research work, but I determined that, as far as possible, this should not be the case.

Still, having faith in the results I had obtained, I left the College of Chemistry and continued my experiments, and found that not only aniline, but also toluidine, xylidine, and cumidine gave a purple colouring matter when oxidised.

The following is a copy of the principal part of the complete specification of the patent I took out at this time:—

DYEING FABRICS

The nature of my invention consists in producing a new colouring matter for dyeing with a lilac or purple colour stuffs or silk, cotton, wool, and other materials in the manner following:—

I take a cold solution of sulphate of aniline, or a cold solution of sulphate of toluidine, or a cold solution of sulphate of xylidine, or a cold solution of sulphate of cumidine, or a mixture of any one of such solutions with any others or other of them, and as much of a cold solution of a soluble bichromate as contains base enough to convert the sulphuric acid in any of the above-mentioned solutions into a neutral sulphate. I then mix the solutions and allow them to stand for 10 or 12 hours, when the mixture will consist of a black powder and a solution of a neutral sulphate. I then throw this mixture upon a fine filter, and wash it with water till free from the neutral sulphate. I then dry the substance thus obtained at a temperature of 100° C., or 212° F., and digest it repeatedly with coal-tar naphtha, until it is free from a brown substance which is extracted by the naphtha. Any other substance than coal-tar naphtha may be used in which the brown substance is soluble and the colouring matter is not soluble. I then free the residue from the naphtha by evaporation, and digest it with methylated spirit, or any other liquid in which the colouring matter is soluble, which dissolves out the new colouring matter. I then separate the methylated spirit from the colouring matter by distillation, at a temperature of 100° C. or 212° F.

Fresh quantities of colouring matter were prepared and taken to Scotland, and, although the method of applying it by means of lacterine (casein) was then found to give very good results, yet the printers who tried it did not show any great enthusiasm; and even Messrs. Pullar began to fluctuate in their opinion as to the advisability of erecting plant for its manufacture, and wrote:—'Should it appear that it will not be of service to printers, it will be questionable whether it would be wise to erect works for the quantity dyers alone will require.' In January, 1867, Mr. R. Pullar, however, advised me to see Mr. Thos. Keith, a silk dyer of Bethnal Green, London, and, after making a few experiments with the colouring matter, and exposing the specimens he dyed to the light for some time, he was much pleased with the result, and encouraged me to go on with its production.

I was then joined in the undertaking by my father—who was a builder, and had sufficient faith in the project to risk the necessary capital—and also by my brother, who also had a good knowledge of building, and, as he had taken part in the preliminary experiments on the preparation of the dye, his assistance proved most valuable, especially as he was possessed of good business capabilities. Plans were prepared and a site obtained at Greenford Green, near Harrow, and in June, 1857, the building of the works was commenced.

At this time, neither I nor my friends had seen the inside of a chemical works, and whatever knowledge I had was obtained from books. This, however, was not so serious a drawback as at first it might appear to be, as the kind of apparatus required and the character of the operations to be performed were so entirely different from any in use that there was but little to copy from.

In commencing this manufacture, it was absolutely necessary to proceed tentatively, as most of the operations required new kinds of apparatus to be devised and tried before more could be ordered to carry out the work on any scale.

W. H. Perkin, 'The Origin of the Coal-tar Colour Industry, and the Contributions of Hofmann and his Pupils', *J. Chem. Soc.*, 1896, **69**, 596.

4m 'On Saccharine'

Meeting held Tuesday, 2nd February

MR. WATSON SMITH IN THE CHAIR

ON SACCHARINE—A NEW SWEETENING AGENT DERIVED FROM COAL TAR

BY IVAN LEVINSTEIN

I am desirous of bringing under your notice this evening a substance which, known for the last few years only to specialists as a chemical curiosity* on account of the difficulty of its production, is likely, in consequence of further improvements by its discoverer in the process of manufacture, to enter the market as a commercial product.

Modern chemistry has achieved a new triumph—one, indeed, the commercial importance of which cannot yet be fully realised. From the black viscous liquid, gas tar, has been prepared, an article which seems likely to enter into daily consumption. We have scarcely as yet become fully familiar with the ever-increasing number of coal-tar dyes, with the production of aromatic substances and those possessing powerful antiseptic and medicinal properties from the same prolific source, when the fresh and startling announcement is made that, by a novel treatment of one of the components of coal tar, a compound may be obtained possessing very similar sweetening properties—but in a far more intense degree—to the best cane or beetroot sugar. The commercial name given to this singular substance by its discoverer, C. Fahlberg, a German chemist residing in the United States, is 'saccharine.' It is not my intention this evening to diverge into any speculations as to the probable economic effects of this wonderful discovery, great as the temptation may be; I shall confine myself to a brief description of the mode of production, and the general and physiological properties of saccharine, and I will also give you an opportunity of inspecting the substance itself and some of its compounds. The chemical name of saccharine is benzoyl sulphonic imide, and it may be expressed by the chemical formula

$$C_6H_4\left\langle{CO \atop SO}\right\rangle NH.$$

In order to obtain saccharine, Fahlberg† commences with toluene, a hydrocarbon present, as you are probably all aware, in coal tar. This toluene $C_6H_5(CH_3)$ he converts into its two monosulphonic acids

$$C_6H_4\left\langle{CH_3 \atop SO_2}\right.$$

and transforms these by treating them with phosphorus-pentachloride (PCl_5) into the corresponding toluene-sulphonic chlorides

$$C_6H_4\left\langle{CH_3 \atop SO_2Cl.}\right.$$

Of these two toluene-sulphonic chlorides—viz, the para and the ortho—only the ortho compound, which is liquid (the other being solid), is suitable for the production of saccharine. By the introduction of the amido group of ammonia, this ortho-toluene-sulphonic-

* American Chemical Journal, vol. 1, page 170; and vol. 2, page 181, and fol. I, page 425.

† Eng. Pat. 6626, June 1, 1885; specification of James Yate Johnson. This Journal, iv. p. 608.

chloride is then converted into ortho-toluene-sulphamide

$$C_6H_4\langle\begin{matrix}CH_3\\SO_2NH_2,\end{matrix}$$

which finally yields, by oxidation, the benzoyl sulphonic imide or saccharine.

Saccharine presents the appearance of a white powder, and crystallises from its aqueous solution in thick short prisms, which are with difficulty soluble in cold water, but more easily in warm. Alcohol,* ether, glucose, glycerol, etc., are good solvents of saccharine. It melts at 200° C., with partial decomposition; its taste in diluted solutions is intensely sweet, so much so that one part will give a very sweet taste to 10,000 parts of water. Saccharine forms salts, all of which possess a powerful saccharine taste; it is endowed with moderately strong antiseptic properties, and is not decomposed in the human system, but eliminated from the body without undergoing any change. It is about 230 times sweeter than the best cane or beetroot sugar. According to Dr. Stutzter, of Bonn, who has carefully investigated the physiological properties of this substance, saccharine, taken into the stomach in the quantities in which it has to be added to food as a sweetening material, has no injurious effect whatever on the human system. Stutzter has given to dogs about 5grms. a-day, without observing any ill effects in them, and when we consider that 5grms. are equal in sweetening power to rather more than 2¼lb. of sugar—a quantity far larger than anyone would like to consume in a day—his views seem amply corroborated by this fact alone; but, in addition to this, patients suffering from diabetes have now been treated for several months in one of the principal hospitals in Berlin, as I am informed, without their feeling the least inconvenience by its use. Physicians must be glad to find in saccharine a substance by means of which diabetic persons may enjoy food which has hitherto not been admissible in their case. Saccharine does not belong to the class of carbohydrates, and does not possess nutritious properties. The use of saccharine will, therefore, as indicated by its properties, be not merely as a probable substitute for sugar, but it may even be applied to medicinal purposes where sugar is not permissible. The inventor was fully aware that in order to supply a perfect substitute for cane or beetroot sugar, something selse—viz., a similar substance—was needed for confectionery and similar purposes, besides sweetening properties, and he has also endeavoured to solve this problem. Dr. Fahlberg combines glucose with starch sugar, a substance very similar to cane or beetroot sugar, as you know, but inferior to these in sweetening properties, with saccharine, and thus obtains a compound which he calls 'dextro-saccharine,' which, as far as the taste is concerned, is scarcely distinguishable from the best sugar. The quantity of 'saccharine' used is in the proportion of one part to from 1000 to 2000 parts of glucose. Now, if we look at the price of saccharine, which is, I believe, at present about 50s. a-pound, we shall find that even at this price such a mixture would be very considerably cheaper than real sugar, but we must bear in mind the fact that there is great likelihood of the process of manufacture of saccharine being considerably cheapened; the expensive phosphorus-pentachloride may probably be replaced by a cheaper compound, etc.

It will then be evident, from what you have heard, not only that 'saccharine' is a most interesting compound, but that it may also be destined to become an article of primary commercial importance. I must leave to the future to decide as to the revolutions it may bring about in the coal-tar industry, in the cultivation of the soil now devoted to growing canes or beets, and in the sugar industry generally, and other industries connected with it; but as great and important commercial interests are in question, it is highly advisable to look well into this matter, and not to allow our foreign competitors in this and other markets to secure exclusively for themselves the benefit which this discovery may confer. I have here for your inspection a specimen of pure crystallised 'saccharine,' and also small balls made from starch, to which has been added 0.05 per cent. of saccharine, of which one is sufficient to impart a very sweet taste, very similar to that of the best sugar, to a large cup of black coffee. Further, some liquors sweetened with 'saccharine,' some lozenges made from 'dextro-saccharine,' and also some raspberry syrup sweetened with saccharine; and finally, a small bottle of quinine, the bitter taste of which has been disguised by the addition of a small quantity of 'saccharine.' These samples will give you an idea of the various and manifold uses to which saccharine may be put, and which may result in an enormous consumption of the article.

* The solubility of saccharine in ether might probably be useful in detecting its presence when mixed with sugar.

I. Levinstein, 'On Saccharine—A New Sweetening Agent derived from Coal Tar', *J. Soc. Chem. Ind.*, 1886, **5**, 75.

5 ATOMISM
A Recurring Theme in Nineteenth-century Science

5a Atomic Symbols

By courtesy of the Chemical Society

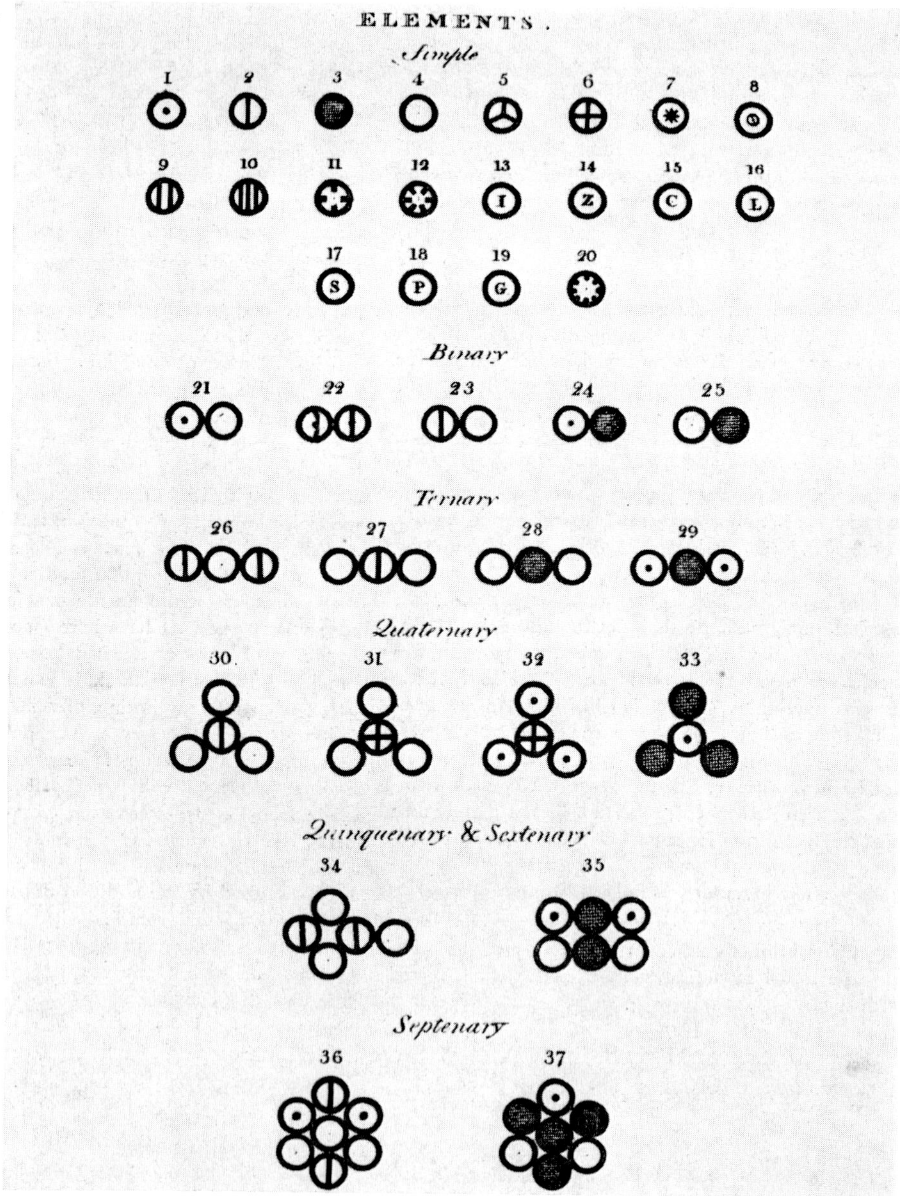

JOHN DALTON, 'On the Constitution of Bodies', *A New System of Chemical Philosophy*, Manchester, 1808, p. 139.

5b Wollaston's 'Scale of Chemical Equivalents'

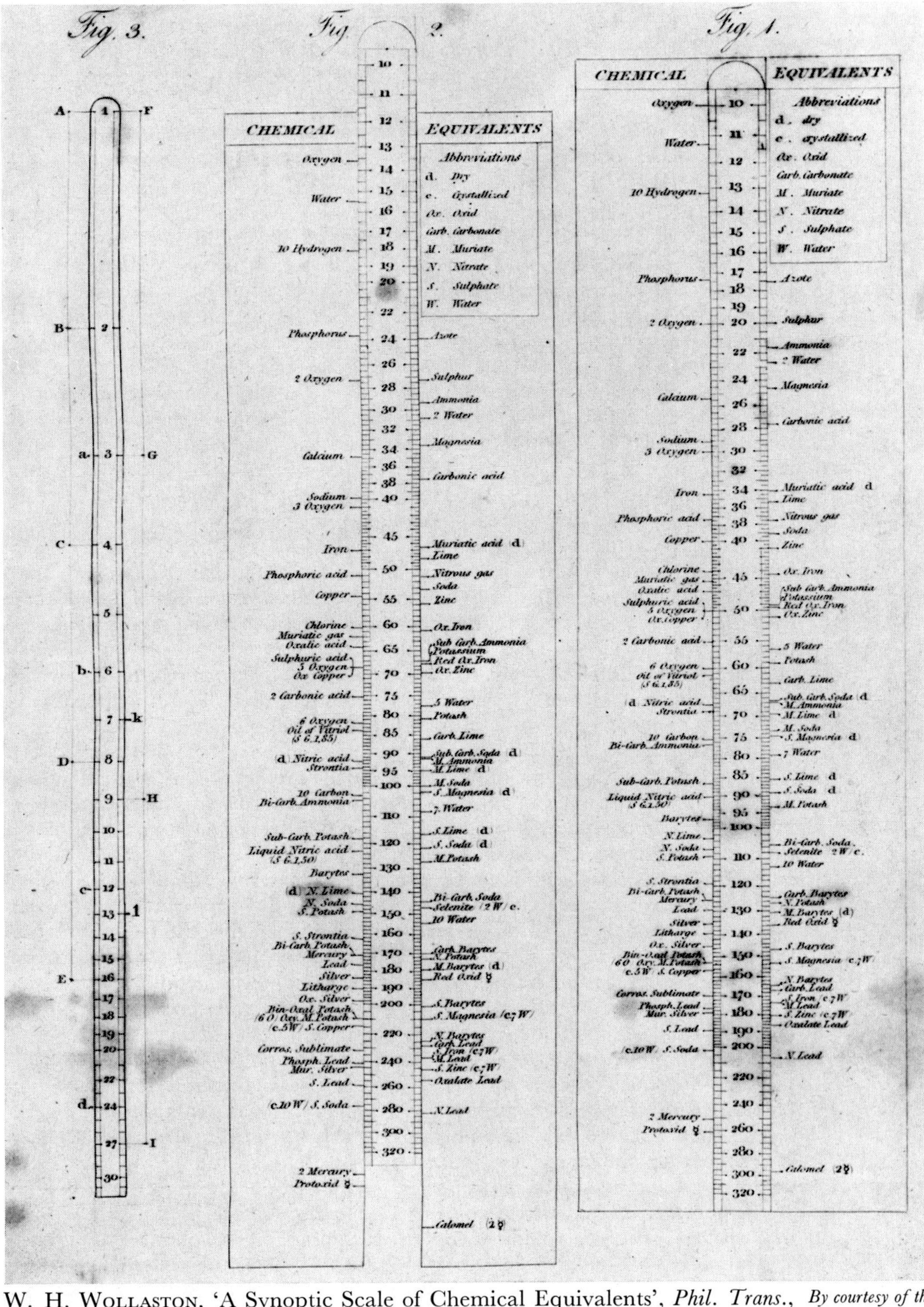

W. H. WOLLASTON, 'A Synoptic Scale of Chemical Equivalents', *Phil. Trans.*, 1814, **104**, plate I, p. 22.

By courtesy of the Chemical Society

5c Berzelius on 'Chemical Proportions'

Essay on the Cause of Chemical Proportions, and on some Circumstances relating to them: together with a short and easy Method of expressing them. By Jacob Berzelius, *M.D., F.R.S. Professor of Chemistry at Stockholm. (Continued from* Vol. II, p. 454)

III. ON THE CHEMICAL SIGNS, AND THE METHOD OF EMPLOYING THEM TO EXPRESS CHEMICAL PROPORTIONS

WHEN we endeavour to express chemical proportions, we find the necessity of chemical signs. Chemistry has always possessed them, though hitherto they have been of very little utility. They owed their origin, no doubt, to the mysterious relation supposed by the alchymists to exist between the metals and the planets, and to the desire which they had of expressing themselves in a manner incomprehensible to the public. The fellow-labourers in the anti-phlogistic revolution published new signs founded on a reasonable principle, the object of which was, that the signs, like the new names, should be definitions of the composition of the substances, and that they should be more easily written than the names of the substances themselves. But, though we must acknowledge that these signs were very well contrived, and very ingenious, they were of no use; because it is easier to write an abbreviated word than to draw a figure, which has but little analogy with letters, and which, to be legible, must be made of a larger size than our ordinary writing. In proposing new chemical signs, I shall endeavour to avoid the inconveniences which rendered the old ones of little utility. I must observe here that the object of the new signs is not that, like the old ones, they should be employed to label vessels in the laboratory: they are destined solely to facilitate the expression of chemical proportions, and to enable us to indicate, without long periphrases, the relative number of volumes of the different constituents contained in each compound body. By determining the weight of the elementary volumes, these figures will enable us to express the numeric result of an analysis as simply, and in a manner as easily remembered, as the algebraic formulas in mechanical philosophy.

The chemical signs ought to be letters, for the greater facility of writing, and not to disfigure a printed book. Though this last circumstance may not appear of any great importance, it ought to be avoided whenever it can be done. I shall take, therefore, for the chemical sign, the *initial letter of the Latin name of each elementary substance:* but as several have the same initial letter, I shall distinguish them in the following manner:—1. In the class which I call *metalloids,* I shall employ the initial letter only, even when this letter is common to the metalloid and to some metal. 2. In the class of metals, I shall distinguish those that have the same initials with another metal, or a metalloid, by writing the first two letters of the word. 3. If the first two letters be common to two metals, I shall, in that case, add to the initial letter the first consonant which they have not in common: for example, S = sulphur, Si = silicium, St = stibium (antimony), Sn = stannum (tin), C = carbonicum, Co = cobaltum (cobalt), Cu = cuprum (copper), O = oxygen, Os = osmium, &c.

The chemical sign expresses always one volume of the substance. When it is necessary to indicate several volumes, it is done by adding the number of volumes: for example, the *oxidum cuprosum* (protoxide of copper) is composed of a volume of oxygen and a volume of metal; therefore its sign is Cu + O. The *oxidum cupricum* (peroxide of copper) is composed of 1 volume of metal and 2 volumes of oxygen; therefore its sign is Cu + 2 O. In like manner, the sign for sulphuric acid is S + 3 O; for carbonic acid, C + 2 O; for water, 2 H + O, &c.

When we express a compound volume of the first order, we throw away the +, and place the number of volumes above the letter: for example, Cu O + S $\overset{3}{\mathrm{O}}$ = sulphate of copper, Cu $\overset{2}{\mathrm{O}}$ + 2 S $\overset{3}{\mathrm{O}}$ = persulphate of copper. These formulas have this advantage, that if we take away the oxygen we see at once the ratio between the combustible radicles. As to the volumes of the second order, it is but rarely of any advantage to express them by formulas as one volume; but if we wish to express them in that way, we may do it by using the parenthesis, as is done in algebraic formulas: for example, alum is composed of 3 volumes of sulphate of alumina and 1 volume of sulphate of potash. Its symbol is 3 (Al $\overset{2}{\mathrm{O}}$ + 2 S $\overset{3}{\mathrm{O}}$) + (P$\overset{2}{\mathrm{o}}$ + 2 S $\overset{3}{\mathrm{O}}$). As to the organic volumes, it is at present very uncertain how far figures can be successfully employed to express their composition. We shall have occasion only in the following pages to express the volume of ammonia. It is 6 H + N + O, or $\overset{6}{\mathrm{H}}$ N O. . . .

EXPLANATION OF SOME COMPOUND CHEMICAL SIGNS

S + 2 O. S + 3 O.—Sulphurous acid, sulphuric acid.
M + 2 O. M + 3 O.—Muriatic acid, oxymuriatic acid.
2 H + O. 6 H + N + O.—Water, ammonia.
N + O. N + 3 O.—Nitrous oxide, nitrous gas.
As + 3 O. As + 6 O.—Arsenic oxide, arsenic acid.
Fe + 2 O. Fe + 3 O.—Black oxide of iron, red oxide of iron.
Po + 2 O. Po + 6 O.—Potash, peroxide of potassium.
2 H + S. As + 12 S.—Sulphureted hydrogen, supersulphuret of arsenic.
Sb + 3 S. St + 2 S.—Sulphuret of antimony, sulphuret of tin.
2 N $\overset{6}{\mathrm{O}}$ + Po $\overset{2}{\mathrm{O}}$.—Nitrate of potash.
2 S $\overset{3}{\mathrm{O}}$ + Po $\overset{2}{\mathrm{O}}$.—Sulphate of potash.
S $\overset{3}{\mathrm{O}}$ + Cu O.—Sulphate of copper.
As $\overset{6}{\mathrm{O}}$ + Cu $\overset{2}{\mathrm{O}}$.—Perarseniate of copper.
M $\overset{2}{\mathrm{O}}$ + $\overset{6}{\mathrm{H}}$ N O + $\overset{2}{\mathrm{H}}$ O.—Hydrous muriate of ammonia.
2 Te $\overset{2}{\mathrm{O}}$ + Pb $\overset{2}{\mathrm{O}}$.—Telluriate of lead.
3 M $\overset{2}{\mathrm{O}}$ + Sb $\overset{3}{\mathrm{O}}$.—Muriate of antimony.
N $\overset{6}{\mathrm{O}}$ + 3 Pb $\overset{2}{\mathrm{O}}$.—Subnitrate of lead at a minimum.
M $\overset{2}{\mathrm{O}}$ + 2 Cu $\overset{2}{\mathrm{O}}$.—Persubmuriate of copper.

J. J. BERZELIUS, 'Essay on the Cause of Chemical Proportions . . . Part III', *Ann. Phil.*, 1814, **3**, 51.

5d Whewell on 'The Atomic Theory'

. . . 8. *Arguments for and against Atoms.*—The leading arguments on the two sides of the question, in their most general form, may be stated as follows:— *For* the Atomic Doctrine.—The appearances which nature presents are compounded of many parts, but if we go on resolving the larger parts into smaller, and so on successively, we must at last come to something simple. For that which is compound can be so no otherwise than by composition of what is simple; and if we suppose all composition to be removed, which hypothetically we may do, there can remain nothing but a number of simple substances, capable of composition, but themselves not compounded. That is, matter being dissolved, resolves itself into atoms.

Against the Atomic Doctrine.—Space is divisible without limit, as may be proved by geometry; and matter occupies space, therefore matter is divisible without limit, and no portion of matter is *indivisible*, or an *atom*.

And to the argument on the other side just stated, it is replied that we cannot even hypothetically divest a body of composition, if by composition we mean the relation of point to point in space. However small be a particle, it is compounded of parts having relation in space.

The Atomists urge again, that if matter be infinitely divisible, a finite body consists of an infinite number of parts, which is a contradiction. To this it is replied, that the finite body consists of an infinite number of parts in the same sense in which the parts are infinitely small, which is no contradiction.

But the opponents of the Atomists not only rebut, but retort this argument drawn from the notion of infinity. Your atoms, they say, are indivisible by any finite force; therefore they are infinitely hard; and thus your finite particles possess infinite properties. To this the Atomists are wont to reply, that they do not mean the hardness of their particles to be infinite, but only so great as to resist all usual natural forces. But here it is plain that their position becomes untenable; for, in the first place, their assumption of this precise degree of hardness in the particles is altogether gratuitous; and in the next place, if it were granted, such particles are not atoms, since in the next moment the forces of nature may be augmented so as to divide the particle, though hitherto undivided.

Such are the arguments for and against the Atomic Theory in its original form. But when these atoms are conceived, as they have been by Newton, and commonly by his followers, to be solid, hard particles exerting attractive and repulsive forces, a new set of arguments come into play. Of these, the principal one may be thus stated: According to the Atomic Theory thus modified, the properties of bodies depend upon the attractions and repulsions of the particles. Therefore, among other properties of bodies, their hardness depends upon such forces. But if the hardness *of the bodies* depends upon the forces, the repulsion, for instance, of the particles, upon what does the hardness *of the particles* depend? what progress do we make in explaining the properties of bodies, when we assume the same properties in our explanation? and to what purpose do we assume that the particles *are* hard?

9. *Transition to Boscovich's Theory.*—To this difficulty it does not appear easy to offer any reply. But if the hardness and solidity of the particles be given up as an incongruous and untenable appendage to the Newtonian view of the Atomic Theory, we are led to the theory of Boscovich, according to which matter consists not of solid particles, but of mere mathematical centres of force. According to this theory, each body is composed of a number of geometrical points from which emanate forces, following certain mathematical laws in virtue of which they become, at certain small distances attractive, at certain other distances repulsive, and at greater distances attractive again. From these forces of the points arise the cohesion of the parts of the same body, the resistance which it exerts against the pressure of another body, and finally the attraction of gravitation which it exerts upon bodies at a distance.

This theory is at least a homogeneous and consistent mechanical theory, and it is probable that it may be used as an instrument for investigating and expressing true laws of nature; although, as we have already said, the attempt to identify the forces by which the particles of bodies are bound together with mechanical attraction appears to be a confusion of two separate ideas.

10. *Use of the Molecular Hypothesis.*—In this form, representing matter as a collection of molecules or centres of force, the Atomic Theory has been abundantly employed in modern times as an hypothesis on which calculations respecting the elementary forces of bodies might be conducted. When thus employed it is to be considered as expressing the principle that the properties of bodies depend upon forces emanating from immovable points of their mass. This view of the way in which the properties of bodies are to be treated by the mechanical philosopher was introduced by Newton, and was a natural sequel to the success which he had obtained by reasoning concerning central forces on a large scale. I have already quoted his Preface to the *Principia*, in which he says, 'Many things induce me to believe that the rest of the phenomena of nature, as well as those of astronomy, may depend upon certain forces by which the particles of bodies, in virtue of causes not yet known, are urged towards each other and cohere in regular figures, or are mutually repelled and recede; and philosophers, knowing nothing of these forces, have hitherto failed in their examination of nature.' Since the time of Newton, this line of speculation has been followed with great assiduity, and by some mathematicians with great success. In particular Laplace has shown that it may, in many instances, be made a much closer representation of nature, if we suppose the forces exerted by the particles to decrease so rapidly with the increasing distance from them, that the force is finite only at distances imperceptible to our senses, and vanishes at all remoter points. He has taught the method of expressing and calculating such forces, and he and other mathematicians of his school have applied this method to many of the most important questions of physics; as a capillary action, the elasticity of solids, the conduction and radiation of heat. The explanation of many apparently unconnected and curious observed facts by these mathematical theories gives us a strong assurance that its essential principles are true. But it must be observed that the actual constitution of bodies as composed of distinct and separate particles is by no means proved by these coincidences. The assumption, in the reasoning, of certain centres of force acting at a distance, is to be considered as nothing more than a method of reducing to calculation that view of the constitution of bodies, which supposes that they exert force at *every* point. It is a mathematical artifice of the same kind as the hypothetical division of a body into infinitesimal parts, in order to find its centre of gravity; and no more implies a physical reality than that hypothesis does.

W. WHEWELL, *Philosophy of the Inductive Sciences*, London, 1840, vol. i, p. 414.

5e The Chemical Society's Atomic Debate

Discussion on Dr. Williamson's Lecture on The Atomic Theory, November 4th, 1869. Sir Benjamin C. Brodie, Bart., *F.R.S., in the Chair*

THE Chairman said the business before the Society was to consider a paper 'On the Atomic Theory,' which had already been read before the Society by Professor Williamson. This theory was not to be regarded as a fixed and definite system of ideas, but one that had undergone great variation and change. There was the atomic theory of Dalton; then that of Berzelius, and also, that of Laurent and Gerhardt; and now they had come to a further form of this theory, in which new ideas had arisen in regard to the nature of atoms and their properties; and he (the Chairman) thought it would greatly facilitate the discussion if Dr. Williamson would explain what was the precise form of the atomic theory which he was prepared to maintain and defend, and how they might discriminate that from other forms of the theory, some of which Dr. Williamson himself would admit to be neither reasonable nor rational. First, there was the question whether there were atoms at all; then, being atoms, what those atoms were, and whether they were endowed with the properties assigned to them in regard to their replacing power, and to what was termed their value. These were the questions raised by the lecture of Dr. Williamson, and which they were now to discuss.

Professor Williamson said that he did not admit that, among chemists, there had been several atomic theories. If there had been, he had yet to learn of them. There had prevailed, he believed, from time immemorial, a belief that matter was built up of small particles; and he might also say that no contrary opinion had ever been embodied in any definite form. Dalton doubtless looked for evidences of what he believed to be true, and saw, in the simple multiple proportions which he specially alluded to, facts which were in accordance with that preconceived conviction; but his was the mere germ of an atomic theory, and the changes which have taken place in it since are consistent with the original notions. The very rapid changes had, in the main, been additions to the starting-point given by Dalton. The line of work so vigorously pursued of late years by many chemists, regarding the order in which atoms are arranged in their compounds, was mentioned by the speaker. His endeavour had been to put together more fully than had been done of late, what is known of the chemical evidence: firstly, respecting the limits to the divisibility of matter, and, secondly, with regard to the properties of those finite particles which constitute the limits of divisibility, solely from the point of view habitually employed by chemists in their ordinary working. He had also endeavoured to separate those conclusions which appeared to him warranted by facts from everything else. Whether the smallest particles of matter have a spherical form or not, whether they are in their nature indivisible, whether they are in reality the ultimate atoms of matter, or like the planets of this system, he knew not, nor did such questions exist for him as a chemist. He therefore thought it wise to exclude them, important as they were, from the actually existing atomic theory. The work of Berzelius was perhaps the most significant transition from the original notion of Dalton. Silently these atoms had been accepted by chemists, and the whole course of investigation had been one grand confirmation of the assumption, that compounds must have molecular weights corresponding to at least the smallest atomic proportion which would represent the actual numbers of analysis. The perfectly independent observations which had been made with reference to the boiling points of homologous liquids, the phenomena of diffusion, and the equality of volumes of masses containing an equal number of these molecules under similar conditions in the gaseous state: all concurred to corroborate the conclusions necessitated by the atomic theory. Thus, independent workmen hewed stones which, when hewn, were found to fit in exactly with the others, forming a perfectly homogeneous whole. Dr. Williamson concluded his remarks by referring to the atomic values, a term which, he thought, represented in a more precise way than some others what is called atomicity; and he also referred to the difference between direct and indirect combination, a fact which he conceived to be essentially atomic in its nature.

Dr. Frankland said that he was not present when Professor Williamson delivered his lecture, and he had only cursorily looked over the printed report;

but, so far as he could gather, the object of the author had been to establish the atomic theory as an absolute truth. There had been, he thought, so little, if any opposition to the application of the atomic theory to the phenomena of chemistry, that a discussion could scarcely be raised upon any other feature of the question than that of its absoluteness. He considered it impossible to get at the truth, as to whether matter was composed of small and indivisible particles, or whether it was continuous—the question belonged to what metaphysicians termed 'the unknowable;' but he acknowledged the importance of the fullest use of the theory as a kind of ladder to assist the chemist in progressing from one position to another in his science. He was, however, averse to accepting the theory as an absolute truth. Any attempt to realise by its help the action of attractive or repulsive forces upon matter was excessively difficult; indeed to realise such an action through a perfectly void space was to him quite impossible. The same difficulty presented itself to Faraday, who was often obliged to throw this atomic theory on one side, as an obstacle to his progress. He said that if matter be assumed to be thus composed of solid particles separated by a void space, then, in considering the phenomena of electricity in connection with that view of matter, this space existing between the atoms must be either a conductor or a non-conductor of electricity. If a conductor, such a thing as an insulator was obviously an impossibility; if it was a non-conductor, such a thing as a conductor was equally inconceivable; so that, apart from chemistry, there were considerations which made it very undesirable that this theory should be represented as an absolute truth. The speaker then referred to the combination which, in certain cases, is effected between gases by the application of heat, and said that, according to the atomic hypothesis, they must assume that the gases, when heated, have their particles driven further asunder; but how, then, could they more readily enter into combination when heated, as in the case of oxygen and hydrogen, and in many other similar instances? No such difficulty presented itself if the gases were regarded as continuous matter. He admired the atomic theory as much as Professor Williamson, and he thought no one could blame him for not making a sufficient use of it; but he did not wish to be considered a blind believer in the theory, or as unwilling to renounce it if anything better presented itself to assist him in his work.

Dr. Odling said that, as a chemist, he was not particularly interested in the question whether matter is infinitely divisible or not, but he did not think that Dr. Williamson's argument had established that conclusion. Dalton was always spoken of as the discoverer of the law, or doctrine, or theory of combination in definite and multiple proportions, to account for which his invention or adoption of the atomic theory was founded; but he conceived that Dalton's great discovery might be better expressed as the law of combination in *reciprocal* and multiple, instead of in *definite* and multiple, proportions. Dalton, in fact, discovered that, if we have three different kinds of matter, A, B, and C, and we find that x times A unites with y times B, and with z times C, then if B and C can combine with one another, it is in the proportion of y and z, or some simple multiples thereof. These were the great facts which Dalton discovered; he showed that they were applicable to all the cases known in his day, and they have proved applicable to all the cases that have been discovered since. Dr. Williamson had said of chemists, that all their modes of thought, and all the government of their actions, are based upon the atomic theory; be [*sic*] he (Dr. Odling) maintained that they are based upon the observed fact that certain bodies combine in certain proportions. Dr. Williamson also argued that the atomic theory is based upon the existence of molecules, and that molecules have no *locus standi* in the absence of the atomic theory. He (Dr. Odling) disputed that position; for the fact that the hydrogen existing in marsh-gas is divisible into four parts, and the hydrogen existing in ammonia only into three parts, is, he conceived, a fact quite independent of the atomic theory. The laws of combination, as far as they go, were compared by Sir Humphry Davy to Kepler's laws of planetary motion, which were simply a general expression of observed facts; and, in the same way, these laws of chemical combination are general expressions of observed facts upon which the atomic hypothesis is super-induced.

Professor Miller thought that Dr. Odling's arguments had not disproved the atomic theory. If we have not the atomic hypothesis, it may fairly be asked what explanation we possess of the laws of combination? Certain facts are admitted on all hands, as determined by experiment, and when an important and extensive series of facts has been ascertained, the endeavour to explain them by some supposition which embraces the facts, and enables us to anticipate new ones if possible, is a necessary result of our mental constitution. Unless some hypothesis is accepted, it is absolutely impossible to reason upon the facts; and, even if we cannot adopt this hypothesis as absolutely true, such a view of the constitution of matter at least explains all the chemical facts that have hitherto been presented. It appeared to him that those who deny the atomic hypothesis are bound to supply something to enable them to interpret phenomena with the same regularity and order as that hypothesis, the accuracy of which they deny and which they desire to displace. No one would suppose that the absence of an undulatory theory for light would conduce to the explanation of optical science; and he held that what the undulatory theory is to the phenomena of light, the atomic theory is to the phenomena of chemistry.

Professor Foster thought the question before the Society was not so much the utility of the atomic theory as its truth. To say that the atomic theory has been very useful, and that, if we refuse to admit it, we have nothing to put in its place, is not exactly the point. A false theory may for the time be an exceedingly useful one, but the present question is whether the atomic theory is true or false, not whether it is a useful or useless theory. With the notions which we are in the habit of entertaining as to the nature of combination, we can indeed scarcely avoid the atomic theory. Naumann has stated, in his 'Relations of Heat to Chemistry,' that, if we assume that in any compound body the components really occupy distinct portions of space, we must assume that body to be incapable of infinite divisibility. To take a concrete case, he speaks of vermilion. Assuming that in this body the sulphur and mercury occupy separate portions of space, and do not interpenetrate, then we know that so far as any actual division goes, we can separate the portion of vermilion into smaller and smaller bits, but each bit is in itself still a perfect bit of vermilion, containing sulphur and mercury. If, however, we could carry the division further, we must ultimately come to the point at which any further division would give us dissimilar portions of matter, at which, indeed, we should actually have broken off the sulphur from the mercury, and thus have arrived at an atom of this compound body vermilion. He (Professor Foster) thought, however, that the ideas generally entertained of the existence of the constituents of bodies in compounds are not absolutely necessary ideas.

The fundamental phenomenon of every chemical change may perhaps be stated in some such terms as these. We begin with some kind or kinds of matter, say oxygen and hydrogen; we perform a certain operation; these kinds of matter disappear, and another kind of matter, viz., water, begins to exist. But it is, perhaps, after all, a conventional mode of expressing the fact, to say that the oxygen and hydrogen are contained in the water. We know that between the bodies which disappear and the body which appears, there are certain relations, not only qualitative but quantitative, the total mass of the disappearing substances being equal to that of the appearing substances; but we may perhaps return, sometimes, at any rate, with great benefit, to the notion that one portion of matter is actually transmuted into another; that it ceases to exist as such, but something else comes in place of it. From such ideas the existence of atoms would not follow as a necessity, but with our present mode of stating and reasoning about chemical changes, an atomic hypothesis or basis appears to be inevitable.

Professor Tyndall sympathized with the ideas of Professor Williamson; his thoughts had, to some extent, been expressed by Dr. Miller, who had made a strong point in referring to the undulatory theory. It would be quite possible, to detach the phenomena of light from the physical theory of an ether; to say, indeed, that the undulatory theory is something superinduced upon the facts, and a mere imagination. 'Still,' said Dr. Tyndall, 'I do not think we should suffer ourselves to be fettered in that way, and I did certainly expect that Dr. Odling would have gone one step further when he referred to Davy's allusion to the laws of Kepler. Those laws stated the facts of planetary motion as clearly as we know them at the present time; but facts are not sufficient to satisfy the human mind. Another man followed Kepler, and superinduced something upon the facts—a something which we now know as the theory of gravitation. I say it is the nature of the human mind, whenever it attains to any degree of depth—it is the nature of a profound thinker when he ponders upon phenomena of this nature, to seek for some underlying principle—something altogether outside the region of facts—from which he deduces the facts as consequences. Professor Foster has very justly remarked that the usefulness of a theory is not a justification for holding what Professor Frankland has called an absolute truth, and that a bad theory is very often useful. Undoubtedly it is. The emission theory of light was useful in the hands of Newton, Malus, Laplace, and especially of Brewster; and still the emission theory fell. Why did it fall? Because in the progress of investigation phenomena were observed which were perfectly inexplicable upon the emission theory, but which have been completely explained by the undulatory theory. The undulatory theory will stand in front of all opposition as long as it is competent to explain the facts of optics; and I apprehend that the atomic theory will stand as long as it is competent to explain the facts of chemistry.'

Dr. Mills observed that to prove the truth of the atomic theory, one of the most important points to establish would be that the general analogy of nature is in favour of indivisibles, whereas the very contrary is the fact. The great kingdoms of nature, animal, vegetable, and mineral, have no positive lines, but pass into one another by imperceptible gradations, as likewise do the forms of nature, geographical, physiological, and crystallographical. In comparing the atomic theory with the undulatory theory of light, it should be remembered that the latter has this advantage, that we have often seen the phenomena of waves, and therefore if an ether exists, it may have a wave-like motion. The existence of waves is a fact; but the existence of atoms, or, in short, of a limit of any sort in nature, is not a fact. No one has ever seen an atom, and in this respect the atomic theory has an exact parallel in the theory of phlogiston. Dr. Mills further remarked that it seems to be forgotten by many writers that matter might be infinitely divisible, and yet that definite proportions might exist; for between two infinites there may be a finite ratio, so that the atomic theory is not perfectly necessary to chemistry. He believed that motion is the highest generalization of modern science, and it therefore affords the only criterion by which all theories, including those of chemistry, must be judged, and the notion of a limit is practically inconsistent with the idea of motion.

Dr. Williamson, in his reply, remarked that the benefit accruing from looking at the question from another point of view would be incalculable, and if another view were developed to supersede the present one, he would be among the first to rejoice at it, but he had not to do with the future of this matter. He had endeavoured to put together the actual evidence of the present view, and none of its opponents had been able to find a fallacy in that evidence. He thought it most important to keep the theory, at each period of its growth, in the modest position which it at present occupies, that of generalizing the best ascertained relations of the elements in their chemical changes.

The Chairman said that he liked to have clear and definite ideas, as far he was able to realize them, on scientific subjects, and he was unable to consider even an atomic theory in the abstract. The atomic theory, as he understood it, was—that any given portion of matter is made up of a number of finite particles, the aggregate of which could be divided and subdivided until at last only one indivisible particle would remain, and this view might be derived from the perusal and common assent of every work on chemistry that defined the atomic theory. He thought that such a view as the physical indivisibility of matter must be separated from the facts and basis of chemistry. Dr. Williamson seemed to think the theory and facts were one, but he (the Chairman) asserted that they are two distinct things. He must also express his dissent from the views expressed by Drs. Frankland and Miller, with reference to the use of a doctrine or theory in which they do not believe. He could not understand using a theory and denying it at the same time. This theory had, he thought, been often the means of deluding chemists into the belief that they understood things about which they knew nothing. He found the works of Kekulé and Naquet scribbled over with pictures of molecules and atoms, arranged in all imaginable ways, for which no adequate reason was given, and if there was no reason for this, it was a mischievous thing to do, for it led to a confusion of ideas, and to mixing up fictions with facts. Many students, he believed, thought the atomic pictures in Naquet's work were the fundamental things to be studied in chemistry; they did not draw the distinction between the facts and theory of the science. His view was that the ultimate constitution of this material universe was one of those questions upon which no light had yet been thrown, and he agreed with Dr. Odling when he said that the science of chemistry did not require or prove the atomic theory. Chemists were bound to have some real and adequate means of working at the science of chemistry, and discovering its laws, and they could effect this, either through the investigation of the laws of gaseous combination, or the study of the capacity of bodies for heat. When they approached the science through the study of gaseous combinations, they discovered that the combining proportions of bodies are capable of being represented by integral numbers, and they also found an analogous fact in studying the capacity of bodies for heat. Either of these methods might be made the basis of the science, and lead to the same general system of ideas, in which we retain what is valuable in the atomic theory, without committing ourselves to a number of assertions incapable of proof.

BENJAMIN C. BRODIE, 'Discussion on Dr. Williamson's Lecture on the Atomic Theory', *J. Chem. Soc.*, 1869, **22**, 433.

5f Wright on 'The Atomic Theory'

THE ATOMIC THEORY

By C. R. A. WRIGHT, D.Sc.

A CORRESPONDENT adopting the signature 'Atom' distinguishes, in your columns, between the atomic theories of Leucippus and Dalton, stating that 'the first is open to discussion, and will long be, ... the second . . . . is not. We must learn much concerning matter before we can say with certainty that it consists *primarily* of indivisible particles or atoms.' He also states that, in writing his paper, he seeks 'only to make an arrangement in discussing the Daltonian chemistry.'

From the tenour of his note, it would seem that 'Atom' does not look upon the views propounded by Dalton, and subsequently extended by others in quite the same light as other writers. Dalton himself ('New System of Chemical Philosophy,' Manchester, 1808, p. 213) states that 'it is one great object of this book to show the importance and advantage of ascertaining *the relative weights of the ultimate particles both of simple and compound bodies*;' and throughout the work the terms 'atom' and 'ultimate particle' are used perfectly interchangeably. On page 141 Dalton puts forth his hypothesis thus: 'All bodies of sensible magnitude, whether liquid or solid, are constituted of a vast number of extremely small particles, or atoms, of matter.'

Later writers have almost uniformly attributed to Dalton the credit of adopting Leucippus's hypothesis, and making it account for the facts (first brought to light by him, *i.e.*, Dalton) of combination in multiple proportion. Gmelin ('Handbook,' vol. i., p. 145) states that, 'according to the *atomic, corpuscular theory*, matter is an original essence, and consists of certain very small parts, called atoms, molecules, or particles.' . . . . In the ancient atomic theory 'the atoms are supposed to be actuated, not by any attractive force, but by a motion existing from all eternity (Leucippus, Epicurus, Democritus, Lucretius, Lesage);' according to the modern atomic theory (of Dalton and his disciples), 'the atoms are supposed to be impressed with innate forces which give rise to the mutual attraction.'

The difference between these theories, *quoad* the final indivisibility of matter, is *nil*. On the other hand, a meaning has been latterly attached to the term atom wholly different from that applied to it by Leucippus, Dalton, and modern physicists, *e.g.*, Clerk-Maxwell, Lecture on Molecules to the British Association, Bradford Meeting (*Nature*, Sept. 25, 1873, p. 437). The transition is thus indicated by Dr. Odling (Watts's 'Dictionary,' article on 'Atomic Weights'): 'At the present day, the word "atom" is most generally employed by chemists [in contradistinction to the term "proportion" used by Davy]; but, while some use it in its strict Daltonian materialistic sense, others use it in an abstract sense, only to express the smallest indivisible combining proportion of a body, and consider the proportional number of a body as an ultimate or unexplained property pertaining to it.' From the general tenour of 'Atom's' note, it would seem as though he attributed to Dalton the views really propounded by later chemists, and thus expressly stated to be entirely different from the ideas conceived by Dalton himself as to the meaning and application of the term *atom*.

A careful comparison of the writings of modern chemists on this subject leads to the conclusion that many of these chemists admit that the question of the actual existence of the Leucippus- Epicurus- Dalton-Clerk-Maxwell-ian ultimate particle or atom is in no way raised in chemistry; it may, indeed, be assumed, as when Kekulé accounts for the identity of 1·2 and 2·3 benzin derivatives by assuming a peculiar kind of vibratory motion amongst the hydrogen and carbon atoms constituting the benzin ring, or when the chemist attempts to account for the fact of combination in multiple proportions. But it is not a necessary condition; on the contrary, numerous arguments tend to show (and not the least, Norman Lockyer's most recent researches) that the hypothesis that sixty-five or more essentially different kinds of atoms exist is not in accordance with the whole range of known facts. To bring this forward somewhat pointedly has been the sole object of the writer's papers on this subject heretofore.

The *atom* now referred to by many chemists is not the same thing as the atom of Leucippus, Dalton, and other physicists; the term connotes wholly different ideas. What is described by these chemists as the 'atomic theory' is in fact neither a theory nor is it atomic; *i.e.*, it does not involve the assumption of any proposition incapable of experimental proof (such as the assumption that matter is constituted in a particular way), nor does it refer to the special proposition of the existence of ultimate indivisible particles. It might with more justice be called the *greatest common divisor system*, inasmuch as fundamentally it is a system for co-ordinating chemical facts based upon the ascription to the various elements of numerical values, which are the greatest common divisors of the weights of the various elements obtainable from constant bulks of their gaseous compounds respectively.

The atom of Leucippus and Dalton has no longer a place in this system of chemistry, saving in special cases where it is referred to as a means of explaining the facts collected together in the system. Let the chemists who admit that the question of the actual ultimate indivisibility of matter is not raised in chemistry generally, also agree to use no terms which imply the raising of the question and its mental decision in the affirmative. Let the terms atom, molecule, atomic weight, &c., &c., be replaced by words not involving by implication the assumption, so that these terms may revert to their legitimate use as employed by the physicist; then will the reasoning of such chemists be less unphilosophical and illogical than it often is at present, whilst the student will be spared much mental confusion, and there will be less room for controversy based on misapprehensions as to the meanings attached to the words employed.

WRIGHT, C. R. A., 'The Atomic Theory', *Chem. News*, 1873, **28**, 295.

6 THE BIRTH OF ELECTRO-CHEMISTRY

6a The Discovery of Electrolysis

Account of the new Electrical or Galvanic Apparatus of Sig. Alex. Volta, *and Experiments performed with the same.*—W.N.

FROM motives of delicacy to the inventor of the most curious and important combination hereafter to be described, I forbore giving an account of its construction and effects in the last number of this Journal, though it has now been a subject of great attention among philosophers for near two months. It appeared proper to avoid the publication of facts, originally flowing from the liberal communication of the worthy president of the Royal Society, until the paper of the inventor had been read to that learned body; and this could not be done till very lately, because the latter part of his memoir did not arrive till long after the first four pages.

The Right Honourable Sir J. Banks, Bart. F.R.S. having favored my friend Anthony Carlisle, Esq. with the perusal and consideration of these four pages at the latter end of last April, I had the pleasure to look them over with him, immediately after which he constructed an instrument according to Sig. Volta's directions. The experiments made with this will form part of the present communication; but in the first place, I shall endeavour to relate the leading particulars of the communication made to the Royal Society, which no doubt will hereafter appear at large in their Transactions.

The portion of letter which first arrived from Sig. Volta, is dated from Como in the Milanese, March 20, 1800. This, together with the subsequent parts, contains a detailed account of the instrument, of which the following is one of the most convenient forms.

Take any number of plates of copper, or which is better of silver, and an equal number of tin, or which is much better, zinc, and a like number of discs, or pieces of card or leather, or cloth*, or any porous substance capable of retaining moisture. Let these last be soaked in pure water, or which is better, salt and water, or alkaline lees. The silver or copper may be pieces of money†. Build up a pile of these pieces; namely, a piece of silver, a piece of zinc, and a piece of wet card: then another piece of silver, a piece of zinc, and a piece of wet card: and so forth, in the same order (or any other order, provided the pieces succeed each other in their turn) till the whole number intended to be made use of is builded up. The instrument is then completed.

In this state it will afford a perpetual current of electricity, through any conductor communicating between its upper and lower plates; and if this conductor be an animal, it will receive an electrical shock as often as the touch is made, by which the circuit is completed. Thus if one hand be applied to the

* Woollen or linen cloth appear to be more durable, and more speedily soaked than card.

† Most of our philosophers have used half crowns for the silver plates. The zinc may be bought at 8d. per lb. at the White Lion in Foster Lane, and cast in moulds of stone or chalk. A pound makes twenty thick pieces of the diameter of half a crown, or 1.3 inches diameter.

lower plate, and the other to the upper, the operator will receive a shock, and that as often as he pleases to lift his finger and put it down again.

This shock resembles the weak charge of a battery of immense surface, and its intensity is so low, that it cannot make its way through the dry skin. It is, therefore, necessary that a large surface of each hand should be well wetted, and a piece of metal be grasped in each, in order to make the touch, or else that the two extremities of the pile should communicate with separate vessels of water, in which the hands may be plunged.

The commotion is stronger the more numerous the pieces. Twenty pieces will give a shock in the arms, if the above precautions be attended to. One hundred pieces may be felt to the shoulders. The current of electricity acts on the animal system while the circuit is complete, as well as during the instant of commotion, and the action is abominably painful at any place where the skin is broken. . . .

On the 30th of April, Mr. Carlisle had provided a pile consisting of 17 half crowns, with a like number of pieces of zinc, and of pasteboard, soaked in salt water. These were arranged in the order of silver, zinc card, &c. which order I shall denote by saying, that the silver was undermost, that is to say, under the zinc; and I make this remark because some philosophers have used the expression that the silver was undermost when they used the order of silver, card zinc, &c. which, as the reader will easily perceive, is contrary to the order here spoken of. This is of no consequence to the effect, though it is material to a clear understanding of the terms we use. This pile gave us the shock as before described, and a very acute sensation wherever the skin was broken. Our first research was directed to ascertain that the shock we felt was really an electrical phenomenon. For this purpose the pile was placed upon Bennett's gold leaf electrometer, and a wire was then made to communicate from the top of the pile to the metallic stand or foot of the instrument. So that the circuit of the shock would have been through the leaves, if they had diverged. But no signs of electricity appeared. Recourse was then had to the revolving doubler, described at page 95 of our present volume. The plate A was connected with the top of the electrometer and the silver end of the pile; and the plate B and ball were made to touch the top of the system by an uninsulated brass wire. The doubler had been previously cleared of electricity by twenty turns in connection with the earth. The negative divergence was produced in the electrometer. Repeated experiments of this kind shewed that the silver end was in the minus, and the zinc end in the plus state.

In all these experiments it was observed, that the action of the instrument was freely transmitted through the usual conductors of electricity, but stopped by glass and other non-conductors. Very early in this course, the contacts being made sure by placing a drop of water upon the upper plate, Mr. Carlisle observed a disengagement of gas round the touching wire. This gas, though very minute in quantity, evidently seemed to me to have the smell afforded by hydrogen when the wire of communication was steel. This, with some other facts, led me to propose to break the circuit by the substitution of a tube of water between two wires. On the 2nd of May we, therefore, inserted a brass wire through each of two corks inserted in a glass tube of half an inch internal diameter. The tube was filled with New river water, and the distance between the points of the wires in the water was one inch and three quarters. This compound discharger was applied so that the external ends of its wire were in contact with the two extreme plates of a pile of thirty-six half crowns with the correspondent pieces of zinc and pasteboard. A fine stream of minute bubbles immediately began to flow from the point of the lower wire in the tube, which communicated with the silver, and the opposite point of the upper wire became tarnished, first deep orange, and then black. On reversing the tube, the gas came from the other point, which was now lowest, while the upper in its turn became tarnished and black. Reversing the tube again, the phenomena again changed their order. In this state the whole was left for two hours and a half. The upper wire gradually emitted whitish filmy clouds, which, towards the end of the process, became of a pea green colour, and hung in perpendicular threads from the extreme half inch of the wire, the water being rendered semi-opaque by what fell off, and in a great part lay, of a pale green, on the lower surface of the tube, which, in this disposition of the apparatus, was inclined about forty degrees to the horizon. The lower wire of three quarters of an inch long, constantly emitted gas, except when another circuit, or complete wire, was applied to the apparatus; during which time the emission of gas was suspended. When this last mentioned wire was removed, the gas re-appeared as before, not instantly, but after the lapse of four beats of a half second clock standing in the room. The product of gas, during the whole two hours and a half, was two-thirtieths of a cubic inch. It was then mixed with an equal quantity of common air, and exploded by the application of a lighted waxed thread.

It might seem almost unnecessary to have reversed the order of the pile in building up, as reversing the tube must have answered exactly the same purpose. We chose, however, to do this, and found that when the zinc was at the bottom, its effects were reversed, that is to say, the gas still came from the wire communicating with the silver, &c.

We had been led by our reasoning on the first appearance of hydrogen to expect a decomposition of the water; but it was with no little surprize that we found the hydrogen extricated at the contact with one wire, while the oxigen fixed itself in combination with the other wire at the distance of almost two inches. This new fact still remains to be explained, and seems to point at some general law of the agency of electricity in chemical operations. As the distance between the wires formed a striking feature in this result, it became desirable to ascertain whether it would take place to greater distances. When a tube three quarters of an inch in diameter, and thirty-six inches long, was made use of, the effect failed, though the very same wires, inserted into a shorter tube, operated very briskly. The solicitation of other objects of enquiry prevented trial being made of all the various intermediate distances; but from the general tenor of experiments, it appears to be established, that this decomposition is more effectual the less the distance between the wires, but that it ceases altogether when the wires come into contact. . . .

The simple decomposition of water by platina wires without oxidation, offered a means of obtaining the gases separate from each other. With this intention, Mr. Carlisle's pile of thirty-six was combined with my two sets of sixteen repetitions. His pile was built with the zinc uppermost, and mine in the reverse order; so that by connecting the upper plates the whole constituted one range, and the communications could be made from the bottom of the one to the bottom of the other. The two platina wires were made to protrude out of two separate tubes, each containing a little water, and through the opposite corks of each were passed copper wires of communication. These tubes were slightly greased on the outside to prevent their becoming damp; and in this state the extremities, armed with the platina, were plunged in a shallow glass vessel of water, in which two small inverted vessels, quite full of water, were so disposed, that the platina of one tube was beneath one vessel, and the platina of the other tube was beneath the other, the distance between their extremities being about two inches. The copper wires of these tubes respectively were made to communicate with the extremities of the intire pile of sixty-eight sets. A cloud of gas arose from each wire, but most from the silver, or minus side. Bubbles were extricated from all parts of the water, and adhered to the whole internal surface of the vessels. The process was continued for thirteen hours, after which the wires were disengaged, and the gases decanted into separate bottles. On measuring the quantities, which was done by weighing the bottles, it was found, that the quantitites of water displaced by the gases, were respectively, 72 grains by the gas from the zinc side, and 142 grains by the gas from the silver side; so that the whole volume of gas was 1.17 cubic inches, or near an inch and a quarter. These are nearly the proportions in bulk, of what are stated to be the component parts of water. The gas from the zinc side, being tried with one measure of nitrous gas, contracted to 1.25, and did not contract more by the addition of another measure; the gas from the silver side by the same treatment contracted to 1.6. The air of the room, on trial, contracted to 1.28. From the smallness of the quantity no attempt was made to detonate the air from the zinc side, but a portion of that from the silver side, being mixed with one third of atmospheric air, gave a loud detonation.

W. Nicholson, 'Account of the New Electrical or Galvanic Apparatus of Sig. Alex. Volta, and Experiments performed with the Same', *Nicholson's Journal*, 1801, **4**, 179.

6b A Page from Davy's Laboratory Notebook, 6 October, 1807 *(see facing page)*

Humphry Davy, *Laboratory Notebook*, 1805–1809, p. 56. (Now at the Royal Institution.)

6c Davy's Electro-chemical Theory

The effect of the electrization by contact belongs, in a high degree, to zinc and copper, but it may likewise be exhibited, in a greater or less degree, by combinations of all other metals; and it is not at all limited to the metallic kingdom, for it seems to belong to bodies in general, to fluid and aëriform substances as well as to solids, to imperfect and non-conductors as well as to perfect conductors.

A plate of glass, laid upon a surface of dry quicksilver, is found positive after contact—the quicksilver is negative. The case is similar, if the plate be laid upon an amalgam of quicksilver, or upon a piece of silk; and the experiment of the excitation of electricity by the contact of glass with other substances is made upon a great scale in the electrical machine, in the continued impression produced by friction.

Liver of sulphur, or an alkaline solution poured into a metallic cup, renders it negative; but an acid solution makes it positive. This effect may be made evident to the electrometer by several combinations, but a single one is sufficient to exhibit the agency upon the legs of frogs.

When a metallic plate is made to touch crystals of the benzoic acid, it is found positive after contact.

By courtesy of the Royal Institution

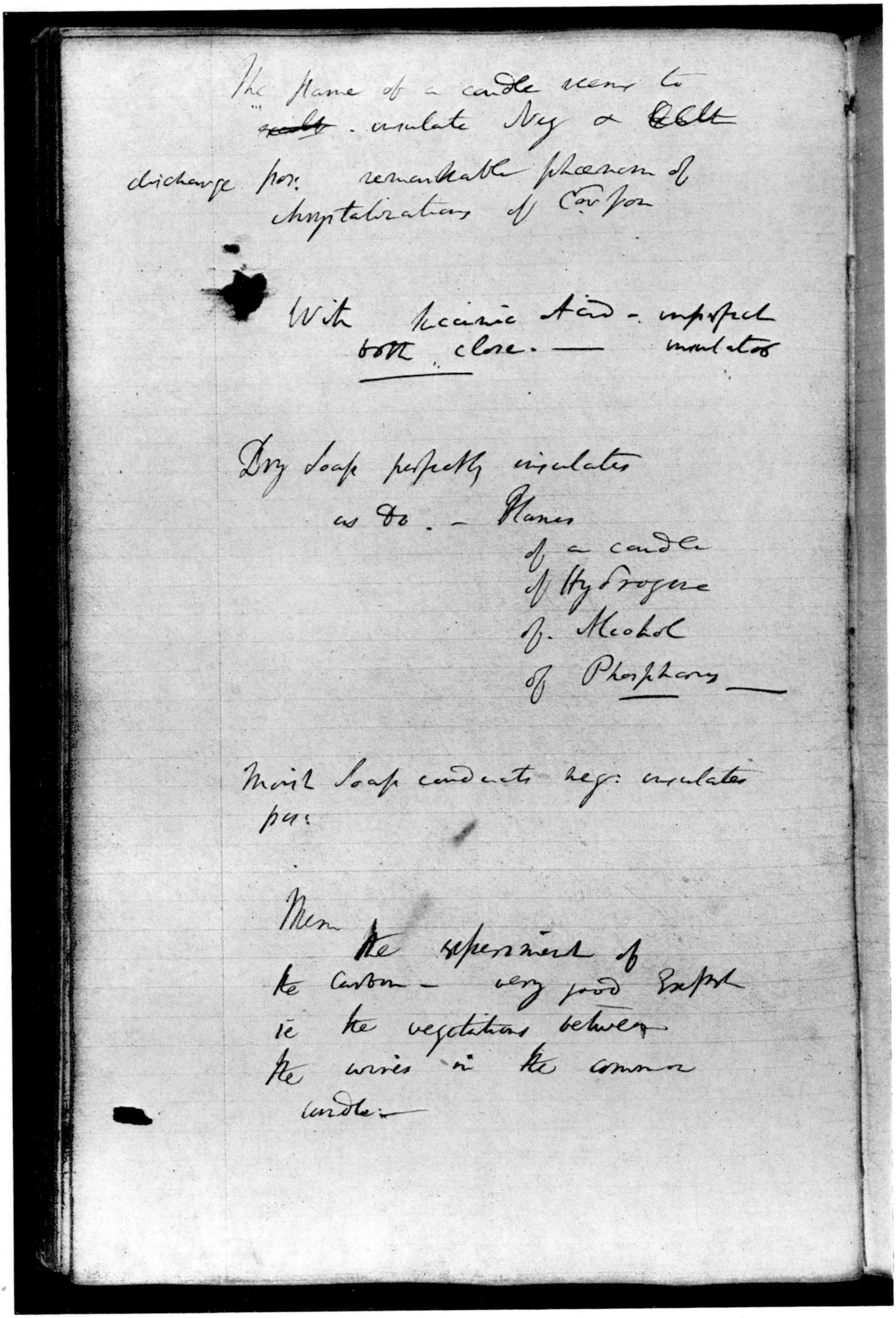

The flame of a candle seems to
~~[illegible]~~ insulate Neg & ~~[illegible]~~
discharge pos: remarkable phenom of
chrystalizations of Cov. pos

With Acetic Acid – imperfect
both close. — insulators

Dry soap perfectly insulates
as do. – flames
of a candle
of Hydrogene
of Alcohol
of Phosphorus —

Moist soap conducts neg: insulates
pos:

Mem
the experiment of
the carbon – very good Expt
ie the vegetations between
the wires in the common
candle. —

But if it were made to touch dry quicklime, it would be found negative.

And a piece of dry quicklime placed upon the acid would be rendered positive, and the acid negative.

In the general arrangement of bodies as affected by contact, zinc is positive with regard to all other metals, manganese, iron, tin, lead, arsenic, antimony, copper, nickel, silver, gold, and platina, form a series in which every metal is negative with regard to those placed before it, and positive with regard to those placed after it.

The metals in general seem to be positive with respect to all bodies except alkaline, and perhaps earthy substances, sulphur and phosphorus; and the acids, as far as they have been examined, are negative with regard to all other substances; and amongst the gasses those that contain oxygen seem to be negative with respect to all that contain inflammable matter.

As the electromotive power, or electrical energy of contact, so generally belongs to bodies differing in species, it must necessarily be in continued operation in all the different kingdoms of nature. The particles of the air cannot move upon the surface of the earth, or upon the waters, without producing some electrical change. The motions of the fluids of the globe must be connected with similar effects, and even in the solid strata, beneath the surface, the contact of metallic ores, by the rocks with which they are enveloped, and the association of the different earths, in a moist state, may be sources of alteration in those parts of the system which appear most fixed and immutable.

In common and obvious facts, which are constantly before our eyes, there are many exhibitions of the electrical effects of contact. In our culinary vessels composed of more than one metal, or containing solder, the greatest effect of corrosion is always at the points of contact of the different substances. Iron nails soon wear out when used to attach copper sheeting to ships, and iron pins employed to attach lead to roofs of buildings rust with great rapidity, which is owing to the chemical operations being increased by the electrical energy of contact. Fermented liquors, which usually contain either the carbonic acid, or some other acid matter, are said to produce some little difference of effect on the palate when taken out of metallic vessels or vessels of glass, and this may naturally be referred to the electricity excited by the contact of the metal and the fluid; and the civilized ancients, who, having no glass, chiefly drank their wines from goblets of metal, and often either silver or gold, must have had constant experience of this electrical taste. Many other instances might be given; and this principle of explanation has been sometimes carried even to a ludicrous extent; thus I have heard it asserted, that a pinch of snuff has a different flavour or odour when taken out of a metallic box, than when it is taken out of a box of wood or shell; which, if it be true, cannot depend on electricity, for no electrical circle is formed in the operation of snuff-taking.

After all that has been stated with regard to the electrical energy or electromotive power of contact, it may perhaps be still asked, is it an ultimate principle?—cannot it be referred to some more general law?

To this the simple reply is, it possibly may not be an ultimate principle, it possibly may, in some advanced age of science, be even referred to a more general law.

But we must reason in natural philosophy not from what we hope, or even expect, but from what we perceive.

The production of positive and negative electricity, by the contact of two bodies, cannot be referred to a chemical agency, for the electrical effects cease the moment chemical combinations begin. Thus zinc and quicksilver, or an alkali and an acid, as long as they are in contact without uniting, give strong marks of electricity, but at the moment they begin to enter into union this electricity disappears.

The energy of contact cannot be owing to a mechanical cause, for it is exhibited without impulse and independent of pressure, and remains permanent, without any relation to motion.

We can no more explain why bodies should differ in specific weight, than why they should differ in their relations to the states of electricity; yet we acknowledge and admire the laws of gravitation. In all our theoretical discussions we must rest somewhere, and that resting place is the most secure where the imagination is controlled by facts and guided by experiments.

The idea of something analogous to electrical energy in bodies was conceived at a very early period, and the same pre-eminent genius which anticipated so many of the results of modern research, appears to have had an indistinct glimpse of this discovery. Newton, in the third book of his Optics, has said, 'May not the electrical attractions be exercised by bodies in an imperceptible manner and constantly?'

The first experiments of electrization by contact were made by Bennett in 1787, but it is to Volta that we owe the development of it as a general property belonging to conductors. He has at once the glory of having discovered the most wonderful philosophical instrument existing, and of having ascertained the great principle of its mode of operation. The fact of the electrical energies of non-conductors is a later work, but after what had been done by the Philosopher of Pavia, it followed as an easy consequence.

I shall resume the subject of excitement by electrical energy, and by other modes in the next lecture. By degrees it will be found that these general facts, the developments of which I am convinced must be very tedious to you, will unfold some brilliant applications, and they are absolutely necessary. The most irksome and difficult business in the progress of scientific discussion, as in building is to lay the foundation. The materials must be of different kinds, and from different sources; they cannot always be adapted to each other, and they usually appear without elegance or beauty, but the permanency and utility of the structure depend upon the labour and care with which they are arranged.

HUMPHRY DAVY, *Collected Works*, London, 1840, vol. viii, p. 300.

6d The Recognition of Ohm's Law

THE Council has awarded the Copley Medal for the present year to Dr. G. S. Ohm, of Nuremberg, for his researches into the laws of Electric Currents, contained in various memoirs published in Schweigger's Journal*, Poggendorff's Annalen, and also in a separate work, entitled *Die Galvanische Kette*

* 1. On the electric conductibility of the metals. (Schweigger's Journal, second series, vol. xiv.)

2. Experiments to discover the power of electro-magnetic multipliers. (Ibid. vol. xxv.)
3. Researches to ascertain the nature of unipolar conductors. (Ibid. vol. xxix.)
4. On hydro-electric currents. (Ibid. third series, vol. iii.)
5. Statement of facts destroying the relations which have been confusedly established between several galvanic properties, and particularly hydro-electric conductors. (Ibid. vol. v.)
6. Theory of galvanic currents. (Ibid. vol. vii.)

Mathematisch Bearbitet, published at Berlin in the year 1827. In these works, Dr. Ohm has established, for the first time, the laws of the electric circuit; a subject of vast importance, and hitherto involved in the greatest uncertainty. He has shown that the usual vague distinctions of intensity and quantity have no foundation, and that all the explanations derived from these considerations are utterly erroneous. He has demonstrated, both theoretically and experimentally, that the action of a circuit is equal to the sum of the electro-motive forces divided by the sum of the resistances; and that whatever be the nature of the current, whether voltaic or thermo-electric, if this quotient be equal, the effect is the same. He has also shown the means of determining with accuracy the values of the separate resistances and electro-motive forces in the circuit. The light which these investigations has thrown on the theory of current electricity is very considerable; and although the labours of Ohm were, for more than ten years, neglected, (Fischner being the only author who, within that time, admitted and confirmed his views,) within the last five years, Gauss, Leng, Jacobi, Poggendorff, Henry, and many other eminent philosophers, have acknowledged the great value of his researches, and their obligations to him in conducting their own investigations. Had the works of Ohm been earlier known, and their value recognised, the industry of experimentalists would have been better rewarded. In this country those who have had most experience in researches in which voltaic agency is concerned, have borne the strongest testimony to the assistance they have derived from this source, and to the invariable accuracy with which the observed phenomena have corresponded with the theory of Ohm. This accordance, it may be observed, is altogether independent of the particular hypothesis which may be adopted as to the origin of electro-motive force; and obtains equally, whether that force is regarded as being derived from the contact of dissimilar metals, or as referable to chemical agency.

Proceedings of the Royal Society, 1837–43, **4**, 336.

7 FARADAY AND ELECTROMAGNETISM

7a The Magnetic Effects of an Electric Current

*ARTICLE IV. Experiments on the Effect of a Current of Electricity on the Magnetic Needle.** By John Christian Oersted, Knight of the Order of Danneborg, Professor of Natural Philosophy, and Secretary to the Royal Society of Copenhagen

THE first experiments respecting the subject which I mean at present to explain, were made by me last winter, while lecturing on electricity, galvanism, and magnetism, in the University. It seemed demonstrated by these experiments that the magnetic needle was moved from its position by the galvanic apparatus, but that the galvanic circle must be complete, and not open, which last method was tried in vain some years ago by very celebrated philosophers. But as these experiments were made with a feeble apparatus, and were not, therefore, sufficiently conclusive, considering the importance of the subject, I associated myself with my friend Esmarck to repeat and extend them by means of a very powerful galvanic battery, provided by us in common. Mr. Wleugel, a Knight of the Order of Danneborg, and at the head of the Pilots, was present at, and assisted in, the experiments. There were present likewise, Mr. Hauch, a man very well skilled in the Natural Sciences, Mr. Reinhardt, Professor of Natural History, Mr. Jacobsen, Professor of Medicine, and that very skilful chemist, Mr. Zeise, Doctor of Philosophy. I had often made experiments by myself; but every fact which I had observed was repeated in the presence of these gentlemen.

The galvanic apparatus which we employed consists of 20 copper troughs, the length and height of each of which was 12 inches; but the breadth scarcely exceeded $2\frac{1}{2}$ inches. Every trough is supplied with two plates of copper, so bent that they could carry a copper rod, which supports the zinc plate in the water of the next trough. The water of the troughs contained $\frac{1}{60}$th of its weight of sulphuric acid, and an equal quantity of nitric acid. The portion of each zinc plate sunk in the water is a square whose side is about 10 inches in length. A smaller apparatus will answer provided it be strong enough to heat a metallic wire red hot.

The opposite ends of the galvanic battery were joined by a metallic wire, which, for shortness sake, we shall call the *uniting conductor*, or the *uniting wire*. To the effect which takes place in this conductor and in the surrounding space, we shall give the name of the *conflict of electricity*.

Let the straight part of this wire be placed horizontally above the magnetic needle, properly suspended, and parallel to it. If necessary, the uniting wire is bent so as to assume a proper position for the experiment. Things being in this state, the needle will be moved, and the end of it next the negative side of the battery will go westward.

* Translated from a printed account drawn up in Latin by the author, and transmitted by him to the Editor of the *Annals of Philosophy*.

If the distance of the uniting wire does not exceed three-quarters of an inch from the needle, the declination of the needle makes an angle of about 45°. If the distance is increased, the angle diminishes proportionally. The declination likewise varies with the power of the battery.

The uniting wire may change its place, either towards the east or west, provided it continue parallel to the needle, without any other change of the effect than in respect to its quantity. Hence the effect cannot be ascribed to attraction; for the same pole of the magnetic needle, which approaches the uniting wire, while placed on its east side, ought to recede from it when on the west side, if these declinations depended on attractions and repulsions. The uniting conductor may consist of several wires, or metallic ribbons, connected together. The nature of the metal does not alter the effect, but merely the quantity. Wires of platinum, gold, silver, brass, iron, ribbons of lead and tin, a mass of mercury, were employed with equal success. The conductor does not lose its effect, though interrupted by water, unless the interruption amounts to several inches in length.

The effects of the uniting wire passes to the needle through glass, metals, wood, water, resin, stoneware, stones; for it is not taken away by interposing plates of glass, metal or wood. Even glass, metal, and wood, interposed at once, do not destroy, and indeed scarcely diminish the effect. The disc of the electrophorus, plates of porphyry, a stone-ware vessel, even filled with water, were interposed with the same result. We found the effects unchanged when the needle was included in a brass box filled with water. It is needless to observe that the transmission of effects through all these matters has never before been observed in electricity and galvanism. The effects, therefore, which take place in the conflict of electricity are very different from the effects of either of the electricities.

If the uniting wire be placed in a horizontal plane under the magnetic needle, all the effects are the same as when it is above the needle, only they are in an opposite direction; for the pole of the magnetic needle next the negative end of the battery declines to the east.

That these facts may be the more easily retained, we may use this formula— the pole *above* which the *negative* electricity enters is turned to the *west; under* which, to the *east*.

If the uniting wire is so turned in a horizontal plane as to form a gradually increasing angle with the magnetic meridian, the declination of the needle *increases*, if the motion of the wire is towards the place of the disturbed needle; but it *diminishes* if the wire moves further from that place.

When the uniting wire is situated in the same horizontal plane in which the needle moves by means of the counterpoise, and parallel to it, no declination is produced either to the east or west; but an *inclination* takes place, so that the pole, next which the negative electricity enters the wire, is *depressed* when the wire is situated on the *west* side, and *elevated* when situated on the *east* side.

If the uniting wire be placed perpendicularly to the plane of the magnetic meridian, whether above or below it, the needle remains at rest, unless it be very near the pole; in that case the pole is *elevated* when the entrance is from the *west* side of the wire, and *depressed*, when from the *east* side.

When the uniting wire is placed perpendicularly opposite to the pole of the magnetic needle, and the upper extremity of the wire receives the negative electricity, the pole is moved towards the east; but when the wire is opposite to a point between the pole and the middle of the needle, the pole is most towards the west. When the upper end of the wire receives positive electricity, the phenomena are reversed.

If the uniting wire is bent so as to form two legs parallel to each other, it repels or attracts the magnetic poles according to the different conditions of the case. Suppose the wire placed opposite to either pole of the needle, so that the plane of the parallel legs is perpendicular to the magnetic meridian, and let the eastern leg be united with the negative end, the western leg with the positive end of the battery: in that case the nearest pole will be repelled either to the east or west, according to the position of the plane of the legs. The eastmost leg being united with the positive, and the westmost with the negative side of the battery, the nearest pole will be attracted. When the plane of the legs is placed perpendicular to the place between the pole and the middle of the needle, the same effects recur, but reversed.

A brass needle, suspended like a magnetic needle, is not moved by the effect of the uniting wire. Likewise needles of glass and of gum lac remain unacted on.

We may now make a few observations towards explaining these phenomena.

The electric conflict acts only on the magnetic particles of matter. All non-magnetic bodies appear penetrable by the electric conflict, while magnetic bodies, or rather their magnetic particles, resist the passage of this conflict. Hence they can be moved by the impetus of the contending powers.

It is sufficiently evident from the preceding facts that the electric conflict is not confined to the conductor, but dispersed pretty widely in the circumjacent space.

From the preceding facts we may likewise collect that this conflict performs circles; for without this condition, it seems impossible that the one part of the uniting wire, when placed below the magnetic pole, should drive it towards the east, and when placed above it towards the west; for it is the nature of a circle that the motions in opposite parts should have an opposite direction. Besides, a motion in circles, joined with a progressive motion, according to the length of the conductor, ought to form a conchoidal or spiral line; but this, unless I am mistaken, contributes nothing to explain the phenomena hitherto observed.

All the effects on the north pole above-mentioned are easily understood by supposing that negative electricity moves in a spiral line bent towards the right, and propels the north pole, but does not act on the south pole. The effects on the south pole are explained in a similar manner, if we ascribe to positive electricity a contrary motion and power of acting on the south pole, but not upon the north. The agreement of this law with nature will be better seen by a repetition of the experiments than by a long explanation. The mode of judging of the experiments will be much facilitated if the course of the electricities in the uniting wire be pointed out by marks or figures.

I shall merely add to the above that I have demonstrated in a book published five years ago that heat and light consist of the conflict of the electricities. From the observations now stated, we may conclude that a circular motion likewise occurs in these effects. This I think will contribute very much to illustrate the phenomena to which the appellation of polarization of light has been given.

JOHN CHRISTIAN OERSTED
Copenhagen, July 21, 1820

J. C. OERSTED, 'Experiments on the Effect of a Current of Electricity on the Magnetic Needle', *Ann. Phil.*, 1820, **16**, 273.

7b Faraday's 'Electro-magnetical Apparatus'

ART. VIII. Description of an Electro-magnetical Apparatus for the Exhibition of Rotatory Motion. By M. Faraday, *Chem. Assist. in the Royal Institution.*

THE account given in the *Miscellanea* of the last Journal, of the apparatus invented in illustration of the paper in the body of that Number, being short and imperfect; a plate is given in the present Number, presenting a section of that apparatus, and a view of a smaller apparatus, illustrative of the motions of the wire and the pole round each other. The larger apparatus is delineated, Fig. 1. Plate VII. on a scale of one-half. It consists of two glass vessels, placed side by side with their appendages. In that on the left of the plate the motion of a magnetic pole round the connecting wire of the voltaic battery is produced. That a current of voltaic electricity may be established through this cup, a hole is drilled at the bottom, and into this a copper pin is ground tight, which projects upwards a little way into the cup, and below is rivetted to a small round plate of copper, forming part of the foot of the vessel. A similar plate of copper is fixed to the turned wooden base on which the cup is intended to stand, and a piece of strong copper wire, which is attached to it beneath, after proceeding downwards a little way, turns horizontally to the left hand, and forms one of the connexions. The surfaces of these two plates intended to come together, are tinned and amalgamated, that they may remain longer clean and bright, and afford better contact. A small cylindrical and powerful magnet has one of its poles fastened to a piece of thread, which, at the other end, is attached to the copper pin at the bottom of the cup; and the height of the magnet and length of the thread is so adjusted, that when the cup is nearly filled with clean mercury, the free pole shall float almost upright on its surface.

A small brass pillar rises from the stand behind the glass-vessels: an arm comes forward from the top of it, supporting at its extremity a cross wire, which at the place on the left hand, where it is perpendicularly over the cup just described, bends downwards, and is continued till it just dips into the centre of the mercurial surface. The wire is diminished in size for a short distance above the surface of the mercury, and its lower extremity amalgamated, for the purpose of ensuring good contact; and so also is the copper pin at the bottom of the cup. When the poles of a voltaic apparatus are connected with the brass pillar, and with the lateral copper wire, the upper pole of the magnet immediately rotates round the wire which dips into the mercury; and in one direction or the other, according as the connexions are made.

The other vessel is of the form delineated in the plate. The stem is hollow and tubular; but, instead of being filled by a plug, as is the aperture in the other vessel, a small copper socket is placed in it, and retained there by being fastened to a circular plate below, which is cemented to the glass foot, so that no mercury shall pass out by it. This plate is tinned and amalgamated on its lower surface, and stands on another plate and wire, just as in the former instance. A small circular bar magnet is placed in the socket, at any convenient height, and then mercury poured in until it rises so high that nothing but the projecting pole of the magnet is left above its surface at the centre. The forms and relative positions of the magnet, socket, plate, &c. are seen in fig. 2.

The cross wire supported by the brass pillar is also prolonged on the right hand, until over the centre of the vessel just described; it then turns downwards, and descends about half an inch: it has its lower extremity hollowed out into a cup, the inner surface of which is well amalgamated. A smaller piece of copper wire has a spherical head fixed on to it, of such a size that it may play in the cup in the manner of a ball and socket-joint, and being well amalgamated, it, when in the cup, retains sufficient fluid mercury by capillary attraction to form an excellent contact with freedom of motion. The ball is prevented from falling out of the socket by a piece of fine thread, which, being fastened to it at the top, passes through a small hole at the summit of the cup, and is made fast on the outside of the thick wire. This is more minutely explained by Fig. 3, and 4. The small wire is of such a length that it may dip a little way into the mercury,

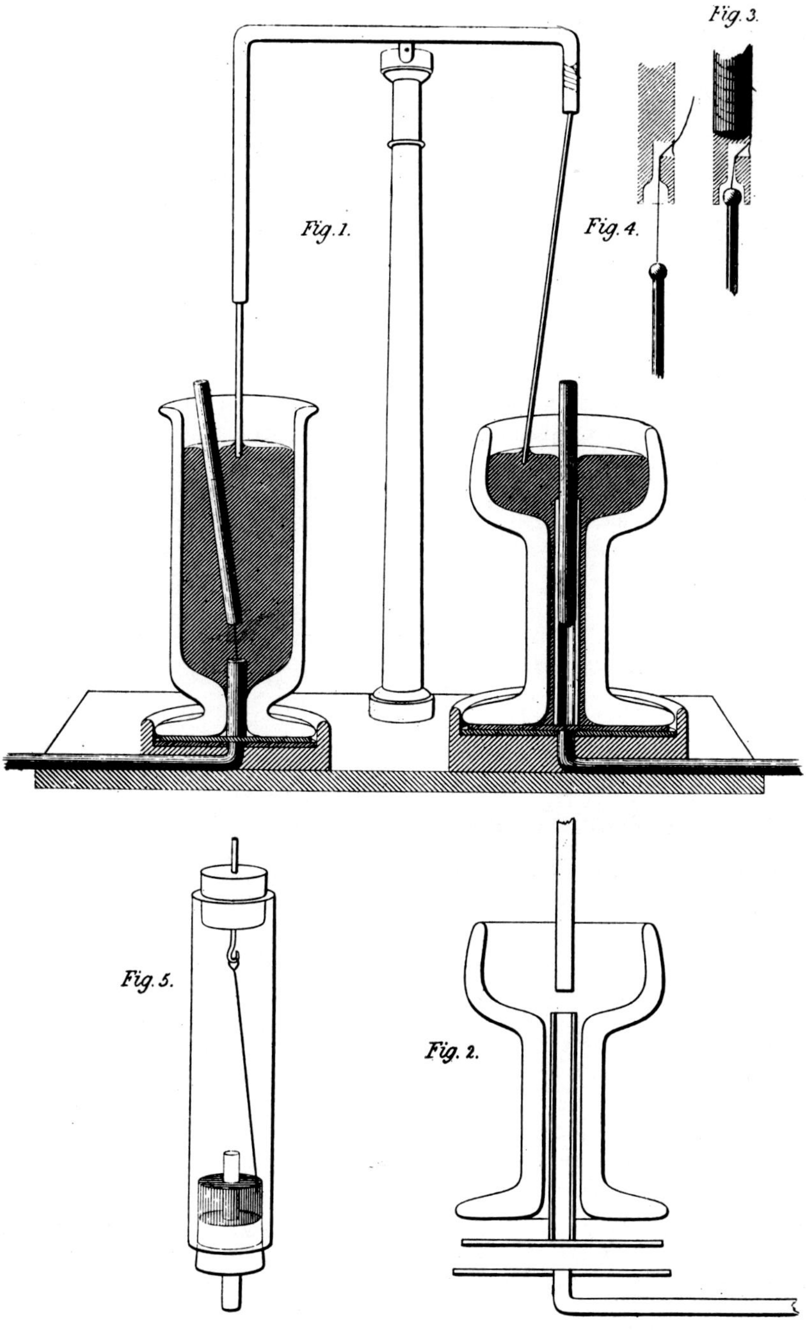

By courtesy of the Royal Institution

and its lower end is amalgamated. When the connexions are so made with the pillar and right hand wire, that the current of electricity shall pass through this moveable wire, it immediately revolves round the pole of the magnet, in a direction dependant on the pole used, and the manner in which the connexions are made.

Fig. 5, is the delineation of a small apparatus, the wire in which revolves rapidly, with very little voltaic power. It consists of a piece of glass tube, the bottom part of which is closed by a cork, through which a small piece of soft iron wire passes, so as to project above and below the cork. A little mercury is then poured in, to form a channel between the iron wire and the glass tube. The upper orifice is also closed by a cork, through which a piece of platinum wire passes, which is terminated within by a loop; another piece of wire hangs from this by a loop, and its lower end, which dips a very little way into the mercury, being amalgamated, it is preserved from adhering either to the iron wire or the glass. When a very minute voltaic combination is connected with the upper and lower ends of this apparatus, and the pole of a magnet is placed in contact with the external end of the iron wire, the moveable wire within rapidly rotates round the magnet thus formed at the moment; and by changing either the connexion, or the pole of the magnet in contact with the iron, the direction of the motion itself is changed.

The small apparatus in the plate is not drawn to any scale. It has been made so small as to produce rapid revolutions, by the action of two plates of zinc and copper, containing not more than a square inch of surface each.

In place of the ball and socket-joint, (fig. 3, and 4,) loops may be used; or the fixed wire may terminate in a small cup containing mercury, with its aperture upwards, and the moveable wire may be bent into the form of a hook, of which the extremity should be sharpened, and the rest in the mercury on the bottom of the cup.

M. Faraday, 'Description of an Electro-magnetical Apparatus for the Exhibition of Rotary Motion', *Quarterly Journal of Science*, 1822, **12**, 283.

7c Sturgeon on Electro-magnets

XXV. On Electro-Magnets. By W. Sturgeon, *Lecturer on Experimental Philosophy at the Honourable East India Company's Military Academy, Addiscombe**

When first I showed that the magnetic energies of a galvanic conducting wire are more conspicuously exhibited by exercising them on *soft iron* than on *hard steel*, my experiments were limited to small masses, generally to a few inches of rod iron about half an inch in diameter. Some of those pieces were employed whilst straight, and others were bent into the form of a horse-shoe magnet, each piece being encompassed by a spiral conductor of copper wire†. The magnetic energies developed by these simple arrangements are of a very distinguished and exalted character, as is conspicuously manifested by the suspension

* Communicated by the Author.

† The pieces of soft iron which were first magnetized in this way are members of an electro-magnetic apparatus, which I had the honour to present to the 'Society of Arts, &c.' in the spring of 1825; for which the Society presented me with their large silver medal, and a purse of thirty guineas.

of a considerable weight at the poles during the period of excitation by the electric influence.

An unparalleled transiliency of magnetic action is also displayed in soft iron, by an instantaneous transition from a state of total inactivity to that of vigorous polarity, and also by a simultaneous reciprocity of polarity in the extremities of the bar;—versatilities in this branch of physics to which soft iron is preeminently prone, and which by the agency of electricity become demonstrable with the celerity of thought, and illustrated by experiments the most splendid in magnetics.

It is, moreover, abundantly manifested by ample experiments, that galvanic electricity exercises a superlative degree of excitation on the latent magnetism of soft iron, and calls forth its recondite powers with astonishing promptitude, to an intensity of action far surpassing anything which can be accomplished by any known application of the most vigorous permanent magnet, or by any other mode of experimenting hitherto discovered. It has been observed, however, by experimenting on different pieces selected from various sources, that notwithstanding the greatest care be observed in preparing them of an uniform figure and dimensions, there appears a considerable difference in the susceptibility which they individually possess of developing the magnetic powers; much of which depends upon the manner of treatment at the forge, as well as upon the natural character of the iron itself*.

The superlative intensity of electro-magnets, and the facility and promptitude with which their energies can be brought into play, are qualifications, admirably adapted for their introduction into a variety of arrangements in which powerful magnets so essentially operate, and perform a distinguished part in the production of electro-magnetic rotations; whilst the versatilities of polarity, of which they are susceptible, are eminently calculated to give a pleasing diversity in the exhibition of that highly interesting class of phænomena, and lead to the production of others, inimitable by any other means.

An experiment of this character is noticed in the Phil. Mag. for January 1825; but as the arrangement by which it is accomplished has not yet been published, a description of it in this place may perhaps still be interesting, especially as it affords a clue to several others which may be exhibited in this curious branch of science.

Experiment.—Fig. 1. is a representation of the apparatus complete. It consists of a cylindrical rod of soft iron, supported in a vertical position by a round wooden foot, into which its inferior extremity is inserted. The superior extremity of the

* I have made a number of experiments on small pieces, from the results of which it appears that *much hammering* is highly detrimental to the development of magnetism in soft iron, whether the exciting cause be galvanic or any other. And although good annealing is always essential, and facilitates to a considerable extent the display of polarity, that process is very far from restoring to the iron that degree of *susceptibility* which it frequently loses by the operation of the hammer.

Cylindric rod iron of small dimensions may very easily be bent into the required form, without any hammering whatever; and I have found that small electro-magnets made in this way display the magnetic powers in a very exalted degree.

An electro-magnet of the above description, weighing 3 ounces, and furnished with one coil of wire, supported 14 pounds. The poles were afterwards made to expose a larger surface by welding to each end of the cylindric bar a square piece of good soft iron: with this alteration only, the lifting power was reduced to about 5 pounds, although the magnet was annealed as much as possible.

It appears to me that the superior magnetic energies displayed by these cylindric rods of iron whilst subjected to the electric influence, are owing in a great measure to the peculiar fibrous texture which the metal is made to assume by the process which brings it to their particular shape.

iron rod passes through the centre of a shallow wooden dish, which is kept firmly in its place by means of cement. The inside of this dish, and also the iron rod, is covered with sealing-wax varnish. A copper wire, one end of which passes through the bottom of the dish and appears at the upper surface, is wound several times round the iron cylinder between the dish at its upper extremity, and the wooden foot in which it is supported. The other extremity of the wire terminates in a small cup, as is seen in the figure.

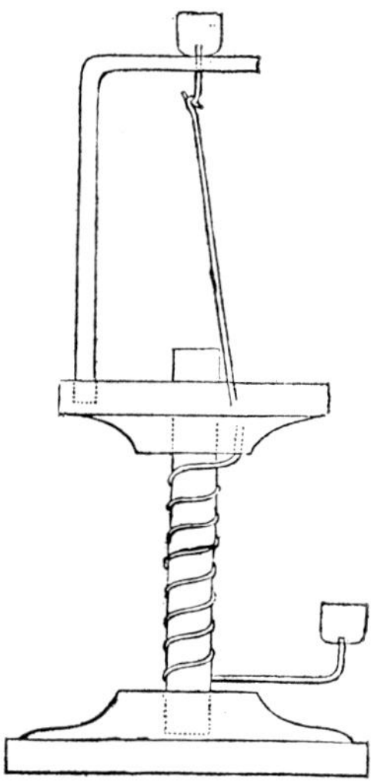

Fig. 1
By courtesy of the Royal Institution

A stout brass wire, one end of which is screwed firmly into the upper edge of the wooden dish, rises vertically about five inches of its length, at which place it is bent at right angles, and continued in a horizontal direction over the axis of the iron rod. A small brass wire screw, the upper end of which is inserted in a small cup, passes through the horizontal arm, and terminates with a hook at its lower end. A light copper wire, having a hook at one end, is by this means suspended to the hook of the screw. The inferior extremity of the pendent wire is pointed, and reaches into the wooden vessel, but not sufficiently low as to touch its inner surface.

When an experiment is to be made with this apparatus, the extremities of the helical wire which pass through the dish and the lower cup, and also the extremity of the short wire which passes through the bottom of the upper cup, are to be well amalgamated. Mercury is now to be placed in the two cups for the convenience of connection, and also in the wooden dish, until the point of the pendent wire (which must also be amalgamated) dips slightly into it. By this means there will be formed a complete metallic connection between the mercury in the upper cup, and that which is placed in the lower.

If the connecting wire from the copper plate of a single galvanic pair be now permitted to enter the upper cup, and that from the zinc to enter the lower, then the direction of the electric current through the system of conductors will be from the upper to the lower cup. The cylindrical rod of iron, inclosed in the spiral, will become highly magnetic, and the suspended moveable wire will perform its revolutions round the upper pole.

With this arrangement the *direction* in which the pendent wire performs its revolutions will entirely depend upon the character of the spiral, or upon the direction in which that part of the conductor passes round the iron rod from the upper to the lower extremity, and not upon any other circumstance whatever. For, it being an established *law* in electro-magnetics that *the direction of its revolutions is not altered* by simultaneously reversing the magnetic polarity and electric current; and as, in the present arrangement, the character of the magnetic poles in the extremities of the bar will at all times be determined by the direction of the electric stream, and that no vicissitude can possibly take place in the one without a corresponding and simultaneous vicissitude in the character of the other,—it follows that the revolving wire will uniformly proceed in *one* and the *same* direction, whatever may be the nature of the electric stream; and that no vicissitudes in the direction of motion can possibly be accomplished by any similar arrangement.

Now, as the direction in which the wire revolves will be determined by the direction in which the spirals encompass the iron, it is plain that the apparatus may be so constructed as to perform its revolutions in any direction the experimenter may think proper to select: but the selection once made, and the standard spiral determined on, the apparatus becomes incapable of exhibiting that beautiful diversity of revolving motions which proceed from various combinations of the electric and magnetic forces.

If, however, the electric current were not to be continuous from one of the extreme cups to the other, the apparatus might then be made to perform the usual variety of electromagnetic rotations; because in that case one current might be employed to give polarity to the iron, whilst another could be transmitted in any required direction through the pendent revolving wire.

Upon the same principle a bar of soft iron, properly connected, will rotate on its axis with an astonishing velocity; the direction of motion being always determined by the character of the helical conductor. The apparatus which I use for the exhibition of this interesting phænomenon is similar to that by means of which I showed the rotation of a steel magnetic bar on its axis, by the influence of two electric currents (Phil. Mag. vol. lxiv. p. 246); the transient electro-magnet in this case being substituted for the permanent steel magnet in the other experiment. The manner in which the spiral is arranged will be understood by contemplating fig. 2. The ends of the spiral are soldered to the cylindrical bar of iron at the points *a a*, and to the centre is soldered a short wire with a descending point for the purpose of maintaining an uninterrupted connection with an annular mass of mercury in which it revolves. The direction of motion is constantly the same whatever may be the nature of the galvanic connections.

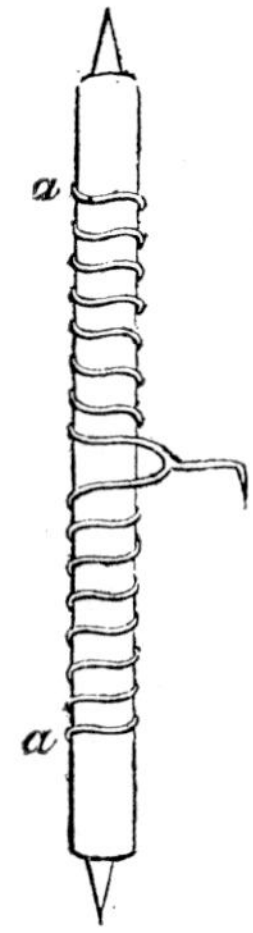

Fig. 2
By courtesy of the Royal Institution

It does not appear that any very extensive experiments were attempted to improve the lifting powers of electro-magnets from the time that my experiments were published in the Transactions of the Society of Arts, &c. for 1825, till the latter part of 1828. Mr. Watkins, Philosophical Instrument-maker, Charing Cross, had, however, made them of a much larger size than any which I had employed; but I am not aware to what extent he pursued the experiment.

In the year 1828 Professor Moll of Utrecht, being on a visit to London, purchased of Mr. Watkins an electro-magnet weighing about five pounds; at that time I believe the largest which had been made. It was of round iron, about one inch in diameter, and furnished with a single copper wire twisted round it 83 times. When this magnet was excited by a large galvanic surface, it supported about 75 pounds.

Professor Moll afterwards prepared another electro-magnet, which when bent was 12½ inches high, 2½ inches in diameter, and weighed about 26 pounds; prepared like the former with a single spiral conducting wire. With an acting galvanic surface of 11 square feet, this magnet would support 154 pounds, but would not lift an anvil which weighed 200 pounds.

The success of these experiments, which established the first grand step in exalting the attractive powers of electro-magnets, gave a new impulse to the inquiry, which the American philosophers have pursued to an extent that will not be very easily surpassed. By dividing about 800 feet of conducting wire into 26 strands, and forming it into as many separate coils round a bar of soft iron about 60 pounds in weight, and properly bent into the horse-shoe form, Professor Henry has been enabled to produce a magnetic force which completely eclipses every other in the whole annals of magnetism; and no parallel is to be found since the miraculous suspension of the celebrated Oriental impostor in his iron coffin*.

This electro-magnet is said to have supported nearly two tons when excited by about five square feet of galvanic surface, an extent comparatively trifling, when compared to the prodigious magnetic force which it is capable of calling into action.

The largest electro-magnet which I have as yet exhibited in my lectures weighs about 16 pounds. It is formed of a square bar of soft iron, 1½ inches across each side; the poles are separate from each other 1¼ inch; the cross piece

* Silliman's Journal.

which joins the poles is from the same rod of iron, and about $3\frac{3}{4}$ inches long.

Twenty separate strands of copper wire, each strand about 50 feet in length, are coiled round the iron one above another, from pole to pole, and separated from each other by intervening cases of silk; the first coil is only the thickness of one ply of silk from the iron, the twentieth or outermost about half an inch distant from it. By this means the wires are completely insulated from each other without the trouble of covering them with thread or varnish. The ends of the wires project about two feet, for the convenience of connection.

With one of my small cylindrical batteries, exposing about 150 square inches of total surface, this electro-magnet supports 400 pounds. I have tried it with a larger battery, but its energies do not appear to be so materially exalted as might have been expected by increasing the extent of galvanic surface. Much depends upon a proper acid solution: good nitric or nitrous acid, with about six or eight times its quantity of water, answers very well. With a new battery of the above dimensions, and a strong solution of salt and water at a temperature of about 190° Fahr., the electro-magnet supported between 70 and 80 pounds, when the first 17 coils only were in the circuit. With the three exterior coils alone in the circuit, it would just support the lifter or cross piece. When the temperature of the solution was between 40° and 50°, the magnetic force excited was comparatively very feeble. With the innermost coil alone, and a strong acid solution, this electro-magnet supports about 100 pounds: with the four innermost wires, about 250 pounds. It improves in power with every additional coil until about the twelfth, but not perceptibly any further: therefore the remaining eight coils appear to be entirely useless; although the last three of them, independent of the innermost seventeen, and at the distance of half an inch from the iron, produce in it a lifting power of 75 pounds.

It is evident from these results, that the exciting power of this miniature battery becomes improved by multiplying the number of conducting wires as far as twelve at least, although the greater part of them be at some considerable distance from the iron; and it is highly probable, although the experiments which I have made on this bar do not satisfactorily prove the fact, that by employing a larger galvanic surface, a much further addition to the number of conducting wires may be advantageously introduced into the circuit for the excitation of the magnetic energies of soft iron. Perhaps the best arrangement would be to have a separate small battery to each wire.

Mr. Marsh has fitted up a bar of iron much larger than mine, with a similar distribution of the conducting wires to that devised, and so successfully employed, by Professor Henry. Mr. Marsh's electro-magnet will support about 560 pounds when excited by a galvanic battery similar to mine*. These two, I believe, are the most powerful electro-magnets yet produced in this country.

A small electro-magnet, which I also employ on the lecture-table, and the manner of its suspension, is represented by fig. 3. The magnet is of cylindric rod iron, and weighs four ounces: its poles are about a quarter of an inch asunder. It is furnished with six coils of wire in the same manner as the large electro-magnet before described, and will support upwards of 50 pounds.

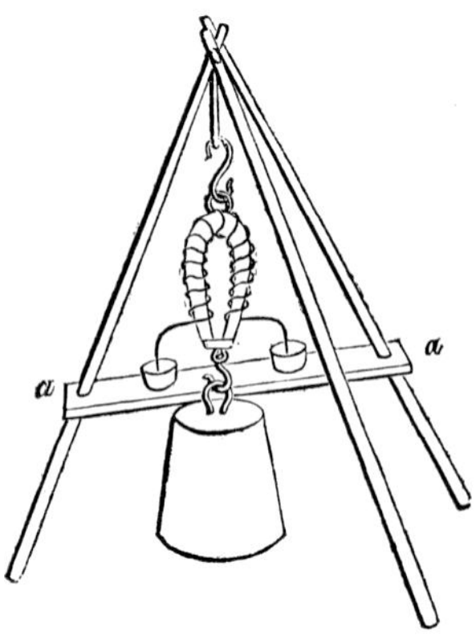

Fig. 3
By courtesy of the Royal Institution

I find a triangular gin very convenient for the suspension of the magnet in these experiments. A stage *a a* of thin board, supporting two wooden dishes, is fastened at a proper height to two of the legs of the gin. Mercury is placed in these vessels, and the dependent amalgamated extremities of the conducting wires dip into it; one into each portion. The vessels are sufficiently wide to admit of considerable play for the wires in the mercury without interruption of

* See the Report of Proceedings at the Royal Institution, in our present Number.—Edit.

contact, which is sometimes occasioned by the swinging of the magnet and attached weight: the circuit is completed by other wires, which connect the battery with these two portions of mercury. When the weight is supported as in the figure, if an interruption be made by removing either of the connecting wires, the weight instantaneously drops on the table. The large magnet I suspend in the same way on a larger gin; the weights which it supports are placed one after another on a square board, suspended by means of a cord at each corner from a hook in the cross piece, which joins the poles of the magnet.

With a new battery, and a solution of salt and water at a temperature of 190° Fahr., the small electro-magnet supports 16 pounds.

I noticed in some of my earliest experiments, that a bar of soft iron which had been intensely magnetized by galvanic action, retained a considerable degree of polarity when the exciting cause had been long removed,—a phænomenon now more conspicuously displayed by the employment of larger masses; nor does it appear to be an easy matter to subdue, entirely, this residual polarity. The poles may be reversed as frequently as we please; but still some, and frequently a considerable degree of polarity remains unneutralized. If the cross piece be permitted to remain attached to the poles when the galvanic connection is broken, the residual polarity will still keep them together with an astonishing force; as is manifest by the very great weight which the magnet continues to support before the cross piece is disengaged from its poles. A residual polarity, however, still remains; but so enfeebled are its energies by the slightest interruption of polar contact, that the cross piece will seldom be supported a second time without a renewed excitation by the battery.

The vigorous residuum of polarity which retains the cross piece to the magnet arises from a continued mutual attraction between the two; for whilst the battery operates upon the magnet, and excites it to action, the latter in its turn also excites the magnetism of the iron connecting its poles, which iron becomes as decidedly polar as the magnet itself; a *north pole* being determined in that end of it which is in connection with the south pole of the magnet, and a *south pole* in the other end, which is connected with the north pole of the magnet. The four magnetic poles thus brought into play, and in vigorous operation on each other, will, if not separated, retain their positions at the points of connection, even though the first exciting cause be entirely withdrawn. But if the connection of the two pieces of iron be in the least interrupted, their magnetism immediately recedes from the extremities, and becomes equally distributed in the metal; the vigour of the poles vanishes, and their magnetism becomes totally incapable of keeping them attached to each other.

Precisely the same kind of reasoning will explain the cause of that well known deterioration of magnetic force which invariably takes place by removing the cross piece from a highly excited steel magnet. The excitation in this case, whether it be performed by the operation of a magnet, or by the more gradual process of adding small weights to those already suspended, is carried on whilst the poles and the cross piece are in contact; and both become excited at the same time. The magnet and cross piece now operate on each other with a gradually increasing vigour; the poles of the one becoming more and more energetic as the power of the opposing poles of the other become further exalted. And this reciprocal increase of action will be exercised on each other till a maximum of mutual attraction is obtained, at which time the magnet will support its greatest load. If more weight be added, the mutual attractive forces of the magnet and cross piece will be overpowered, and the two will separate. The magnetism of the iron no longer displays polarity, and that of the steel partially recedes from its extremities; but in consequence of the character of the metal, a considerable polarity is still displayed by the magnet: its energies, however, are very much diminished; and when the cross piece is replaced at the poles, the effort to stimulate it to polarity is proportionably diminished. The reciprocal attractions now operate with impaired forces, and consequently the load which can be supported is much less than before.

A compound steel magnet in my possession weighs 9 pounds, which, when well magnetized will support 120 pounds; but if weights be added till the cross piece falls off, the power is reduced to about 75 pounds.

Fig. 4. is a representation of an electro-magnetic sphere, mounted on a mahogany frame, consisting of a stout base board and two upright pillars, to

By courtesy of the Royal Institution

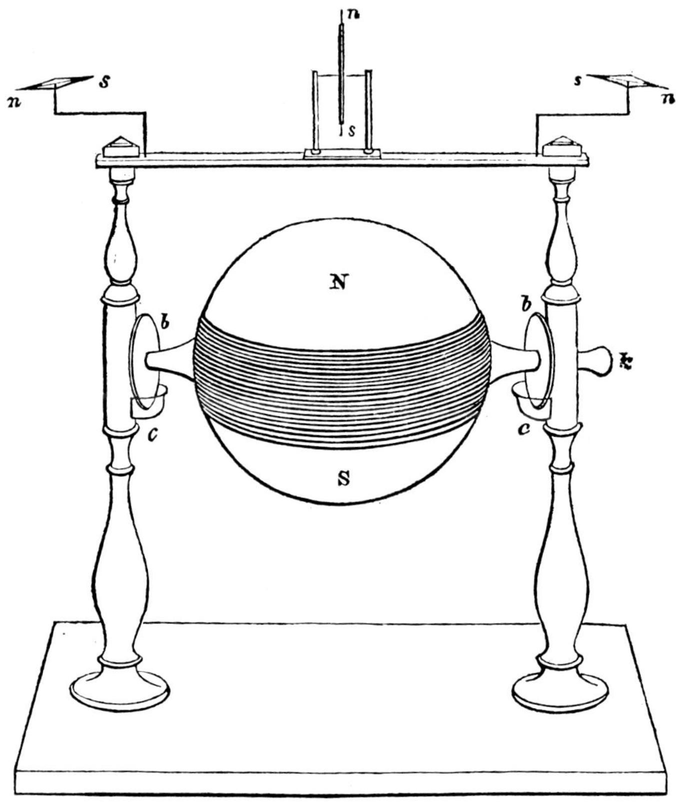

Fig. 4.

the upper extremities of which is fixed a cross piece or stage. On the centre of the stage, and directly over the sphere, is placed a dipping needle; and near to the extremities are inserted the lower ends of two wires bent twice at right angles, as seen in the figure; the upper extremities of these wires are finely pointed, and support two horizontal needles.

The sphere N S is a cast-iron shell, about 8 inches in diameter, and weighs about 68 pounds. Its surface is divided into three parts, which for convenience may be called *tropical* and *polar*, having one of the former and two of the latter. The tropical region is covered with four coils of copper wire in separate strata, insulated from each other, and also from the metallic sphere, by means of slips of silk; the exterior coil, being uncovered, is seen in the figure. The coils are prevented from slipping towards the poles by means of stout iron rings soldered to the sphere, one at each tropical circle.

The two sets of extremities of the conducting wires are soldered, one to each of the two copper discs or wheels *b b*, through the centres of which the extremities of the horizontal axis of the sphere pass. The lower edges of the wheels dip into portions of mercury placed in two semicircular vessels *c c*, which slide into the side of the pillars a little below the holes which receive the axis of the sphere.

By this arrangement the extremities of the wires which encircle the globe will constantly be in connection with the mercury in the two cups, in whatever position the poles may be placed. The connections with the battery are accomplished by two wires joining its plates, and the two portions of mercury in the vessels *c c*.

If this apparatus be so placed, before the battery be attached, that the needles stand at right angles to the vertical plane of the frame, and the dip of the central needle is counterbalanced, they will be arranged nearly parallel to each other, as their distance prevents them being affected by the ball.

If the axis joining the poles of the sphere be now brought into a horizontal position, by turning the key *k*, which enters the horizontal axle, the equator will be directly under the dipping needle. With this position of the apparatus, make the galvanic connections so that the pole of the sphere which faces the north may assume north polarity, or that species of polarity which is natural to the northern regions of the earth. The two horizontal needles will deviate from their former positions, having their north ends drawn towards the sphere: the central needle being placed on the equator will remain undisturbed. If the north pole of the sphere be now gradually turned upwards, the horizontal needles will be directed still more towards this pole, and the north end of the central needle will incline. The angles of deviation and dip will increase more and more as the north pole of the sphere advances towards the zenith; and when it has arrived at that point, those angles will have arrived at their maximum. The central needle will stand vertically, exhibiting a dip of 90° over the pole; and the two horizontal needles will point almost directly towards it. By permitting the sphere to resume its first position as gradually as it left it, the needles which are influenced by it will be observed to recede as gradually to their former positions.

If now the galvanic connections be reversed, the whole of the needles become reversed also, because of a change in the polarity of the sphere; and the dipping needle, after a few oscillations, will settle in a horizontal position.

By experimenting in this way with various positions of the sphere, it will be found to operate on the needle in a manner highly imitative of the earth's magnetism on different parts of its surface.

The two polar regions of the sphere exhibit a diffused polarity, the centres of which are nearly, perhaps exactly, in the poles of its equator. There are, however, in the polar regions several points which exhibit distinct polarity; and although of the same character as the general pole in which they are situated, will draw a delicate needle held near to them, from the direction which it takes when held at a greater distance from the action of the general or aggregate pole of that particular region. These local poles, I imagine, arise from a want of uniformity in the character of the iron. There are on the earth's surface aberrations of this kind arising from the local attractions of certain islands, and from causes not easily determined; which, as it happens, these accidental poles may serve to imitate.

W. STURGEON, 'On Electro-magnets', *Phil. Mag.*, 1832 (N.S.), **2**, 199.

7d Insulated Wire

To the Editor of the Annals of Electricity, &c.

Sir,

THE enormous price charged by the opticians for covered copper wire, and a desire to diminish as much as possible the expense of electro-magnetic studies, induces me to inform your readers how they may get it cheaper.

Those who cover bonnet wire charge about 6*d. per pound*: it is only necessary then to discover a wire coverer, and supply him with the copper wire, as, thus, it will not cost one third the opticians' charge. A man of this description, whose name is Green, and who lives near the western end of Quaker Street, Spitalfields, covered lately 2½ pounds of copper wire with silk, for 2*s*. 6*d*. The first cost of the wire, which was No. 16, was 1*s*. 3*d*. per pound; thus I had 2½ pounds of well covered wire for 5*s*. 7½*d*. Smaller wire in proportion. When covered with cotton it is about 4*d*. per pound less. A full description of the apparatus by which wires are covered may be seen in the Mechanics' Magazine, No. 717.

With best wishes for the continued success of your excellent Annals,

I remain,
Yours respectfully,
G. FRANCIS
55, *Great Prescott Street*

G. FRANCIS, Letter to the Editor, *Annals of Electricity*, 1838, **2**, 396.

7e Clerk Maxwell on 'Faraday's Lines of Force'

Dec. 10.—*A paper was read by Mr. Maxwell on Faraday's Lines of Force*

THE method pursued in this paper is a modification of that mode of viewing electrical phænomena in relation to the theory of the uniform conduction of heat, which was first pointed out by Professor W. Thomson in the Cambridge and Dublin Mathematical Journal, vol. iii. Instead of using the analogy of heat, a fluid, the properties of which are entirely at our disposal, is assumed as the vehicle of mathematical reasoning. A method is given by which two series of surfaces may be drawn in the fluid so as to define its motion completely. The uniform motion of an imponderable and incompressible fluid permeating a medium, whose resistance is directly as the velocity, is then discussed, and it is shown how a system of surfaces of equal pressure may be drawn, which, with the two former systems of surfaces, divides the medium into cells, in each of which the same amount of work is done in overcoming resistance. It is then shown that if the fluid be supposed to emanate from certain centres, and to be absorbed at others, the position of these centres can be found when the pressure at any point is known; and that when the centres are known, the distribution of pressures may be found. Methods are then given by which the motion of the fluid out of one medium into another, the resistance of which is different, may be conceived and calculated, and the theory of motion in a medium in which the resistance is different in different directions is stated.

The mathematical ideas obtained from the fluid are then applied to various parts of electrical science. It is shown that the expression for the electrical potential at any point is identical with that of the pressure in the fluid, provided that 'sources' of fluid are put instead of positive electrical 'matter,' and centres of absorption or 'sinks' for negative 'matter.'

The theory of Faraday with respect to the effect of dielectrics in modifying electric induction, is illustrated by the case of different media having different conducting power; and it is shown, that, in order to calculate the effects by the ordinary formulæ of attractions, we must alter in a certain proportion the quantities of electricity within the dielectric, and conceive an imaginary distribution of electricity over the surface which separates it from the surrounding medium.

The theory of magnets and of the phænomena of paramagnetic and diamagnetic bodies is expressed with reference to the 'lines of inductive magnetic action;' and elementary proofs of the tendency of paramagnetic bodies toward places of stronger magnetic action and of diamagnetic bodies toward places of weaker action, are given. This distinction of paramagnetic and diamagnetic is not here used absolutely, but indicates a greater or less conductivity for the lines of inductive action than that of the surrounding medium.

The magnetic phænomena of crystals are then examined, and referred to unequal magnetic conductivity in different directions; and the case of a crystalline sphere in a uniform field of force is worked out.

The laws of electric conduction, as laid down by Ohm, are shown to agree with those of the imaginary fluid, and definitions of quantity and intensity are given, which will apply to magnetism as well as galvanism.

The theory of the attractions of closed circuits, as established by Ampère, is shown to lead to the following results:—

1. The total intensity of the magnetizing force estimated along any closed curve embracing the circuit is a measure of the quantity of the current.

2. The quantity of the current, multiplied by the quantity of inductive magnetic action, from whatever source, which passes through it, gives what may be called the potential of the circuit. The tendency of the resultant forces is to increase this potential.

The theory of Faraday with respect to the induction of currents in closed circuits takes the following form:—

When the quantity of inductive magnetic action which passes through a given circuit changes in any way, an electromotive force proportional to the rate of change acts in the circuit, and a current is produced whose quantity is the electromotive force divided by the total resistance of the circuit.

The mathematical discussion of the electro-magnetic laws is reserved for another communication.

J. C. MAXWELL, 'Faraday's Lines of Force', *Phil. Mag.*, 1856, **11** (4), 404.

8 ORGANIZATION OF SCIENCE

8a Towards a British Association

ART. I.—Meeting of the Cultivators of Natural Science and Medicine at Hamburgh, in September 1830. *By* James F. W. Johnston, *M.A. &c. &c. Communicated by the Author*

> Während man in der ältern zeit die naturforschung als eine angenehme aber nutzlose Beschäftigung und als ein harmloses Spielzeug müssiger Köpfe ansah, hat man sich in der neuesten zeit immer mehr von ihrem grossen Einfluss auf den Cultur-Zustand und das Wohl der Völker überzeugt:—und so sehen wir die Lenker der Völker bemüht grossartige Anstalten zu ihrer Beförderung und Erweiterung zu errichten.
>
> Whereas, in former times, men regarded the inquisition of nature as a pleasant but useless employment, and as a harmless pastime for idle heads, they have, of late years, become every day more and more convinced of its influence upon the civilization and welfare of nations, and the leaders of the people are everywhere bestirring themselves for the erection of establishments to promote its advancement and extension.
> *Tiedeman's Address to the Meeting at Heidelberg in* 1829.

MANY illustrations of the fact stated in this sentence of Tiedeman's oration are to be found in our own country, in which the change of public opinion in regard to scientific pursuits has been nearly as great as in any other; but no single illustration of it to be met with in any country is more strikingly instructive, than what is contained in the history of the Society of German Scientific Men. Commencing at its outset with the trifling concourse of some twenty lovers of natural science, it has every year augmented and grown,—despite of the openly avowed hostility of some governments, and the secret espionage of others,—till in the short space of nine years, it has attained to the character of a great national congregation, of which the most distinguished naturalists of the age are proud to be members, and which kings vie with each other in honouring. At first a few cities only were open to them, and the dread of political associations shut many gates against them;—now their task is to choose among many rival claimants, each of which would gladly entertain them. They are borne along now by the tide of public opinion, directing at once and directed by it; and the honour formerly bestowed willingly on individuals, from a knowledge and appreciation, in some degree, or their labours, is now bestowed with equal cheerfulness, and with great increase, on the whole body of indefatigable men, of which individual philosophers are but members.

The Society of German Naturalists owes its origin to Professor Oken of Munich. This indefatigable and free-minded man was formerly Professor of Natural History in the University of Jena in Weimar, to which chair he was appointed in 1807. He was already favourably known for five or six volumes in natural science, especially zoology; and, amidst his professional labours, he found leisure for adding every year one or two to their number. Among these was *A System of Natural Science*, *A Treatise on Light and Heat*, and *A System of Natural History*. In 1817 he commenced at Jena a monthly journal of Literature and Science in a quarto form, of which he still continues to be the editor. But the times were critical, or men in power, at least, thought them so. The principles of the Holy Alliance demanded a strict surveillance of the periodical literature; and it was dangerous for small states to give countenance to liberal men, or to permit political treatises to be published in their dominions. Oken cared little for men in power. He acted independently, and admitted into his Journal some articles of a political nature, which gave high offence. The consequence was the intimation, 'either you must discontinue the *Isis*, or give up your chair.'—'I told them,' said Oken, 'I cared nothing for their chair, and I would go on with the *Isis* in spite of them.' Of course, he lost his chair; and, though he was allowed to remain, the *Isis* was forbidden to be published in Weimar. The publication therefore was transferred to Leipzig, while Oken continued to reside at Jena. In 1827 he obtained a chair from the King of Bavaria in the university of Munich, where he is now professor of physiology.

Oken is a little man, probably near fifty years of age, of dark, yet sanguine complexion, and features whose habitual, if not natural, expression, is severity and determination. His dark eye and compressed lips have a forbidding and distance-keeping expression, for one can read upon them our own national motto, 'Nemo me impune lacessit.' I do not know how far his power of saying severe things corresponds with his apparent inclination; but, if the one equal in amount only half of the other, I should be very sorry indeed to come under his lash.

In conversation Oken is nevertheless pleasant and communicative; and I shall not soon forget the buzz and general sensation,—the turning of eyes, and moving of feet, in the rooms of the Apollo Saal, on the occasion of our first public soirée, when the words, 'Oken is come!' were passed along the assembly. His friends crowded first to greet him, after which foreigners and other strangers were severally introduced; and one could easily forgive the slight air of patronage with which he, though the last comer, *welcomed* them to the meeting, when we considered how goodly an assemblage his efforts had brought together.

It was in the *Isis*, and while still at Jena, that Oken proposed the plan of a great yearly meeting of the cultivators of natural science and medicine, from all parts of the German Fatherland. It was a noble idea, and nobly has it at length been brought about. But in 1821 Oken was still a tainted man,—the remembrance of his political sins was still fresh,—and his proceedings were consequently regarded with suspicion. Societies of all sorts were dreaded by the German governments, and they feared some hidden and dangerous design under the guise of a concourse of philosophers. But open influence could not be exerted against that which as yet had no being; and, in 1822, the first meeting took place at Leipzig, Dr Schwägrichen, Professor of Botany, in the chair. But what a contrast did this first meeting present to those of the last three or four years. At Leipzig there were about twenty came together from the city, and these were joined by about a dozen strangers. It was, however, a beginning. In 1823 they met in Halle in greater force, the celebrated botanist Sprengel being president, and Schweigger, well known for his Journal, which he has edited so indefatigably for twenty years, being secretary. Wurtzburg, famed as a medical school, was the seat of the third assemblage, D'Outrepont and Schönlein, of the medical faculty of that University, holding office. They now began to muster strong, both in numbers and in talent; and here the meeting first obtained a consistency and fixed establishment. Frankfort received them hospitably in 1825; and the scientific men, and the authorities of the city, united in showing attention to the strangers. This place had the honour of first adding public respect to the private entertainments got up by the inhabitants to gratify their visitors; and, among these private entertainments, that of Banker Bathman deserves especial commemoration. If not the first, it was the greatest yet paid to the entire body since their assembling commenced. Thus Frankfort, though the seat of no university, has a claim to much merit as a patron of scientific men. At Dresden, in the following year, preparations were also made for their reception; and the learned men connected with that seat of the fine arts, exerted themselves to make their visit a pleasant one. Seiler, the Director of the Surgical Academy, was president, and Carus the Anatomist was secretary. The sixth meeting in 1827 was held at Munich, the seat of a flourishing University, opened only the preceding year under the favouring auspices of Louis Maximilian of Bavaria. This city also deserves well of the Society, and the attentions of the King were such as it had not hitherto experienced. Besides general attention to the comfort and accommodation of the whole body, particular attentions were paid to the individual members; and each person, during the period of his stay, had an invitation to dine at least once in the palace. They began now to reckon their number by hundreds; and the amount and variety of subjects brought forward at their public meetings have increased beyond expectation, it was found necessary to break themselves up into sections, of which the botanists, an amiable and enthusiastic race of men, first set the example. Thus time was gained; men of like tastes and pursuits brought more frequently and more closely together; and every one spared the infliction of dissertations and discussions upon the thousand and one subjects in which he felt no earthly interest; for, though all cultivators of natural science rejoice in the advancement, and admire those who successfully cultivate any one department, yet each one has his own favourite branch or branches, beyond which he has little anxiety to roam, and unconnected with which discussions, however learned, are often only tiresome. It was a judicious plan, then, to make the separation into sections, and thus to permit the shell and fly men to discuss the mysteries of their several ologies, without scandalizing the more *grave* and *weighty* pursuits of medicine and oryctognosy. This practice, begun at Munich, assumed a more extended and definite form at Berlin, and was finally arranged and consolidated at Heidelberg.

Berlin gave a powerful impulse to the rising distinction of the Deutscher Naturforscher Versammlung. Every thing was done by the Prussian government, and under the immediate superintendence of the distinguished president, Baron Humboldt, for the convenience and accommodation of the strangers; and arrangements were there first entered into by which the comfort of those from a distance was materially increased. Purses, even, as well as persons were attended to; and as living in hotels was considered too expensive for many, who, nevertheless, in a strange place, would be unable to provide themselves with private lodgings, several hundreds of the latter were secured, chiefly in situations which gave convenient access to the places of meeting; and thus, the most complete strangers found themselves at once economically and conveniently situated. Fetes and excursions were also got up, and concerts given, which the royal family, and even the king himself, graced with his presence; while poetry and music lent their aid to welcome and eulogize the votaries of science. The celebrated Humboldt presided, and Lichtenstein, the well known zoologist and South African traveller, held the office of secretary. The number of strangers who came from various parts of Germany and the northern countries, was 269 and the total amount of members enrolled was 464. In this numerous assemblage of learned men, England had but one representative. Yet one man may be worth a host, and the science of England suffered no derogation in the person of 'Charles Babbage, London.' This meeting at Berlin was by far the most splendid that has yet taken place, not from the number of strangers who attended it, for in that respect it was nearly equalled by the late meeting at Hamburgh, but from the circumstance of its being held in the capital of a powerful kingdom, where the government had shown a disposition to honour, and pay attention to scientific men,—in Berlin, the seat of the first university in Germany, where the professors of every science are among the most eminent men, and the collections in every department of natural history of the most splendid description to be found in the whole empire. It remains to see what Vienna can do, in the ensuing September, to rival the more northern metropolis.

The beautiful and romantic city of Heidelberg was the seat of the following anniversary: Tiedeman and Gmelin, whose names have so long illustrated the university of Heidelberg, and whose labours have so much increased and diffused the knowledge of the anatomical and chemical sciences, held on this occasion the two official situations. At this meeting the number of strangers amounted only to 193, and the total amount of members to 273, but it proved, nevertheless, a very interesting and satisfactory assemblage. . . .

The first object of these meetings is to promote this acquaintance and friendly personal intercourse among men of science; but other great and perhaps more important benefits grow spontaneously out of them. They draw public attention to science and scientific men, and make people inquire concerning both them and their pursuits. They exalt science in general estimation, and with it those who devote themselves to its advancement; and, above all, they spur on the governments of the different states to examine into and ameliorate the condition of their scientific institutions; and to seek for men of true science to fill the chairs of public instruction. Such and similar benefits have already resulted from the meetings in Germany. Might not similar results in our own country be looked for from a similar institution?

Portobello, 23d February 1831

J. F. W. Johnston, 'Meeting of the Cultivators of Natural Science and Medicine at Hamburgh, in September 1830', *Edinburgh Journal of Science*, 1831, **4**, 189.

8b Foundation of the Chemical Society

Old Burlington House, Piccadilly, in which the Society had Rooms from 1857 *to* 1873

By courtesy of the Chemical Society

First Page of the Obligation Book

By courtesy of the Chemical Society

We the undersigned, do hereby engage, that we will endeavour to promote the interests and welfare of the Chemical Society of London, and observe its Orders and Bye-laws, so long as we shall continue Members thereof.

T. S. Moore and J. C. Philip, *The Chemical Society 1841–1941*, London, 1947, plates 4, 5.

8c Babbage on 'The Position of Science' (1851)

Science in England is not a profession: its cultivators are scarcely recognised even as a class. Our language itself contains no *single* term by which their occupation can be expressed. We borrow a foreign word [*Savant*] from another country whose high ambition it is to advance science, and whose deeper policy, in accord with more generous feelings, gives to the intellectual labourer reward and honour, in return for services which crown the nation with imperishable renown, and ultimately enrich the human race.

The first question which presents itself to a government desirous of advancing science, is to consider what departments of knowledge it is important that it should reward. This is a point upon which much misunderstanding prevails, and with regard to which interested parties have studiously endeavoured to delude the public.

As the fund which can be applied to this purpose even by a generous nation is moderate, the first limitation of its application ought naturally to be,—to confine it to those discoveries which are from their very nature not immediately capable of becoming a source of profit.

One of the most common errors, is to reward persons who have merely acquired an extensive knowledge of various departments of science, but who have neither extended its boundaries by new methods, nor added new principles to its theories.

An analogous mistake often occurs to wealthy and benevolent persons residing in the country, who, finding in the son of their village blacksmith or other artificer, some great aptitude for figures, immediately conclude that if properly trained and then sent to College, he will turn out a great mathematician. Now although in very rare instances such cases may have occurred, the general result is quite different. The lad thus selected, if as is usually the case he is somewhat above the average intellect, will under such favourable circumstances probably acquire a considerable knowledge of science, and become a very respectable member of society. But if the benevolent person who thus totally changed the position in life of this young man, had first made inquiries at our national schools, he would probably have found several out of every hundred scholars, capable under similar treatment of acquiring a still larger amount of that knowledge.

With the increasing extension of science the labour of some of its details becomes excessive, and those who are able to afford the expense, gladly employ computers to relieve them from the more irksome portions of their toil. The reduction of astronomical and meteorological observations are of this kind. When once the formulæ to be used are decided upon, and a skeleton form is ruled or printed and a system of checks is devised, the remaining work may be executed by persons of very moderate attainments. This may be extended to the computation of the orbits of planets, of comets, and of double stars, and such assistance may usually be had on very moderate terms. In more extensive operations, the liability to error from the want of sufficient checks, and the great tediousness and even uncertainty of the result must remain, until mechanism shall entirely relieve the mind from these difficulties.

Let us now consider what is the present situation of men of science in England.

The estimate which is formed of the social position of any class of society, depends mainly upon the answer to these two questions:—

What are the salaries of the highest offices to which the most successful may aspire?

What are the honorary distinctions which the most eminent can attain?

Offices of a strictly scientific nature are few, and their salaries are generally of small amount: amongst these there are—

A few of the professorships at our universities.

The Astronomer Royal.

The Astronomers of some of our Colonial Observatories.

The Master of Mechanics to the Queen.

The Conductor of the Nautical Almanac.

The Director of the Museum of Economical Geology and of the Geological Survey.

Various officers of the same institution.

Some of the officers in the Natural History department of the British Museum.

The most valuable of these, that of Astronomer Royal, receives about 1,300*l.* a-year, including a pension of 300*l.*

Thus there is amongst this class one solitary prize of at the utmost 1,300*l.* a-year, and that is confined to one department of science.

Offices for which men of science are at least as fit as any other persons, are numerous, though they are very rarely attained by those who pursue it.

It may, perhaps, have been expected that the recent appointment of Sir John Herschel to the Mastership of the Mint, should have been noticed in the previous list. But until the motives which dictated it are known, I have no observation to make, except that it is gratifying to me to find that the great principle of the 'claims of science,' for which I have all my life been contending, has been thus as it were, unconsciously admitted by the minister: and had the accident of birth placed me in his position, the appointment would have been the same, although the motives for it might have been different.

Let us now turn to the *honorary distinctions* which await science. During the eleven years of the present reign, one solitary instance is to be found of a baronetcy given for science, and that too occurred only at a festival (the coronation) at which baronetages and peerages were showered upon those whose sole claim was founded on the mere support of party.

During the same interval, about half a dozen of those who cultivate science, have been knighted.

It appears then that the highest position a man of science can attain, and that but very rarely, is a baronetcy; that the highest salary is about 1,000*l.* a-year. When this is compared with the most successful prizes in the army, the navy, the church, or the bar, it shows at once the inferior position occupied by science.

Connected with the navy is an office which ought to be held by a person eminently uniting science with practical skill. The Surveyor-General of the Navy has to decide upon questions of the greatest difficulty. The mathematical theories and inquiries on which the various qualities of sailing vessels and steamers depend, are of the most complicated kind, and are not even yet sufficiently advanced to serve as secure and absolute guides. Yet without a knowledge of their present state, and a power of advancing those theories, it is hopeless to expect the greatest and most valuable additions to the science of naval architecture. This can only be accomplished by one who combines a great facility in applying such portions of them as admit of it, to the practical facts which experience is continually bringing to light.

The talent for commanding a fleet is by no means rare: the most successful in that line may attain fortune, the peerage, and a large pension. The talent for investigating the laws regulating the forms of ships, is of the very rarest order. Even if its possessor should happen to be of the naval profession, his greatest reward could only extend to knighthood, and a thousand a-year during the tenure of an office of great labour. Of course, naval men having the requisite talents, would never turn them into so unprofitable a direction: yet it would be difficult to say how many millions of money have been, and continue to be, uselessly expended for want of that knowledge.

Amongst those situations in the appointment of the government, there are many in which a knowledge of various branches of science is highly useful. A considerable number of these are filled by officers of engineers, artillery, and other corps of the army and of the navy. Thus those whose service is already paid for by the country, are excused from doing their ordinary duty, and are paid again for doing another and perhaps a more agreeable duty.

Under the delusive plea that *military* and *civil* engineering are the same science, military engineers have been placed in situations for which they were unfit, and civil engineers have been excluded, to the injury of that profession, and to the much greater damage of the country. The Ordnance Magnetical Observatories will furnish an example of the *economy* which, it is pretended, results from such arrangements.

Some ten or twelve years ago, it was proposed by Humboldt that various governments should establish magnetical observatories at different points on the earth's surface, so chosen that by the united information thus obtained, we might arrive at more accurate and correct ideas of the state of the earth's magnetism. That plan has been pursued with great advantage to science. A magnetical observatory was built at Greenwich, and continuous observations were made which have been reduced and published annually under the direction of the Astronomer Royal. The expense* of the Magnetic and Meteoric Observatory, excluding that portion of the Astronomer Royal's salary which may be considered due to his services in the direction of this department, but *including the whole of the making and recording the observations themselves*, is 720*l*. annually.

There are other magnetical observatories in several of our colonies in which observations are made. These observations appear to be sent for reduction to an establishment at Woolwich, under the superintendence of Colonel Sabine.

Now the first and most obvious course would have been to have employed an additional number of computers at Greenwich, who should use the same formulæ and methods of reduction. This would ensure perfect uniformity, and would apparently be the most economical plan.

The course that is actually pursued is to have a separate establishment at Woolwich, with an officer, and several non-commissioned officers on extra pay, so that the account stands thus:—†

	£	s.	d.
One officer, extra pay	182	10	0
One non-commissioned officer, ditto	27	7	6
Three non-commissioned officers, ditto	68	8	9
Contingent, not exceeding	200	0	0
Apparent expense	£478	6	3

But to this must be added—

	£	s.	d.
The full pay of Lieut.-Colonel	300	0	5
His extra pay	273	15	0
Carried forward	£1,052	1	8

* See App. to Rept. of Select Com. on Misc. Expenditure, p. 222.

† See p. 221 App. to Rept. on Misc. Expenditure, p. 848 (543) II.

	£	s.	d.
Brought forward	1,052	1	8
Full pay of one officer, if a Captain	192	16	3
Ditto one non-commissioned ditto*	20	0	0
Ditto three ditto ditto	50	0	0
Real expense	£1,314	17	11

In the estimate for civil service for 1850† the following items occur:—

	£	s.	d.
Extra pay to Colonel Sabine, Royal Artillery, for services in connexion with the Magnetic and Meteorological Observations, for ten years, from 7th May, 1839, to 7th May, 1849, at 15s. a-day	2,739	15	0
Deduct 3s. 4d. per day granted him from 1st June, 1841, to 7th May, 1848, as compensation for loss of command pay	434	8	0
	£2,305	7	0

This certainly requires an explanation. Here is an officer not doing the services of his profession, who it seems has been allowed a compensation for what he *might* have received if he *had* rendered those services: notwithstanding which, at the end of ten years, he claims and is allowed the above sum of £2,305 7s. for services the payment of which it would seem by this account was never contemplated during those ten years.

It is also to be remarked that Colonel Sabine does not reside at Woolwich, where the only effective portion of the work is carried on.

But to return to our argument: it is singular that even the principles on which science ought to be rewarded, are not entirely settled.

Should all equally great discoveries be rewarded in the same way, without regard to the different positions in society which the discoverers occupy? If this principle were admitted, the rewards must be very large, or there would be none for the higher classes of society.

Of all steps in the social scale, that which first elevates a man into the class of Gentlemen is by far the greatest. In this country, where the differences of rank are great, there is fortunately, until we approach royalty, no absolute line of demarcation between any classes, except the one alluded to; even the peerage to a private gentleman is not so great an advance.

It is without doubt very desirable that all classes should contribute to the intellectual advancement of the country. But unless different advantages are proposed to different classes, it is not possible to apply any general stimulus to all.

Those who maintain that science is its own reward, cannot have remarked the vicious circle in which they reason. The delight derived from discovery is indeed a high intellectual reward, but the force of this maxim is only known practically to those who have already advanced in the career of discovery: it can, therefore, never direct the inquirer into that line. All men are subject to the same feelings and passions. It is assuredly true that men of wealth and rank will be happier if they cultivate their faculties, and add to the amount of human knowledge: but they cannot be aware of this truth until they are considerably advanced, consequently it cannot have induced them to commence this cultivation.

* The pay of the non-commissioned officers has been assumed as somewhat less by ten per cent. than their extra pay.

† See p. 41, App. to Rept. on Misc. Expenditure, p. 848, (268) IV.

But it is for the interest of those who are the consumers of knowledge, that all other minds should be induced to advance it: therefore it is our interest to place even before the highest classes, at the commencement of their career, motives for its pursuit. Having raised such expectations, justice requires us to fulfil them; nor can we regret that the advantages derived from the course into which we have invited them, should have proved beneficial to them beyond even the limits of our prediction.

It is of the very nature of knowledge that the recondite and apparently useless acquisition of to-day, becomes part of the popular food of a succeeding generation. Thus the nobleman who spends his wealth in constructing unrivalled instruments, and his nights in scrutinizing with them the remotest boundaries of space into which human vision has yet penetrated, is preparing a source of pleasure and happiness for the descendants of those very peasants whom his practical skill in engineering has raised by his own instructions above the ranks in which he originally found them.

Another question has been raised, but not yet answered, respecting those pensions which have been awarded for scientific discoveries. A certain definite limit has been fixed by practice, which has never yet been exceeded in pensions assigned to science. The sum of three hundred a-year, the maximum of reward to science, is almost the minimum of reward for other services.

The most important question is, Whether these pensions are given as the reward of scientific services rendered to the country, or as charity to enlightened and studious persons who happen to be poor? In the one case, they are an honour which a philosopher may be proud of receiving from his country: in the other, they are no more than a higher order of pauper relief, which an independent gentleman can scarcely condescend to accept.

Another important question, though of a different nature, also arises here. Are these pensions, thus small in amount, fit to be offered to those who, in order to arrive at their discoveries, have themselves in some cases spent out of their own private fortune, sums far larger than the fee simple of the rewards thus offered to them?

Is it just that the *same rewards* should be given to persons filling well-endowed scientific offices, supplied with all the means of discovery which the most perfect art can produce, as to other philosophers, who, at the expense of their own personal comfort and perhaps of the interests of their family, have purchased the costly means by which they have succeeded in *equally* improving their several departments of science?

For the honour and the advancement of science, it is necessary that these questions should be distinctly answered. It is to be hoped that some independent member of parliament will at last press them in a manner which no ministerial shuffling can evade.

C. Babbage, *The Exhibition of 1851*, London, 1851, p. 189.

8d 'The X-Club'

In the year 1863 a small number of earnest scientific workers met each other at the Athenæum, Royal Institution, Royal Society, and elsewhere. But these meetings were, as a rule, quite accidental and fragmentary as regards numbers;

only two or three being, perhaps, gathered together on each occasion. These meetings, although casual and usually of short duration, were so pleasant and congenial that the desirability of periodical gatherings of longer duration occurred to some of us, and it was decided that we should dine together monthly during the nine months of the scientific session—namely, from October to June inclusive.

Our little party thus formed consisted of nine members representing the Sciences of Mathematics, Physics, Chemistry, and Biology. It has never been added to and has only been diminished by death, which has reduced our numbers to four (1897).

The Club originally consisted of G. Busk, T. A. Hirst, J. D. Hooker, T. H. Huxley, J. Lubbock, Herbert Spencer, W. Spottiswoode, John Tyndall, and myself. Three of our numbers have since been Presidents of the Royal Society, three have received the Copley medal—the highest honour the Royal Society can confer—I have been president of other learned societies, one is a baronet, one a member of H.M. Most Honourable Privy Council, one is granted a cross of the Most Honourable Order of the Star of India, one K.C.B., and one who repudiates and refuses all honours and distinctions will be admitted by all to belong to the most eminent of living moral philosophers.

We agreed to meet and dine together on the first Thursday of each of the months already mentioned, except in the month of June, when the meeting was to be on one of the Saturdays in June. It was to be held somewhere in the country, to extend from Saturday to Monday, and to be attended by the wives of members.

The ordinary meetings were held successively at St. George's Hotel, Albemarle Street, Almond's Hotel, Clifford Street, and the Athenæum Club. The country meetings were most held at Maidenhead, but sometimes near the foot of Leith Hill, at Oxford, and at Oatlands Park. The ordinary meetings were announced upon a postcard thus—

$$X = 2 \text{ (the day of the month),}$$

whilst the country meetings were similarly announced thus—

$$X \times uyvs = 5.$$

At the first meeting the subject of the rules for the Club was discussed, and it was resolved that there should be only one rule, namely, 'That there shall be no rules.'

Each of the members held in turn the offices of treasurer and secretary, his duties being to collect subscriptions, send off notes of meetings, order the dinners, pay the bills, and keep rough minutes of the proceedings.

The ordinary meetings of the Club were very well attended. There was rarely an absentee, and still more rarely two or more. The country meetings were also well attended, but, owing to deaths, they fell into abeyance in the year 1874 and were never renewed.

In illustration of the character of the meetings, the minutes (kept by Hirst) of the first eight are here given.

First Meeting

At St. George's Hotel, Albemarle Street, November 3, 1864

Present: Busk (in the chair), Hirst, Hooker, Frankland, Lubbock, Spencer, Tyndall, Huxley.

Resolved that meetings be held on the first Thursday of every month, from October to June (inclusive), and that the June meeting be held in the country.

The re-organisation of the 'Reader' was discussed.

Dinner Bill . . . *6l.* 3*s.* (15*s.* 4½*d.* each).

December 8th, SECOND MEETING, 1864
(*Postponed a week in consequence of Royal Society Anniversary Dinner*)

Present: Hirst, Hooker, Huxley, Lubbock, Spencer, Tyndall.

Absent: Busk, Frankland.

Spottiswoode was elected a member.

Resolved that the Bishop of Natal be invited as a guest.

Dinner Bill . . . 4*l.* 2*s.* (13*s.* 8*d.* each).

January 5th, THIRD MEETING, 1865

Present: Hooker, Frankland, Hirst, Huxley, Lubbock, Spencer, Spottiswoode, Tyndall.

Absent: Busk.

Hirst elected Treasurer.

Resolved that each member, whether present or absent, shall bear his share of the evening's expenses; and that he shall, when possible, give previous notice of his absence to the treasurer.

Fergusson was suggested as a tenth member.

Dinner bill . . . 5*l.* 6*s.* 9*d.*
Stationery . . . 0*l.* 1*s.* 7*d.* } (13*s.* 6½*d.* each).

February 2nd, FOURTH MEETING, 1865

Present: Huxley, Busk, Frankland, Hirst, Hooker, Lubbock, Spencer, Spottiswoode, Tyndall.

The claims of several candidates now proposed for admission to the Athenæum and Royal Society formed one of the subjects of conversation.

Dinner bill . . . 5*l.* 14*s.* 6*d.*
Postage . . . 0*l.* 0*s.* 3*d.* } (12*s.* 9*d.* each).

March 2nd, FIFTH MEETING, 1865

Present: Spencer, Busk, Frankland, Hirst, Hooker, Huxley, Lubbock, Spottiswoode, Tyndall.

It was agreed that Busk should invite Fergusson to join us, provided the latter should wish to do so. The desirability of again proposing Colenso as a visitor to the Athenæum was discussed.

Dinner bill . . . 6*l.* 0*s.* 6*d.*
Postage . . . 0*l.* 0*s.* 3*d.* } (13*s.* 5*d.* each).

April 6th, SIXTH MEETING, 1865

Present: Busk, Frankland, Hooker, Huxley, Lubbock, Spencer, *Spottiswoode*.

Absent: Tyndall, Hirst.

The Bishop of Natal and Mr. Wilson were present as guests.

Dinner bill . . . 5*l.* 16*s.* 0*d.*
Postage . . . 0*l.* 0*s.* 3*d.* } (12*s.* 11*d.* each).

May 4th, SEVENTH MEETING, 1865

Present: Busk, Frankland, Hirst, Hooker, Lubbock, Spencer, Spottiswoode, Tyndall.

Absent: Huxley.

It was resolved that the next meeting should be held in the country, and that the wives of members should be invited. Spottiswoode was requested to make enquiries with a view of deciding upon the place and time of meeting.

Dinner bill . . . 5*l.* 15*s.* 3*d.*
Postage . . . 0*l.* 0*s.* 3*d.* } (12*s.* 10*d.* each).

June 10th and 11th, EIGHTH MEETING, 1865
Held at Maidenhead

Present: Mr. and Mrs. *Busk*, Professor and Mrs. Frankland, Professor and Mrs. Huxley, Mr. and Mrs. Lubbock, Mr. and Mrs. Spottiswoode, Professor Tyndall, Mr. H. Spencer, Mr. T. A. Hirst.

Absent: Dr. and Mrs. Hooker.

Rooms were secured for the party at the Orkney Arms, Maidenhead. After dinner on Saturday the 10th, an excursion on the Thames was made. On Sunday the party proceeded in carriages to spend the day at Burnham Beeches. On their return Max Müller, Esq., dined with them, and at 8.30 they returned to town.

	£	*s.*	*d.*
Hotel expenses	21	18	0
Railway fares, carriages, boat, &c. . .	8	8	0
	30	6	0

To be divided by eight (3*l.* 15*s.* 9*d.* each).

The topics of conversation at these meetings may, perhaps, be best recorded by the following brief extract from the minute book:

The following four names were suggested as new members of the Council of the Royal Society: Roscoe, Rolleston, B. Stewart, and Dr. Thomson. Lubbock's proposition was discussed of the foundation of a Christie Lectureship at the Royal Institution.

The advisability of Tyndall's acceptance of the Professorship of Physics at Oxford was a subject of conversation. The present method of nominating members of the Council of the Royal Society was discussed, and it was resolved that some effort to alter it should be made before next year.

Spottiswoode informed us that the Liberal party at Oxford were about to try to utilise the present movement for university extension, originated by the theological party. The former would be glad to receive support from the friends of Science outside the university. A conversation ensued relative to the changes which ought now to be introduced into school education generally.

Huxley's forthcoming Sunday lecture at St. Martin's Hall, and the Sunday League generally, were subjects of conversation. Spencer spoke of some of the results of his late botanical inquiries.

Frankland proposed that we should take into consideration some method of hastening the publication of papers in the 'Phil. Trans.' Hirst read a letter from the Secretary of the Sunday League in reference to the late suppression of the Sunday lectures by the Sabbatarians.

The present unsatisfactory mode of election of the Council of the Royal Society was discussed. Frankland, Hirst, and Spottiswoode expressed their intention of bringing the subject before the Council as soon as possible.

Autobiographical Sketches from the Life of Edward Frankland, London, 1902, p. 148.

8e 'The B-Hive Club'

By courtesy of Bristol Central Library

A. Scott, 'The B-Hive Club', *J. Chem. Soc.*, 1913, 343.

8f J. C. Brough on 'Modern Chemistry'

MODERN CHEMISTRY

I'm all in a flutter; I scarcely dare utter
  The words I have set to a jingle;
For I see at this table philosophers able,
  Whose ears at my verses will tingle.
Still, I don't mind confessing, I'm fond of expressing
  My notions and thoughts, in defiance
Of every great gun who can't see the fun
  Of winnowing chaff out of science.

I've read till I'm weary books weighty and dreary
  In which certain chemists seem aiming
To prove to outsiders they're excellent riders
  Of hobbies in theory and naming.
With 'monads' and 'diads', and 'pentads' and 'triads,'
  My brain has been addled completely;
And what's really meant by 'something-valent,'
  Is a question I give up discreetly.

Though Frankland's notation commands admiration,
  As something exceedingly clever,
And Mr. Kay Shuttleworth praises its subtle worth,
  I give it up sadly for ever:

Its brackets and braces, and dashes and spaces,
And letters decreased and augmented
Are grimly suggestive of lunes to make restive
A chemical printer demented.

I've tried hard, but vainly, to realise plainly
Those bonds of atomic connexion,
Which Crum Brown's clear vision discerns with
Projecting in every direction. [precision
In fine, I'm confounded with doctrines expounded
By writers on chemical statics,
Whom jokers unruly may designate truly
As modern atomic fanatics.

I turn for instruction to Brodie's production,
But stick at the famous equations
Which make chlorine fare as 'alpha ki square,'
Or the product of three operations.
It may be the case that the 'unit of space'
Requires symbolic expression;
But I cannot extract any notions exact
From Sir Benjamin's daring aggression.

For years I received the doctrines believed
About acids with much satisfaction,
And constantly swore H_2SO_4
Was an acid above all detraction;
But Williamson's views my notions confuse,
And make me once more undecided
Whether old SO_3 the acid should be,
Or merely a fragment divided.

When Odling with unction dilates on a function,
I sink out of sheer inanition;
For I find his 'aplones' and 'diamerones'
Indigestible mental nutrition.
In fact, I am dazed with the systems upraised
By each master of chemical knowledge,
Who seems to suppose that truth only grows
In the shadows of one little college.

J. C. B.

March, 1868.

A. Scott, 'Presidential Address', *J. Chem. Soc.*, 1913, 346.

9 SCIENTIFIC AND TECHNICAL COMMUNICATION

9a 'Conversations on Chemistry'

CONVERSATION I: On The General Principles Of Chemistry

MRS. B.

HAVING now acquired some elementary notions of NATURAL PHILOSOPHY, I am going to propose to you another branch of science, to which I am particularly anxious that you should devote a share of your attention. This is CHEMISTRY, which is so closely connected with Natural Philosophy, that the study of the one must be incomplete without some knowledge of the other; for it is obvious that we can derive but a very imperfect idea of bodies from the study of the general laws by which they are governed, if we remain totally ignorant of their intimate nature.

CAROLINE

To confess the truth, Mrs. B., I am not disposed to form a very favourable idea of chemistry, nor do I expect to derive much entertainment from it. I prefer those sciences that exhibit nature on a grand scale, to those which are confined to the minutiæ of petty details. Can the studies which we have lately pursued, the general properties of matter, or the revolutions of the heavenly bodies, be compared to the mixing up of a few insignificant drugs?

MRS. B.

I rather imagine that your want of taste for chemistry proceeds from the very limited idea you entertain of its object. You confine the chemist's laboratory to the narrow precincts of the apothecary's shop, whilst it is subservient to an immense variety of other useful purposes. Besides, my dear, chemistry is by no means confined to works of art. Nature also has her laboratory, which is the universe, and there she is incessantly employed in chemical operations. You are surprised, Caroline; but I assure you that the most wonderful and the most interesting phenomena of nature are almost all of them produced by chemical powers. Without entering therefore into the minute details of practical chemistry, a woman may obtain such a knowledge of the science, as will not only throw an interest on the common occurrence of life, but will enlarge the sphere of her ideas, and render the contemplation of nature a source of delightful instruction.

CAROLINE

If this is the case, I have certainly been much mistaken in the notion I had formed of chemistry. I own that I thought it was chiefly confined to the knowledge and preparation of medicines.

MRS. B.

That is only a branch of chemistry, which is called Pharmacy; and though the study of it is certainly of great importance to the world at large, it properly belongs to professional men, and is therefore the last that I should advise you to study.

EMILY

But did not the chemists formerly employ themselves in search of the philosopher's stone, or the secret of making gold?

MRS. B.

These were a particular set of misguided philosophers, who dignified themselves with the name of Alchemists, to distinguish their pursuits from those of the common chemists, whose studies were confined to the knowledge of medicines.

But, since that period, chemistry has undergone so complete a revolution, that, from an obscure and mysterious art, it is now become a regular and beautiful science, to which art is entirely subservient. It is true, however, that we are indebted to the alchemists for many very useful discoveries, which sprung from their fruitless attempts to make gold, and which undoubtedly have proved of infinitely greater advantage to mankind than all their chimerical pursuits.

The modern chemists, far from directing their ambition to the imitation of one of the least useful productions of inanimate nature, aim at copying almost all her operations, and sometimes even form combinations, the model of which is not to be found in her own productions. They have little reason to regret their inability to make gold (which is often but a false representation of riches), whilst by their innumerable inventions and discoveries, they have so greatly stimulated industry and facilitated labour, as prodigiously to increase the luxuries as well as the necessaries of life.

EMILY

But I do not understand by what means chemistry can facilitate labour; is not that rather the province of the mechanic?

MRS. B.

There are many ways by which labour may be rendered more easy, independently of mechanics; but even the machine the most wonderful in its effects, the steam engine, cannot be understood without the assistance of chemistry. In agriculture, a chemical knowledge of the nature of soils, and of vegetation, is highly useful; and in those arts which relate to the comforts and conveniences of life, it would be endless to enumerate the advantages which result from the study of this science.

CAROLINE

But, pray, tell us more precisely in what manner the discoveries of chemists have proved so beneficial to society?

MRS. B.

That would be an unfair anticipation; for you would not comprehend the nature of such discoveries and useful applications, so well as you will do hereafter. Without a due regard to method, we cannot expect to make any progress in chemistry. I wish to direct your observation chiefly to the chemical operations of Nature; but those of Art are certainly of too high importance to pass unnoticed. We shall therefore allow them also some share of our attention.

EMILY

Well, then, let us now set to work regularly. I am very anxious to begin.

MRS. B.

The object of chemistry is to obtain a knowledge of the intimate nature of bodies, and of their mutual action on each other. You find therefore, Caroline, that this is no narrow or confined science, which comprehends every thing material within our sphere.

CAROLINE

On the contrary, it must be inexhaustible; and I am at a loss to conceive how any proficiency can be made in a science whose objects are so numerous.

MRS. B.

If every individual substance was formed of different materials, the study of chemistry would indeed be endless; but you must observe that the various bodies in nature are composed of certain elementary principles, which are not very numerous.

CAROLINE

Yes; I know that all bodies are composed of fire, air, earth, and water; I learnt that many years ago.

MRS. B.

But you must now endeavour to forget it. I have already informed you what a great change chemistry has undergone since it has become a regular science. Within these thirty years especially, it has experienced an entire revolution, and it is now proved that neither fire, air, earth, nor water, can be called elementary bodies. For an elementary body is one that cannot be decomposed, that is to say, separated into other substances; and fire, air, earth, and water, are all of them susceptible of decomposition.

EMILY

I thought that decomposing a body was dividing it into its minutest parts. And if so, I do not understand why an elementary substance is not capable of being decomposed, as well as any other.

MRS. B.

You have misconceived the idea of *decomposition*; it is very different from mere *division*: the latter simply reduces a body into parts, but the former separates it into the various ingredients, or materials, of which it is composed. If we were to take a loaf of bread, and separate the several ingredients of which it is made, the flour, the yeast, the salt, and the water, it would be very different from cutting the loaf into pieces, or crumbling it to atoms.

EMILY

I understand you now very well. To decompose a body is to separate from each other the various elementary substances of which it consists.

CAROLINE

But flour, water, and the other materials of bread, according to your definition, are not elementary substances?

MRS. B.

No, my dear; I mentioned bread rather as a familiar comparison, to illustrate the idea, than as an example.

The elementary substances of which a body is composed, are called the *constituent* parts of that body; in decomposing it, therefore, we separate its constituent parts. If, on the contrary, we divide a body by chopping it to pieces, or even by grinding or pounding it to the finest powder, each of these small particles will still consist of a portion of the several constituent parts of the whole body: these we call the *integrant* parts; do you understand the difference?

EMILY

Yes, I think, perfectly. We *decompose* a body into its *constituent* parts; and *divide* it into its *integrant* parts.

MRS. B.

Exactly so. If therefore a body consist of only one kind of substance, though we may divide it into its integrant parts, it is not possible to decompose it. Such bodies are therefore called *simple* or *elementary*, as they are the elements of which all other bodies are composed. *Compound bodies* are such as consist of more than one of these elementary principles.

CAROLINE

But do not fire, air, earth, and water, consist, each of them, but of one kind of substance?

MRS. B.

No, my dear; they are every one of them susceptible of being separated into various simple bodies. Instead of four, chemists now reckon upwards of forty elementary substances. These we shall first examine separately, and afterwards consider in their combinations with each other.

Their names are as follows:

LIGHT,	(2 *Alkalies.*),	SILEX,	PLATINA,	ZINC,	URANIUM,
CALORIC,	POTASH,	ALUMINE,	SILVER,	BISMUTH,	TELLURIUM,
OXYGEN,	SODA.	YTTRIA,	MERCURY,	ANTIMONY,	TITANIUM,
NITROGEN,	(10 *Earths.*),	GLUCINA,	COPPER,	ARSENIC,	CHROME,
HYDROGEN,	LIME,	ZIRCONIA,	IRON,	COBALT,	OSMIUM
SULPHUR,	MAGNESIA,	AGUSTINA.	TIN,	MANGANESE,	IRIDIUM,
PHOSPHORUS,	STRONTITES,	(25 *Metals.*),	LEAD,	TUNGSTEN,	PALLADIUM,
CARBONE,	BARYTES,	GOLD,	NICKEL,	MOLYBDENUM,	RHODIUM.

CAROLINE

This is, indeed, a formidable list!

Mrs Marcet, *Conversations on Chemistry*, London, 1806, vol. i, 1.

9b 'Science for All'

Until this last quarter of a century, a boy though able to read and write, was still kept back and hindered by the price and technical composition of books; but high minded and noble hearted men have arisen, and joined their gigantic energies to break down the barriers, and fill up the gulph, which separated the poor man from knowledge; and behold! books are put into his hands, at a low price, unincumbered and stripped of their hard and untelligible terms; and, if he will not read them, he has nobody to blame but himself—

Here it would be affectation, if not insult, to pass on without mentioning Brougham—I did not call him Lord Brougham, because *he* is a man upon whom titles are utterly thrown away, for they cannot dignify or ennoble *him*. As a Statesman, we know nothing of him here; but as the firm and mighty friend of the diffusion of knowledge; as the first of those master-minds who have given an irresistable impulse to society, which can end in nothing less than enlightening every class of people; as the foremost and most zealous of the promoters of the general good, he has an irresistable claim upon the admiration and respect of every benevolent mind.

An Association of which he is the head has been lately formed, under the title of 'the Society for the diffusion of useful Knowledge'. It consists of the first and greatest men in their different departments; and these are incessantly labouring for the benefit and information of the working classes. They have adopted and steadily pursued the noble design of making philosophy attractive, easy and cheap. The works they have already produced, (some of which are in our book-case,) will testify how successful they have been in their undertaking. The chief of these works is 'the Library of useful Knowledge,' where the different Arts and Sciences are separately taught, and explained, in a manner accessible to the lightest purse, or the most dull apprehension. The veil of mystery which has so long enveloped the works on Science, is here torn off, and the plain truth exposed that every man may be a Philosopher, if he will.

For sixpence or a shilling a reading Mechanic may now possess a plain, correct, and intelligible treatise upon any branch of knowledge with which he may wish to become acquainted; and with less labour and expense than it formerly required to learn Latin, he may become very tolerably acquainted with the leading principles of every science. To encourage and interest him in the noble pursuit, 'the Library of Entertaining Knowledge' is put into his hands—a work which will, I am sure, afford to a man of common capacity more amusement than all the trash of the novelists and romance writers.

The tendency and effects of such labours must obviously be, practically to shew that knowledge is useful, interesting and beneficial to all men; and that it is in the power of all men to obtain it. They will give men a habit of investigating and thinking for themselves, and thus destroy injurious monopolies, and weaken ill-directed influences. They will send hundreds, nay thousands,—of a new set of labourers into the boundless and half cultivated fields of science; to explore new tracts, find new riches, and add to the heap of existing knowledge. The same pursuit will exalt and improve those engaged in it, in every sense; and a spirit of generous emulation will rise up to stimulate, and push forward this new race of Philosophers. Idleness, Drunkenness, Debauchery, and the low catalogue of degrading vices, which are the companions of ignorance, will be scouted and despised; and the lower classes will become more useful, intelligent, and respectable.

If these should turn out to be the effects of the labours of a small number of men, where shall we fix the bounds of the benefit they have conferred upon their country, and mankind? and into what paltry insignificance shrinks the glory of Cæsars, Alexanders, and Buonapartes, when compared with theirs!—But further—

Although the furnishing of cheap and proper books was absolutely indispensible to the diffusion of knowledge, it was found that men whose habits of thinking and acting were already formed, required some additional encouragement and assistance; and the great advantages of associations, for the purpose of acquiring and cultivating knowledge, were soon felt and appreciated. By means of a combination and subscription among a number of persons, many facilities and opportunities are afforded to every individual which no one, by himself, could command. In this manner the more expensive books and apparatus could be provided; the student could be assisted by the advice and experience of others; and instruction by Lecture and Demonstration could be obtained.

But these benefits, great as they are justly considered, are trifling compared with that derived from the mutual intercourse, assistance, and emulation which such societies foster and encourage. The interchange of ideas and observations among numbers of men upon subjects of a scientific nature, could not fail, to be, and in the highest degree, important and beneficial, as well as to themselves, as to the cause they were united to support: and experience has realized, and even surpassed the most extravagant expectation.

Thus Associations were formed in different places, under different names; but all having the avowed object of acquiring and cultivating useful knowledge. Second to none among these was the first 'Mechanics' Institute' established by Dr. Birkbeck, at Glasgow, in the year 1800; and here it cannot be uninteresting to give you a hasty sketch of the circumstances attendant on the origin of what may be termed our Parent Society.

In the year 1799, Dr. Birkbeck was appointed Lecturer on natural and experimental Philosophy in the Andersonian Institution of Glasgow. It happened that, at this period, there was but one man in that large town who was a regular and established mathematical Instrument maker; and this person had so much employment, or was so dilatory in the execution of his orders, that the Doctor was frequently disappointed of the apparatus with which it was necessary to illustrate his Lectures. He resolved, however, to bear this inconvenience no longer, and, with his usual industry and perseverance, determined to become his own Instrument maker. He now visited the Joiner at his bench, the Smith at his forge, and the Turner at his lathe; and he soon observed the mechanics to be,—not what they have been represented, doltish and stupid as they were ignorant; but, what they really are,—acute observers and curious enquirers. Encouraged by that condescending affability which Dr. Birkbeck possesses in common with most men of generous and cultivated minds, they pressed round him for information with regard to the uses of the Instruments they were assisting him to construct. He answered their numerous questions in his own simple and forcible manner, and soon had the satisfaction to see a spirit of enquiry, and a desire for knowledge awakened among many of them.—He, then, used his influence to procure for some of them admittance to the Lectures he was giving at the College; and the use they made of this privilege, and the dissappointment expressed by others, who from the rules of the College, could not enjoy it, set him upon devising some other means of imparting to them that information and knowledge they were so anxious to acquire;—Pope has remarked,—

> Whene'er to mischief mortals bend their will,
> How soon they find fit instruments of ill!

and, when a man, with a head and a heart like Birkbeck's, is determined to do good to others, the means to serve are as readily found as the means to injure. While on a visit to his native county, (and Yorkshire may be proud of such a man,) he issued a prospectus of Lectures which he would deliver in the following Session. When the time came, he sent circulars to the various manufactories and work-shops of Glasgow, requesting lists of the more sober and intelligent workmen in each; and furnished them with tickets of gratuitous admission—Seventy-five Mechanics attended the first Lecture; the next Lecture (such was the deep impression made upon their minds) was attended by two hundred—Three hundred were present at the third Lecture, and five hundred at the fourth. The Society was now organized by the Doctor under the name of the 'Glasgow Mechanics' Institution' and regulated by laws, the plan of which has been imitated, in most of the large towns in this country. He continued to superintend, and lecture to the Institution, during the few years he remained in Glasgow; and had the gratification of finding that the Mechanics were regular, attentive, and grateful receivers of the benefits he conferred. The fruits of his labours soon blossomed and ripened: one mechanic communicated to him a new plan for an air-pump; another sent him a new method for determining the Sun's distance; and all united in the purchase of a handsome silver cup, which they presented to him, appropriately inscribed, and he received other tokens of their gratitude and esteem upon which he must look with delight far greater than upon 'the costly gifts of those who distribute from their abundance, not always from their hearts'. Well have the lines of the Poet been applied to this man;—

> What justly envied feelings warm the breast,
> Of him whose ardent soul can find no rest,
> But in imparting wisdom's light serene,
> And scattering science o'er life's dreary scene.

Dr. Birkbeck is now the President of the London Mechanics' Institution, which was established and grown up under his fostering care, and which is very prosperous and increasing.

We have now briefly considered the circumstances which, until the nineteenth century, have made learning and knowledge the property of the few, to the exclusion of the many; we have noticed the manner in which serious and effectual obstacles were opposed to the poor man's acquiring it; and how he was, as it were, thrown into and kept in a state of intellectual bondage;—we have said that cultivation and exertion will invigorate and enlarge the powers and capacity of the human mind; and that the more a man reads, studies, and observes, and the more he thus learns, the easier it will be to him to grasp and retain still more difficult and extensive knowledge.

We have seen how a few noble and enterprising men have brought about what may be well called a glorious revolution in science, by throwing open her doors to the humblest and poorest classes; by the establishment of Schools, by furnishing cheap and intelligible works on Philosophy, and by the formation of Associations throughout the kingdom, where men may meet together, and enjoy the best means of acquiring and cultivating useful and practical knowledge. Before this new system of enquiry and improvement, mystery and error are now rapidly declining. The days are gone when nobody was supposed to have any thing to do with Mechanics but the professed Engineer, and that nobody should meddle with Chemistry but the Chemist or the Apothecary.

It is true, indeed, now as ever it was, that the field of knowledge is too boundless in its extent, and too inexhaustible in its treasures, to be thoroughly cultivated, or entirely explored by one person, and it is as necessary now as ever it was, that its particular departments should be separately investigated, for practical purposes, by different classes of men;—but it is no longer practicable to exclude the multitude, or to prevent any man from having a general and useful acquaintance with what is equally important and interesting to all.

ALFRED SMITH, *An Introductory Lecture on the Past and Present Study of Science, as regards the Working Classes*, London, 1831, p. 18.

9c Whewell on 'University Science'

SECT. 6.—On the Sciences as Subjects of University Teaching

FROM what has already been said of the use of mathematics in university education, it will easily be inferred, that we cannot find, in any of the more modern physical sciences, anything that can fitly be substituted for that study. The effect of the clear insight of geometry or mechanics cannot be efficiently replaced by sciences which exhibit a mass of observed facts, and consequent doubtful speculations, as geology; or even by other sciences, as chemistry and natural history, which, though they involve philosophical principles, can only be learnt by presenting numerous facts to the senses. But though such sciences cannot do the work of mental cultivation, they are highly-valuable acquisitions to the student, and may very beneficially engage his attention during the later years of his University career. For although they do not constitute the *culture*, they belong to the *information* of the well-educated man; though his habits of thought must be *formed* among other subjects, they may be well *employed* on these. And it is well for the general sympathy and mutual understanding of the cultivated part of mankind, that they should have many subjects of common interest.

But we may say much more than this. A considerable general knowledge of the modern progressive sciences, is as requisite to connect the educated man with the future, as a thorough acquaintance with ancient literature is, to connect him with the past. Except he know what has been done and is doing, in the way of extending our knowledge of earth, its elements, and its inhabitants, how can he judge what are the probable prospects of our knowledge? And if he be indifferent to this, how can he feel that interest in the future fortunes of his race, which becomes a person of his lofty extraction? Some insight into the progressive sciences is an essential part of a liberal education, in any large sense of the term.

This consideration is in some measure connected with the choice of our course of mathematical reading. For mathematics has been one of the great instruments of the progress of the physical sciences; and the constant use of this instrument, and the efforts to make it more effective, lead to frequent modifications of its form and expansions of its powers; and thus occasion, from time to time, changes in the current system of mathematics. This change, so far as the use of mathematics in education is concerned, is an evil rather than a good: for in our design we do not wish our pupils to possess mathematics principally as information, nor even as an instrument, although in that way it is of great advantage to many persons; but as an intellectual discipline; and this end is best answered by teaching the stable system of a demonstrative science. But at the same time we must recollect, that we cannot give to our mathematical studies their true dignity, without showing the place they hold in the progress of science; and, for the reason which we have mentioned, namely, in order to nourish in our more advanced pupils an interest in that progress, we do not hesitate to introduce to them all new mathematical methods, which have a prominent and permanent importance. But this is salutary to the more advanced pupils only. No deviation from the plain instructive circle of ancient elementary mathematics is fitted for the first stages of the student's University progress. The mode in which his elementary mathematical studies best produce their effect upon him is, when they are presented in the luminous simplicity in which the Greek intellect contemplated them, not when they are disguised and obscured by being translated into the modern language of symbols. To learn this language, (a valuable lesson for the mind, if it be rightly taught,) is best made an ulterior step in the student's advance.

But the physical sciences are useful, not only as belonging to the information of the educated man, but also as supplying him with examples of Inductive Reasoning. The general rules and conditions of such reasoning have hitherto been very imperfectly pointed out; but a knowledge of the sciences gives any one the means of speculating for himself on the subject;—a subject of the strongest interest with reference to the future progress of knowledge. I have already said, that a practical instruction in Inductive Reasoning is not possible, except so far as it depends on the cultivation of the Deductive Faculty. We may lead men to feel the force of demonstration, but we cannot teach them to discover new truths. And I may repeat my observation, that this use of the practical teaching of the reason by means of mathematics may be much enhanced by a proper selection of our mathematical studies;—by not confining ourselves to pure mathematics, and, least of all, to the pure mathematics of symbols. On the subject, considered in this point of view, I, a little while ago, published a few remarks, under the title of 'Thoughts on the Study of Mathematics, as a part of a Liberal Education;' and as these remarks may serve to illustrate further what is here said, I reprint them at the end of the present book.

It will be observed, that I intentionally omit, at present, all reference to professional education; and, consequently, to lectures, examinations, and degrees, in Divinity, Medicine, and Law. All such studies should be subsequent to the intellectual culture of which I now speak; and the professions to which these studies belong will derive the greatest part of their real dignity and refinement, from their being built on such a foundation. I omit, also, all consideration of what is called 'practical knowledge,' such as civil engineering, practical arts and trades, and the like; which are sometimes recommended as useful elements of education. If these are wanted as professional knowledge, they must be learnt, as, in fact, they are learnt, among professional men and practical applications. If they are wished for as information, they stand on the same ground as the higher physical sciences, of which we have spoken above, so far as the arguments there employed are applicable to them. But this 'practical knowledge' can never stand in the place of a really liberal education, nor in the smallest degree supersede the necessity of the studies we have pointed out, if such an education be our object.

W. WHEWELL, *On the Principles of English University Education*, London, 1837, p. 42.

9d Science at Cambridge

7. *Board of Natural Science Studies*

THE sciences included in this group, which are represented by Professors in the University, are Chemistry, Botany, Mineralogy, and Geology. There are no representatives of Zoology, considered apart from Comparative Anatomy, and as a branch of Natural History; nor of Meteorology, a science of rapidly-increasing importance, of which the chief generalizations have only recently been established; nor of Electricity and Magnetism, unless considered as departments of the vast science of Chemistry, or connected by other relations with Natural Philosophy. In our classification we have considered the more exact Natural Sciences as included in the application of Mathematics; but in foreign Universities they are considered, under the name of Physics, as a branch of experimental rather than of mathematical science, though there are very few departments of them where the aid of a mathematical treatment of most of the questions involved is not indispensably necessary.

The arrangements of the Natural Sciences Tripos, adopted in 1848, include Human and Comparative Anatomy and Physiology, which are essential branches of medical instruction, and are more properly referred, therefore, to the group of Medical Sciences.

There have been two examinations for the Natural, as well as the Moral, Sciences Tripos in the past and the present year. The number of candidates for both of them was small, but much more so in the former than in the latter; a result which might naturally have been expected as these studies have been only recently encouraged. It is obvious also that there cannot be continuous instruction given to Students engaged in learning those sciences which are dependent upon the exhibition or manipulation of experiments, or in the examination of specimens, unless the Professors or well-qualified assistants are present to direct or advise them.

Chemistry

The science of Chemistry is rapidly extending its ramifications into all the arts of life, and a knowledge not merely of its principles, but also the practical power of applying them, is becoming daily more and more important as a part of general as well as professional education. The present venerable Professor of this science has lived to see the day, after many years of comparative indifference and neglect, when his lecture-room is again well attended by students; an effect partly produced by the recent legislation in the University, and partly by the interest which the results of chemistry have created amongst all classes of the community. But the science is much too extensive for the teaching of one Professor: and we venture to hope that the University will eventually add a second who may divide with the present Professor the vast range of subjects contained in this extensive science. . . .

Recommendations

It appears that the University contains some noble collections, and that courses of lectures are given in Anatomy, Botany, Geology, and Mineralogy. These lectures have been, during past years, comparatively ineffective from two causes; 1st. The several subjects formed no regular and acknowledged part of the academic course. 2ndly. The Student had not sufficient facilities for making observations, or carrying on a course of direct experiments. The first of these impediments has been removed by the institution of the new Natural Sciences Tripos. The second can only be removed entirely by the erection of several new buildings, such as those suggested by Dr. Clark and Professor Miller, as well as in various other statements of our Report and Evidence.

New Buildings required

Appropriate buildings will, we hope, soon be erected in the new Botanic Garden, and when finished, the whole site of the old Garden will be available for other Academical purposes; and would afford a very appropriate site for Museums and Lecture Rooms. We recommend that measures should be taken for the purpose of obtaining a site for the erection of an ample series of buildings which are greatly needed; such as Museums of Comparative Anatomy and Zoology, with convenient contiguous rooms for dissections, experiments, and books of reference, a Museum of Mineralogy, with requisite rooms for experiments; and a Museum for the deposit of physical and chemical apparatus, with all possible conveniences for their ready use by Students.

For such buildings as these the University has a pressing want; and it is very fortunate that an ample site may be found for the whole of them. The cost of their erection need not be very great. They ought to be erected for use, and not for ornament. The University has not funds for costly architectural decorations; which, besides, are often injurious to the scientific ends of a building.

A new Museum of Comparative Anatomy and Zoology we think of great and immediate importance. The present Museum might then be appropriated to Human Anatomy, and to Pathology. Should the fossils of the Woodwardian Museum increase beyond the capacity of the whole floor on which they are now arranged, including therein the present Mineralogical Museum; some of the larger fossils might, from time to time, find a place in the new Museum of Comparative Anatomy, which ought to be a building of ample dimensions. The two Collections would thus become interlaced; and, in the way of exchange, this might be a great advantage to both of them. . . .

ANSWERS FROM THE REV. J. CUMMING, M.A., PROFESSOR OF CHEMISTRY

Professorship of Chemistry

The Professor of Chemistry was founded by the University in the year 1702. The present Professor was elected in 1815. No duties are annexed to the office, except the delivery of a course of Lectures annually. There has been no modification by subsequent authority.

Duties

A course of about 28 Lectures was given annually by the present Professor and his immediate predecessors until the year 1831. From that time thirty Lectures were given in the Lent, and twenty in the Easter Term. This was continued until the year 1845; but the attendance in the Easter Term was so small, consisting of only four or five medical students, that the original plan was resumed of giving Lectures only in one Term.

Endowment

The original endowment was a grant from the Treasury of 100*l.* per annum, which has not been augmented.

Fees

The extraneous sources of emolument are the fees for attendance on the Lectures.

Present Value

The present Professor has admitted, gratuitously, pupils recommended by their tutors. The fees from others average, at present, about 20*l.* per annum. The present value of the Professorship is about 120*l.* per annum, being 100*l.* granted by Parliament and 20*l.* from fees annually. This is the gross income, subject to deductions, for pay of an assistant, and for expenses for apparatus, &c.

Mode of Appointment

The mode of appointment to the office is by open election, 'More burgensium.' There are no conditions of eligibility or tenability.

Residence, &c.

There is no residence, museum, library, collection, or apparatus attached to the Professorship.

The apparatus is the private property of the Professor, and there are no funds for this purpose.

There are no opportunities afforded to students for instruction in the actual manipulation of instruments, as there is only one room appropriated to the Chemical Professor, and there is no apparatus for the use of students.

Hitherto the study of Chemistry has not only been neglected but discouraged in the University, as diverting the attention of pupils from what have been considered their proper academical studies. The recent establishment of a Tripos for Natural Science will, it is to be expected, remove this objection.

(Signed) J. CUMMING

Report of Her Majesty's Commissioners appointed to inquire into the State, Discipline, Studies and Revenues of the University and Colleges of Cambridge. *B.P.P.*, 1852–3 (44).

9e Science at Public Schools

Report

1. A demand for the introduction of Science into the modern system of education has increased so steadily during the last few years, and has received the approval of so many men of the highest eminence in every rank and profession, and especially of those who have made the theory and practice of education their study, that it is impossible to doubt the existence of a general, and even a national desire to facilitate the acquisition of some scientific knowledge by boys at our Public and other Schools.

2. We would point out that there is already a *general* recognition of Science as an element in liberal education. It is encouraged, to a greater or less degree, by the English, Scotch, and Irish Universities, it is recognised as an optional study by the College of Preceptors, it forms one of the subjects in the Local Examinations of Oxford and Cambridge; and it has even been partially introduced into several Public Schools. We have added an appendix containing information on some of these points; but the means at present adopted in our Schools and Universities for making this teaching effective, are, in our opinion, capable of great improvement.

3. That general education in Schools ought to include some training in Science is an opinion that has been strongly urged on the following grounds:—

As providing the best discipline in observation and collection of facts, in the combination of inductive with deductive reasoning, and in accuracy both of thought and language.

Because it is found in practice to remedy some of the defects of the ordinary school education. Many boys, on whom the ordinary school studies produce very slight effect, are stimulated and improved by instruction in science; and it is found to be a most valuable element in the education of those who show special aptitude for literary culture.

Because the methods and results of Science have so profoundly affected all the philosophical thought of the age, that an educated man is under a very great disadvantage if he is unacquainted with them.

Because very great intellectual pleasure is derived in after life from even a moderate acquaintance with Science.

On grounds of practical utility as materially affecting the present position and future progress of civilization.

This opinion is fully supported by the popular judgment. All who have much to do with the parents of boys in the upper classes of life are aware that, as a rule, they value education in Science on some or all of the grounds above stated.

4. There are difficulties in the way of introducing Science into schools, and we shall make some remarks on them; they will be found, we believe, to be by no means insuperable.

First among these difficulties is the necessary increase of expense; for if science is to be taught, at least one additional master must be appointed, and it will be necessary in some cases to provide him with additional schoolrooms and a fund for the purchase of apparatus. It is obvious that the money which will be requisite for both the initial and current expenses must in general be obtained by increasing the school fees. This difficulty is a real but not a fatal one. In a wealthy country like England, a slight increase in the cost of education will not be allowed (in cases where it is unavoidable) to stand in the way of what is generally looked on as an important educational reform; and parents will not be unwilling to pay a small additional fee if they are satisfied that the instruction in Science is to be made a reality.

Another ground of hesitation is the fear that the teaching of Science will injure the teaching in classics. But we do not think that there need be the slightest apprehension that any one of the valuable results of a classical education will be diminished by the introduction of Science. It is a very general opinion, in which schoolmasters heartily concur, that much more knowledge and intellectual vigour might be obtained by most boys, during the many years they spend at school, than what they do, as a matter of fact, obtain. It should, we think, be frankly acknowledged, and indeed few are found who deny it, that an exclusively classical education, however well it may operate in the case of the very few who distinguish themselves in its curriculum, fails deplorably for the majority of minds. As a general rule the small proportion of boys who leave our schools for the Universities consist undeniably of those who have advanced furthest in classical studies; and judging the existing system of education by these boys alone, we have to confess that it frequently ends in astonishing ignorance. This ignorance, often previously acknowledged and deplored, has been dwelt on with much emphasis, and brought into great prominence by the recent Royal Commission for Inquiry into our Public Schools. We need not fear that we shall do great damage by endeavouring to improve a system which has not been found to yield satisfactory results. And we believe, further, that the philological abilities of the very few who succeed in attaining to a satisfactory knowledge of classics will be rather stimulated than impeded by a more expansive training.

Lastly, it may be objected that an undue strain will be put upon the minds of boys by the introduction of the proposed subjects. We would reply that the same objections were made, and in some schools are still made, to the introduction of Mathematics and Modern Languages, and are found by general experience to have been untenable. A change of studies, involving the play of a new set of faculties, often produces a sense of positive relief; and at a time when it is thought necessary to devote to games so large a proportion of a boy's available time, the danger of a general over-pressure to the intellectual powers is very small, while any such danger in individual cases can always be obviated by special remissions. We do not wish to advocate any addition to the hours of work in schools where it is believed that they are already as numerous as is desirable; but in such schools some hours a week could still be given up to science, by a curtailment of the vastly preponderant time at present devoted to classical studies, and especially to Greek and Latin Composition.

5. To the selection of the subjects that ought to be included in a programme of scientific instruction in public schools we have given our best attention; and we would make the following remarks on the principles by which we have been guided in the selection that we shall propose.

There is an important distinction between scientific *information* and scientific *training*; in other words, between general literary acquaintance with scientific facts and the knowledge of methods that may be gained by studying the facts at first hand under the guidance of a competent teacher. Both of these are valuable; it is very desirable, for example, that boys should have some general information about the ordinary phenomena of nature, such as the simple facts of Astronomy, of Geology, of Physical Geography, and of elementary Phisiology. On the other hand, the scientific habit of mind, which is the principal benefit resulting from scientific training, and which is of incalculable value whatever be the pursuits of after life, can better be attained by a thorough knowledge of the facts and principles of one science than by a general acquaintance with what has been said or written about many. Both of these should co-exist, we think, at any school which professes to offer the highest liberal education; and at *every* school it will be easy to provide at least for giving some scientific information.

I. The subjects that we recommend for scientific *information* as distinguished from training, should comprehend a general description of the solar system; of the form and physical geography of the earth, and of such natural phenomena as tides, currents, winds, and the causes that influence climate; of the broad facts of Geology; of elementary Natural History, with especial reference to the useful plants and animals; and of the rudiments of Physiology. This is a kind of information which requires less preparation on the part of the teacher; and its effectiveness will depend on his knowledge, clearness, method, and sympathy with his pupils. Nothing will be gained by circumscribing these subjects by any general syllabus; they may safely be left to the discretion of the masters who teach them.

II. And for scientific *training* we are decidedly of opinion that the subjects which have paramount claims are Experimental Physics, Elementary Chemistry, and Botany.

i. The science of Experimental Physics deals with subjects which come within the range of every boy's experience; it embraces the phenomena and laws of light, heat, sound, electricity, and magnetism, the elements of mechanics, and the mechanical properties of liquids and gases. The thorough knowledge of these subjects includes the practical mastery of the apparatus employed in their investigation. The study of experimental physics involves the observation and colligation of facts, and the discovery and application of principles. It is both inductive and deductive. It exercises the attention and the memory, but makes both of them subservient to an intellectual discipline higher than either. The teacher can so present his facts as to make them suggest the principles which underlie them, while, once in possession of the principle, the learner may be stimulated to deduce from it results which lie beyond the bounds of his experience. The subsequent verification of his deduction by experiment never fails to excite his interest and awaken his delight. The effects obtained in the classroom will be made the key to the explanation of natural phenomena,—of thunder and lightning, of rain and snow, of dew and hoar-frost, of winds and waves, of atmospheric refraction and reflexion, of the rainbow and the mirage, of meteorites, of terrestrial magnetism, of the pressure and buoyancy of water and of air. Thus the knowledge acquired by the study of experimental physics is, of itself, of the highest value while the acquisition of that knowledge brings into healthful and vigorous play every faculty of the learner's mind. Not only are natural phenomena made the objects of intelligent observation, but they furnish material for thought to wrestle with and to overcome, the growth of intellectual strength being the sure concomitant of the enjoyment of intellectual victory. We do not entertain a doubt that the competent teacher who loves his subject and can sympathise with his pupils, will find in experimental physics a store of knowledge of the most fascinating kind, and an instrument of mental training of exceeding power.

ii. Chemistry is remarkable for the comprehensive character of the training which it affords. Not only does it exercise the memory and the reasoning powers, but it also teaches the student to gather by his own experiments and observations the facts upon which to reason.

It affords a corrective of each of the two extremes against which real educators of youth are constantly struggling. For on the one hand it leads even sluggish or uncultivated minds from simple and interesting observations to general ideas and conclusions, and gives them a taste of intellectual enjoyment and a desire for learning. On the other hand, it checks over-confidence in mere reasoning, and shows the way in which valid extensions of our ideas grow out of a series of more and more rational and accurate observations of external nature.

It must not, however, be supposed that all so-called teaching of chemistry produces results of this kind. Young men do occasionally come up to public examinations with a literary acquaintance with special facts and even principles of chemistry, sufficient to enable them to describe those facts from some one point of view and to enunciate the principles in fluent language, and who yet know nothing of the real meaning of the phrases which they have learnt. Such mere literary acquaintance with scientific facts is in chemistry an incalculable evil to the student if he be allowed to mistake it for science.

Whether the student is to learn much or little of chemistry his very first lessons must be samples of the science. He must see the chief phenomena which are described to him, so that the words of each description may afterwards call up in his mind an image of the thing; he must make simple experiments, and learn to describe accurately what he has done and what he has observed; he must learn to use the knowledge which he has acquired before proceeding to the acquisition of more; and he must rise gradually from well-examined facts to general laws and theories.

Among the commonest non-metallic elements and their simplest compounds the teacher in a school will find abundant scope for his chief exertions.

iii. Botany has also strong claims to be regarded as a subject for scientific training. It has been introduced into the regular school course at Rugby (where it is the first branch of Natural Science which is studied); and the voluntary pursuit of it is encouraged at Harrow and at some other schools with satisfactory results. It only requires observation, attention, and the acquisition of some new words; but it also evolves the powers of comparison and colligation of facts in a remarkable degree. Of all sciences it seems to offer the greatest facilities for observation in the fields and gardens; and to this must be added the fact that boys, from their familiarity with fruits, trees, and flowers, start with a considerable general knowledge of botanical facts. It admits therefore pre-eminently of being taught in the true scientific method. The teaching of Science is made really valuable by training the learner's mind to examine into his present knowledge, to arrange and criticise it, and to look for additional information. The science must be begun when it touches his past experience, and this experience must be converted into scientific knowledge. The discretion of the teacher will best determine the range of Botany at which it is desirable to aim.

6. The modes of giving instruction in the subjects which we have recommended are reducible to two. I. A compulsory system of instruction may be adopted, similar to that which exists at Rugby, where science has now for nearly three years been introduced on precisely the same footing as Mathematics and Modern Languages, and is necessarily taught to all boys. II. A voluntary system may be encouraged, as has been done for many years at Harrow, where scientific instruction on such subjects as have been enumerated above is now given in a systematic series of lectures, on which the attendance of all boys who are interested in them is entirely optional.

Of these systems it is impossible not to feel that the compulsory system is the most complete and satisfactory. The experience of different schools will indicate how it may best be adopted, and what modifications of it may be made to suit the different school arrangements. It will often be very desirable to supplement it by the voluntary system, to enable the boys of higher scientific ability to study those parts of the course of Experimental Physics which will rarely, if ever, be included in the compulsory school system. Lectures may also be occasionally given by some non-resident lecturer with a view of stimulating the attention and interest of the boys. We add appendices containing details of these two systems as worked at Rugby and Harrow, and we believe that a combination of the two would leave little or nothing to be desired.

The thorough teaching of the Physical Sciences at schools will not, however, be possible unless there is a general improvement in the knowledge of Arithmetic. At present, many boys of 13 and 14 are sent to the Public Schools almost totally ignorant of the elements of Arithmetic, and in such cases they gain only the most limited and meagre knowledge of it; and the great majority enter ill-taught. It is a serious and lasting injury to boys so to neglect Arithmetic in their early education; it arises partly from the desire of the masters of preparatory schools to send up their boys fitted to take a good place in the classical school, and from the indifference of the public schools themselves to the evil that has resulted.

7. With a view to the furtherance of this scheme, we make the following suggestions:—

i. That in all schools Natural Science be one of the subjects to be taught, and that in every Public School at least one Natural Science master be appointed for the purpose.

ii. That at least three hours a week be devoted to such scientific instruction.

iii. That Natural Science should be placed on an equal footing with Mathematics and Modern Languages in effecting promotions, and in winning honours and prizes.

iv. That some knowledge of Arithmetic should be required for admission into all Public Schools.

v. That the Universities and Colleges be invited to assist in the introduction of scientific education, by making natural science a subject of examination, either at Matriculation, or at an early period of a University career.

vi. That the importance of appointing Lecturers in Science, and offering Entrance Scholarships, Exhibitions, and Fellowships for the encouragement of scientific attainments be represented to the authorities of the Colleges.

With reference to the last two recommendations, we would observe that, without the co-operation of the Universities, Science can never be effectively introduced into School education. Although not more than 35 per cent., even of the boys at our great Public Schools, proceed to the University, and at the majority of schools a still smaller proportion, yet the curriculum of a public school course is almost exclusively prepared with reference to the requirements of the Universities and the rewards for proficiency that they offer. No more decisive proof could be furnished of the fact that the Universities and Colleges have it in their power to alter and improve the whole higher education of England.

Report of a Committee appointed by the Council of the British Association for the Advancement of Science to consider the best means for promoting Scientific Education in Schools. *B.P.P.*, 1867–8 (54).

9f Technical Education in Europe

Industrial Progress and the Education of the Industrial Classes, in France, Switzerland, Germany, &c.

LETTER OF MR. SAMUELSON, M.P., TO THE RIGHT HONOURABLE LORD R. MONTAGU, THE VICE PRESIDENT OF THE COMMITTEE OF COUNCIL FOR EDUCATION

Bodicote Grange, 16 November 1867

My Lord,

THE progress of continental manufactures and the elements of that progress, including more especially the theoretical and technical instruction of continental nations, are subjects so evidently beyond the scope of any single person, even if ample time be granted, that in complying with your request that I would make known the experience of my recent visit to some of the industrial and educational establishments abroad, I can offer no more than a sketch which at the best may be of some service to more competent observers as a partial outline of the ground to be investigated by them at greater leisure and with more adequate preparation.

I am happy to acknowledge the readiness with which the most detailed information on every point connected with the industry and education of France, Switzerland, Germany, and Belgium, was afforded to me by the various public departments of those countries, and by the proprietors of every factory, and the teachers and managers of every school which I visited. Our own representatives also, including the gentlemen in charge of the British department of the International Exhibition at Paris, appreciated fully the importance of the inquiry in which I was engaged. I received the most cordial assistance from them, and many will, no doubt, contribute valuable information in reply to the questions referring to technical education which have been addressed to them by our public departments.

Before leaving England I revisited several manufacturing towns of the north, and also Nottingham; and on the Continent, after a brief inspection of the lace-works of St. Pierre, near Calais, and some preparatory study of the iron, cotton, and silk manufactures exhibited at the Champs de Mars, and visits to some of the large works in and around Paris, I saw the great iron works of Monsieur Schneider at Creuzot; the schools and some of the factories of Lyons; several important iron-works and ribbon factories at St. Etienne; Mulhouse, in its more important industrial and educational features; the Ecole d'Horlogerie, and other schools of Geneva; the admirable schools of Winterthur, and the great manufactory of spinning machinery in that town; the Polytechnic Institution, and the schools of Stuttgard; the great locomotive factory at Esslingen; several of the coal-pits and metallurgical establishments of Westphalia, including the steel-works of Krupp and of Bochum, and the schools at the latter place;

the trade school at Crefeld; Gladbach as a principal centre of the cotton manufactures of Germany; and Verviers, as representing some of the more important departments of the woollen manufactures of Belgium. The mining districts and iron-works of Belgium have been so recently described by Messrs. Creed and Williams, that I limited my inquiries at Brussels, and through Monsieur Havréz, one of the Government mining engineers at Charleroi, to some questions relating to education and mining statistics, which those gentlemen did not include in their inquiry. On my return to France, I saw several of the primary schools and the combined elementary and trade schools of St. Nicolas, as well as the Ecole Turgot and Collége Chaptal in Paris; and I obtained such detailed information respecting the three Ecoles des Arts et Métiers at Chalons, Angers, and Aix, from former pupils of those schools, and from other sources, that I do not regret so much as I should otherwise have done, my inability to visit one of them. The same remark will apply to the great iron-works of the Moselle. But it was a great disappointment that I was unable to inspect the Polytechnicum of Zürich, which was closed until the middle of this month, in consequence of the prevalence of cholera. Death, attributed to excessive brain excitement during a visit to the International Exhibition, has quite recently deprived that institution of Professor Schroeter, whose celebrity as a teacher of mechanical engineering has been one of the chief causes of its having so rapidly obtained a great European reputation.

Already at Manchester I found reason to rate more highly than I had been inclined to do from the published statistics, the growing importance of continental competition in some of those branches of manufacture which we have been accustomed to regard as peculiarly our own, and I could not refuse to accept the evidence which forced itself upon me, of the injury which has resulted to the interests of our working population from some of the strikes supported by the great trades unions. Powerful as their organisation is for good when judiciously directed, the responsibility of the officers and members of such corporations as that of the Amalgamated Engineers, and others of similar importance, is all the more serious when its resources are employed in sustaining such strikes as that which took place last year in the locomotive factory of Messrs. Beyer and Peacock of Gorton. Until its occurrence more than half of the engines constructed by them were exported, but since that period the foreign trade has been diverted more than ever to the other countries whose natural advantages are far inferior to ours. . . .

I find that in France, with a population of 38,000,000, the two great technical schools, the Ecolé Polytechnique, and the Ecolés des Arts et Manufactures, furnish from 350 to 400 young men yearly for the scientific departments of the army, and for the higher industrial positions. In Great Britain (I quote from the 14th Report of the Science and Art Department), 'The number of students entering the Royal School of Mines for one, two, or three years, with a view to become associates, has, including the exhibitioners, been 13, the total number of entries as occasional students for the different courses, was 108. The chemical laboratory was attended during the three terms by 116, and the metallurgical laboratory by 23 students.' A few young men attended the technical classes of King's College. Add to these the pupils of the school of Naval Architecture, a proportion of those of Sandhurst, and a small number of technical students at the universities of Glasgow, Edinburgh, and Oxford, and we shall have given a fair representation of the higher technical education of this country.

On the other hand, Prussia, Baden, Wurtemberg, Saxony, and Switzerland, numbering together about 32,000,000 of inhabitants, besides the great chemical laboratories of Heidelberg, Giessen, Berlin, and Bonn, the mining academies of Berlin and Freiburg, and the various military technical schools, possess the following polytechnic institutions of the first class:—

Zurich, with	–	–	–	550	regular pupils.
Carlsruhe ,,	–	–	–	800	,,
Stuttgart ,,	–	–	–	450	,,
Dresden ,,	–	–	–	270	,,
Berlin ,,	–	–	–	370	,,
Hanover ,,	–	–	–	430	,,
				2,870	

besides numerous occasional students, educated at a total cost to the states of 40,000*l.* per annum, in addition to school fees, amounting to 9,800*l.*, or at an average annual cost of 15*l.* per pupil to the state, and of 3*l.* 10*s.* to the pupils themselves.

The programmes of all are not before me, but, in stating that in all, with the exception of the preparatory courses of Zurich, the system followed is that of the German universities, namely, absolute freedom to the young men to attend or to neglect the lectures and classes of the professors, I need scarcely add, that the collegiate system, the 'internat' of the Ecole Polytechnique is unknown in the higher schools of Germany and Switzerland. I hope the Government will invest competent persons with authority to visit these magnificent institutions. The buildings of the Swiss Polytechnicum have cost about 100,000*l.*; those of Carlsruhe, probably an equal sum. Wurtemberg, a comparatively poor inland state, with less than 3,000,000 of inhabitants, without minerals, and without navigable rivers, has expended 45,000*l.* on its Polytechnic Institute.

The statistics of the Polytechnicum of Zurich, which stands at the head of all these schools, afford a curious illustration of the appreciation in which the higher technical studies are held by natives of countries less fortunately endowed with similar institutions. Thus, I find, that in the year 1866–67 the students included—

43 Hungarians,
31 Poles,
26 Bohemians,
25 Russians,
4 Norwegians,
6 Dutchmen,
2 Brazilians, and also
2 *Englishmen.*

The Polytechnicum being closed on account of the cholera, I visited the Polytechnic School of Stuttgart, one of those which is rapidly rising in public favour, chiefly on account of the high estimation in which Professor Veith, the director of the division for mechanical construction, is held. Dr. Veith was one of the assistants, at Carlsruhe, of the late Professor Redtenbacher, who is regarded as the founder of the theoretical instruction in mechanical construction in Germany.

The school consists of (first) a mathematical division of two classes, to which a mercantile class is attached, and of (secondly) a technical division, which latter is again subdivided into four schools:—

1. Architecture.
2. Civil engineering.
3. Mechanical construction.
4. Chemical technology.

The minimum age of admission to the mathematical division is 16 years, but 15 years only in the mercantile class. In the technical division (subject to special exceptions) it is 18 years. There are 150 students in the former and 300 in the latter. The entrance examination for the mathematical division proper requires a knowledge of quadratics and familiarity with the use of logarithms, with plane trigonometry, the construction of the French language, ability to translate German into French, and to write an essay in German correctly, and acquaintance with the chief periods and events of history. There is no prescribed entrance examination for the various sub-classes of the technical division. The applicant must satisfy the director under whose superintendence each is placed, that his preliminary knowledge is adequate to enable him to follow the course of instruction with advantage.

The teaching staff consists of 24 professors, 20 assistants, and seven private tutors in modern languages for the youths who attend the mathematical division. It will not be supposed that mathematics end with the mathematical division, which in fact is a species of preparatory school; on the contrary, the study of pure mathematics occupies a prominent position in every sub-division, each of which also includes instruction in the allied subjects. For instance, the chemical division includes a special course of mechanical construction for chemists, the architectural school—a special course of chemistry for architects. What important consequences such a course as this may exercise in the choice of building materials will suggest itself to those who are familiar with the Houses of Parliament and other buildings, old and new, of our great cities. The period of study for each of the four technical sub-divisions is three years, the first year's technical course being general, preparatory and nearly identical for all the branches. One of the great difficulties of a technical education is felt, here as everywhere, to be the difficulty of combining practice with science. As a means of solving it, the pupils are recommended to work two years as engineers, or in the factory after the conclusion of their first year, and then to resume their theoretical studies. The first year's study induces them to observe the phenomena of the workshop; that of the subsequent years, to solve its problems correctly, and by the most direct methods.

B.P.P., 1867–8 (54).

9g Public Schools Commission Report*

XXXVI. Natural Philosophy School

Natural Philosophy became a subject of instruction at Rugby in the year 1849, just at the period when the University of Oxford, while remodelling and expanding its system of instruction, established a School of Physical Science in which examinations for a degree were held, and three scales of honour were awarded to meritorious candidates. To Dr. Tait is due the credit of supporting at Rugby this movement of the Universities, by providing, for the benefit of any boy whose parents might wish him to learn Natural Philosophy, a tutor
Answers III, 11 exclusively devoted to that branch of learning. During the Head Mastership of Dr. Goulburn this office became attached to one of the then existing Classical

* EDITORS' NOTE. This section deals with evidence given to the Commission concerning Rugby School.

Masterships, and is now assigned to one of the Mathematical Masters. In the year 1859 the Trustees of the School with exemplary liberality built a Physical Science Lecture Room and Laboratory, and partially furnished both, at the cost of more than 1,000*l*. withdrawn from the capital belonging to the School. Those on the Foundation receive this instruction without payment, and no fee is paid for them by the Trustees, on the ground that their outlay in building a lecture room and laboratory and partially furnishing them entitles them to take credit for the fees of Foundationers without actual payment.

Evidence, Mr. Wilson, 1364

Boys in general are not admitted to Lectures in Natural Philosophy until they reach the Middle School. The present Teacher has established this practice in the belief that boys before the age at which they commonly reach that point in the School are not well qualified for it. Nor are boys in any part of the School compelled to learn it. It is, in fact, regarded as a substitute for Modern Languages, to which parents may have recourse if they think fit. This alternative, too, is encumbered with the condition that an extra fee of 6*l*. 6*s*. per annum, not required for the teaching of Modern Languages, must be paid for instruction in Physical Science. It is formally permissible, however, to learn both Modern Languages and Physical Science, and some boys at the present moment actually take instruction in both, but the practice is discouraged, probably as being supposed to distract the mind with too many pursuits.

XXXVII. Arrangement of the School of Natural Philosophy

In analogy with the organization of the Schools of Mathematics and Modern Languages the main divisions of the School of Natural Philosophy correspond with those of the Classical School. The sub-schools, however, in this department are few and comprehensive, being only two in number, one of which embraces the Sixth Form and whole Upper School, the other the whole Middle School. Again, they are not subdivided into sets or classes as are the sub-schools in Mathematics and Modern Languages. Each division or sub-school is taught together in one class, in which the boys are arranged in order corresponding with their divisions or classes in the Classical School. The single class of the Sixth and Upper School division contains 29 boys; that of the Middle School contains 12, inclusive of a single boy from the Lower School, whose age and general intelligence seemed to qualify him for the study. It has been the practice of the Lecturer, instead of forming his pupils into smaller classes for class instruction, to draft off into the laboratory the most proficient without making any extra charge for it, while the rest are listening to lectures more suited to less advanced boys.

Table B. Natural Philosophy Mr. J. M. Wilson, 1377

XXXVIII. Subjects and Method of Natural Philosophy Teaching in Class

The instruction given in this School during twelve months ended July 1861 consisted entirely of subjects formerly comprehended under the name of chemistry, *i.e.* chemistry and electricity. Lectures, following the arrangement and explaining the details of some approved text book, such as 'Fownes' Chemistry,' are given twice in the week to each class. They are illustrated by experiments and diagrams, and brought home to individual boys by questions put to test their understanding of the lecture. Notes taken at the time of the lecture are subsequently expanded into reports drawn up by the boys out of school, and containing sketches of the apparatus. These are shown up once in a fortnight at least, and are then corrected by the Lecturer, as a classical exercise might be by a tutor. At the close of every seventh lecture a paper of questions is set on the matter of that and the six previous lectures.

Note to Table B. Natural Philosophy

Note to Table B. Natural Philosophy

XXXIX. Number of Hours spent in Natural Philosophy Classes

Table B. Natural Philosophy

The hours spent by each boy in class are two in the week beside the work already described, which may be termed rather digestion of what has been, than preparation of what is to be taught.

XL. Private Tuition in Natural Philosophy

Every boy learning Physical Science in class may become also, if his parents wish it, a private pupil, or Laboratory pupil. If he is a Foundationer this costs him nothing; if a Non-foundationer, an extra fee of five guineas, beside that paid for class instruction, is taken. This payment seems rather to confer the privilege of using the laboratory at all hours than to involve any very definite kind or amount of extra instruction. The number of such private pupils seems to vary greatly and suddenly at different seasons. It amounted to two only between October and Christmas 1860, and to eight between Christmas and June 1861.

XLI. Encouragements to the Study of Natural Philosophy

1. PROMOTION

Promotion in the Classical School is affected by the examination in Physical Science in the ratio of eight to a hundred marks given in all subjects.

2. PROMOTION IN NATURAL PHILOSOPHY SCHOOL

Table B. Natural Philosophy

It appears possible for boys to gain and lose places in the two Natural Philosophy classes, yet this movement is so restricted in extent, and is so sluggish, that a very marked effect on the industry of the boys can hardly be ascribed to it.

3. DISTINCTIONS AND PRIZES

To boys who distinguish themselves in the Natural Philosophy branch of the Christmas examination in any form, either a first or second class is awarded; the value of which, in contributing to a prize, is equal to the same grade of honour in any other branch except that of pure Classical Scholarship. Seven first classes and four second classes were awarded at the Christmas examination of 1861.

School List, Christmas, 1861 *Evidence, Mr. Wilson,* 1385

A prize for Practical Chemistry is given annually by the Assistant Master in this School to the best analyst in the Rugby Laboratory. He also gives a prize to all boys who, in his papers of questions, gain 75 per cent. of the maximum number of marks. A handsome prize of 10*l.* for Natural Science formerly established has recently been abolished.

4. EXHIBITIONS

Evidence, Dr. Temple, 731–737. *Mr. Wilson,* 1412

Natural Philosophy forms one of the subjects for examination for the Exhibitions. The Classical examiners are responsible for this part of the examination, but are at liberty to obtain assistance in setting papers, and appreciating the work if they think it necessary. Occasionally they avail themselves of this. An Assistant Master of a well known school in the neighbourhood on a late occasion did the work and reported upon the general proficiency of all the boys under examination.

XLII. Immediate Results of the Teaching in Natural Philosophy

One first class has been obtained by a Rugbean in the Physical Science School at Oxford in the last ten years, and one first class in the Natural Science Tripos at Cambridge, and one first 'Honours' in Experimental Science at Dublin. No first class at Oxford has been obtained within the last seven years. The Examiner at the last examination for Exhibitions reported the examinees as fairly well up in the Chemistry of the non-metallic elements. *List of Oxford Honours, Answers III. 20.*

Out of five candidates for direct commissions from Rugby four offered themselves for examination in the natural and experimental sciences, and of these, three failed in them, and one succeeded. The single candidate for Woolwich did not take them up. *Evidence, Mr. Wilson, 1412*

It is impossible to feel that the immediate results are as yet quite proportionate to the place which is now formally given to the study in the arrangement of the School, and to the expenditure which the Trustees have devoted to it.

Report of Her Majesty's Commissioners appointed to inquire into the Revenues and Management of certain Colleges and Schools (the Clarendon Report). *B.P.P.*, 1864 (20).

9h Royal Commission on Scientific Instruction: Sixth Report

Detailed Time Tables.—Sixth Form, 1st Term 1873

	Classical		Mathematical and Modern*	
Day	Time	Scheme	Time	Scheme
Monday	9–10	Grammar Paper (Latin or Greek)	9–10	Grammar Paper (Latin or German)
	10–11	Divinity	10–11	Divinity
	11–12.30	Mathematics	11–12.30	Mathematics
	2–3.30	Greek	2–3.30	Mathematics
	3.30–4.30	French	3.30–4.40	French
	Evening Work	Greek Rep	Evening Work	Mathematics
		Mathematics		German
		Latin Verse		
Tuesday	9	Greek Rep		
	9.15–11	Latin	9–11	Latin
	11–12.30	Greek	11–12.30	Physical Science
	2–3.15	First-sight Trans. (or German)	2–3.15	German
	3.15–4.30	Greek Prose	3.15–4.30	Mathematics
	Evening Work	Latin with Rep	Evening Work	Latin with Rep
		French		French
Wednesday	9	Latin Rep	9	Latin Rep
	9.15–10.30	Latin	9.15–10.30	Latin
	10.30–11	Correction of Paper Work in Class	10.30–11	Correction of Paper Work in Class
	11–1	(Special Classes)	11–1	(Special Classes) Science Mathematics

* *Science Students.*—On Wednesday afternoons (after ordinary school hours) a lesson is given in Analytical Chemistry, with practical work in the laboratory: 1 to 2 hours.

If a Science Student wishes to take up an additional subject, he can do so on Tuesday and Friday afternoons. In this case the necessary time ($2\frac{1}{2}$ hours for instruction, and $1\frac{1}{2}$ hours for preparation) is obtained either by relinquishing German, or by reducing the time devoted to Mathematics.

Day	Classical		Mathematical and Modern	
	Time	Scheme	Time	Scheme
WEDNESDAY (*cont.*)	Evening Work	History Latin Latin Prose	Evening Work	History Latin
THURSDAY	9–10 10–11 11–12.30 2–3.30 3.30–4.30 Evening Work	Latin Mathematics History Greek French English Literature and Rep Latin	9–10 10–11 11–12.30 2–3.30 3.30–4.30 Evening Work	Latin Mathematics History German or Mathematics French English Literature and Rep Latin
FRIDAY	9 9.15–11 10–11.30 11.30–12.30 2–3.15 3.15–4.30 Evening Work	English Rep Latin Greek English Literature Greek Prose (or German) First-sight Translations Greek with re-translation French Greek Verse	9 9.15–10 10–11.30 11.30–12.30 2–3.15 3.15–4.30 Evening Work	English Rep Latin Mathematics English Literature Mathematics German Mathematics German
SATURDAY	9–10.30 10.30–11 11–1	Latin Greek (Special Classes)	9–10.30 10.30–11 11–1	Latin Physical Science (Special Classes) Science Mathematics
	Evening Work	Latin or Greek Grammar Divinity (English Essay)	Evening Work	Grammar (Latin or German) Divinity English Essay

The English Essay is set on Saturday, and is delivered up on the following Saturday.

Essays and Classical Compositions are looked over with each boy separately during Preparation Time.

In the Special Classes a 'Classical' boy will generally take up either Classics entirely, or Classics and German; but he has the option, under the usual restrictions, of any of the other Special Subjects.

In the Special Classes a 'Mathematical' boy will generally take up Science and Mathematics; but he has the option, under the usual restrictions, of any of the other Special Subjects.

College or School	Lessons: Oral Teaching and Demonstration	Lessons: Book Work	Practical Work
Eton - -	All - - - -	Subsidiary and supplementary	Practical work in the laboratory is done entirely out of school, the time devoted to it being from 3 to $3\frac{1}{2}$ hours in each week. The time spent in regular school lessons is about one half of this
Rugby - -	All - - - -	Supplementary - -	Botany in lessons, chemistry in laboratory
Clifton - -	All, when not catechetical, or occupied with working out examples and propositions	Preparatory or supplementary	Botany in lessons, botanical and geological expeditions, chemistry
Marlborough -	All - - - -	Preparatory - -	Various branches of Natural History
Winchester	All in lower classes - - Part in upper classes - -	Nil - - - Part in upper classes	Nil
Westminster -	All - - - -	Preparatory and supplementary	Nil
Cheltenham -	From one fourth to one half -	Varying with subject and with master*	Botany, electricity, chemistry in lessons
Wellington -	Two thirds - - -	Supplementary - -	Electrical measurements in lessons, chemistry in laboratory (voluntary)
Rossall - -	Lectures followed by oral and written examinations	Subsidiary to lectures -	Chemistry in the laboratory
Charterhouse -	All - - - -	Supplementary - -	Nil
Christ's Hospital	Chemistry, one half; natural philosophy, three fourths; elementary mechanics, and examples are set to be worked from books	Chemistry, one half; natural philosophy, one fourth; supplementary and to supply problems and examples	Chemistry in laboratory
City of London School	All - - - -	Preparatory at home -	Chemistry in laboratory. Also, with a select few, Experimental Physics, Physiology, and Botany
Taunton College School	Of physiology, all; of botany, one half; of chemistry, for the Upper School, one third; for the Middle School, all; of mechanics, all	Out of school as an assistance to notes taken at the lectures	Botany, half of each lesson employed in dissections, schedule writing, and classification; chemistry in laboratory
Dulwich College	Each of the classes receives two lessons per week of an hour or an hour and a quarter each. A course of evening lectures (generally four or five lectures) on some branch of science is also given in each term. Attendance at these lectures is voluntary. They are open to all boys, whether pupils in any of the science classes or not	Book work is got up during the 'preparation time,' which, for the more advanced pupils, amounts to about two hours for chemistry and one hour and a half for physics	Chemistry in laboratory, one hour a week: for advanced pupils, two hours; out of ordinary school hours Physics in laboratory, during the school time assigned to the Science lesson

College or School	Lessons: Oral Teaching and Demonstration	Lessons: Book Work	Practical Work
University College School	In chemistry and chemical physics, the subjects are explained solely by means of oral teaching and demonstration, no portion of any lesson being ever devoted to book work. This oral teaching includes the explanation of those portions of science which the boys are expected by the Universities to know, but which are not capable of experimental proof in the class rooms of a school. All the subjects in physics and mechanics are explained by means of oral teaching and demonstration, no books being used. The lessons are alternately devoted to experimental demonstrations of general principles and to the working of illustrative exercises and problems on, or deductions from, these principles	Preparatory at home -	Chemistry in laboratory accompanied by oral teaching; and occasionally there is added such further demonstration as is considered necessary to supplement the experiments made by the boys themselves. [A laboratory for physics has been recently constructed in the new building]
King's College School	Physical classes, one half -	One half - - -	Chemistry in laboratory; physics in laboratory
Manchester Grammar School	All, with very few exceptions	Preparatory, excepting when used in the absence of large diagrams	Chemistry in laboratory. Optical, heat, and electrical experiments

[*NOTE.—Masters and methods differ; subjects differ also. One master would reply, chiefly preparatory; both for first (or demonstrative) lesson, and also for last (or catechetical) lesson. Another master would say, simply preparatory, except so far as useful for revision of work before examination. The Head Master of the Military Department, a man of great experience and great wisdom in teaching, who himself taught science for many years before any special master for science was appointed, says: 'The subject matter of each lesson should generally be explained in advance; it is then learnt from text-books during preparation hours, and tested at the next lesson time. If any points are not clearly understood, they are again explained, and notes made by the boys.']

Chemical Laboratory, Dulwich College

By courtesy of the British Museum

Sixth Report of the Royal Commission on Scientific Instruction and the Advancement of Science (the Devonshire Report). *B.P.P.*, 1875 (c. 1279), (28).

9i A Nineteenth-century Physics Examination

SUBJECT III.—EXPERIMENTAL PHYSICS

Subdivision I.—Acoustics, Light, and Heat

EXAMINER, PROFESSOR TYNDALL, F.R.S.

EXPERIMENTAL PHYSICS.—1ST PAPER

SECTION I

1. State how you conceive sound to be propagated through the atmosphere; how is its velocity affected by the temperature of the air?
2. What are the conditions necessary to the production of a musical note? Wherein does harmony differ physically from noise? What is the cause of pitch; what is the measure of intensity?
3. The theoretic velocity of sound, as found by Newton, was much less than the observed velocity: state fully how Laplace explained the discrepancy.

SECTION II

1. A luminous beam strikes a plane mirror obliquely, and is reflected from it. Take two points, the one in the incident, the other in the reflected beam, and

prove that the sum of the incident and reflected rays is less than that of any other two lines drawn from the two points to the reflecting surface.
2. Two mirrors form with each other an angle of 40°: how many images will be formed of an object placed between them? Apply your knowledge to the explanation of the kaleidoscope.
3. If the intensity of light diminishes as the square of the distance increases, how is it that on a clear night the distant lamps appear as bright as the near ones?

SECTION III

1. Lenses are divided into two great classes: name them, and show by a diagram the action of one of each class.
2. What is meant by the spherical aberration of a lens, and what by its chromatic aberration?
3. What is meant by the dispersion of light? Is it possible to neutralize the dispersion without at the same time neutralizing the refraction? State what you know regarding the construction of achromatic lenses.

SECTION IV

1. Describe the human eye.
2. What are the conditions of distinct vision? Explain the action of spectacles as applied to the remedy of long and short sight.
3. Describe some one form of the stereoscope, and state what you know regarding the principle of the instrument.

SECTION V

1. Explain the mirage of the desert.
2. A bright moon when looked at with the naked eye appears considerably larger than when looked at through a dark glass: explain the fact.
3. Describe and explain the rainbow.

SECTION VI

1. The water at the bottom of the Great Geyser-tube in Iceland has sometimes been found 50° above 212° Fahrenheit, and it existed thus without ebullition: how is this possible?
2. Two bombshells are accurately filled with cold water and closely stopped: one of them is placed in boiling water, the other in a freezing mixture; both of them burst—why?
3. Describe, as far as you know, the influence of pressure on the fusing points of bodies.

SECTION VII

1. The temperature of the human body is nearly constant in all climates; how is this constancy preserved while the external temperature varies so considerably?
2. A block of ice is heated from a temperature of 20° below the freezing point, until it is reduced to a state of vapour; state the indications of a thermometer which shows the temperature from first to last.
3. The specific heat of water is 4 times, and its density is 770 times that of air: supposing a cubic mile of water to yield up one degree of its heat to a cold atmosphere, what quantity would 1,000 cubic miles of the atmosphere be heated by the transference?

SECTION VIII

1. Explain the production of winds.
2. Explain the formation of rain.
3. Explain the formation of hoar frost; and why it collects chiefly on the blades of grass.

Subdivision II.—Magnetism and Electricity
EXPERIMENTAL PHYSICS

SECTION I

1. You are required to prepare a magnetic needle for your class experiments: how will you do it?
2. How would you demonstrate to a class the general laws of magnetic attraction and repulsion?
3. State what you know regarding the earth's magnetism; the position of its poles, the intensity of its force, and the terms used to express its general phenomena.

SECTION II

1. Common soft iron is always attracted by the poles of a magnet; hard steel is sometimes repelled: state the reason.
2. Place an iron sphere in contact with the north pole of a magnet; it clings, a second sphere clings to the first, a third to the second, and so on; what is the magnetic condition of the row of spheres? If I place the north pole of a second magnet underneath the row, what occurs? What happens when a south pole is placed underneath?
3. State what you know regarding the phenomena of diamagnetism.

SECTION III

1. How would you demonstrate to a class the general laws of electric attraction and repulsion?
2. How would you explain the attraction of light bodies by a bit of sealingwax rubbed on the coat sleeve?
3. I give you an insulated sphere charged with positive electricity and require you to charge by means of it a gold-leaf electroscope, with negative electricity: how will you do it?

SECTION IV

1. Give a scientific explanation of thunder and lightning.
2. Explain the action of lightning conductors; and state generally the influence of the shape of a conductor on the distribution of electricity upon its surface.
3. Describe and explain the phenomenon known as the 'return stroke' in a thunder storm.

SECTION V

1. I furnish you with a flat smooth sheet of gutta percha, a metal plate with a glass handle, and a catskin; and require you to charge a Leyden jar by these means: how would you do it?
2. Why in charging a Leyden jar do you connect its outer coating with the earth?
3. Describe the manner in which the arrangement known as the 'cascade battery' is charged with electricity.

SECTION VI

1. Describe accurately the ordinary pile of Volta, and state the conditions of the two ends of the pile.
2. What part does the nitric acid play on the batteries of Grove, and Bunsen?
3. If our object be to overcome a great external resistance, what is the best arrangement of a series of cells? If we wish to send a copious current through a thick wire, what should our arrangement be?

SECTION VII

1. What do you mean by the 'direction' of an electric current? State the direction in which the current flows in Grove's battery.
2. A bar of iron is surrounded by a wire carrying an electric current; assume a direction for the current, and state the consequent polarity that will be excited in the bar.
3. How would you ascertain experimentally the relation between the magnetism of the bar and the strength of the current circulating around it?

SECTION VIII

1. In firing a mine by electricity why is platinum wire preferable to silver?
2. Two pieces of thin platinum wire are introduced at different parts of a circuit, in which they are rendered incandescent by the current: one of them is plunged into cold water, its glow being thereby extinguished; the other immediately glows with greater intensity; why?
3. Describe and explain the ordinary magneto-electric machine used by physicians.

Papers given to Candidates in the Examination for Science Certificates at South Kensington, November, 1860.

9j Berlin Chemical Laboratory

By courtesy of the Chemical Society

W. Ruske, 100 *Jahre Deutsche Chemische Gesellschaft*, Berlin, 1967, p. 35.

9k Huxley's Evidence on Technical Instruction

Eleventh Day, Saturday, 28th October 1882

PRESENT:
Mr Bernhard Samuelson, M.P., F.R.S., in the Chair
Professor H. E. Roscoe, LL.D., F.R.S.
Mr. Philip Magnus
Mr. John Slagg, M.P.
Mr. Swire Smith
Mr. William Woodall, M.P.

... 3000. In what directions do you think the Science and Art Department might advantageously develope its operations?—I have referred to one direction in which I think a good deal might be done, that is to say, by the encouragement of the classes, by going further than it now does in the direction of supplying apparatus, for example, that is wanted for teaching, and by encouraging, or at any rate not impeding the establishment of the Sunday teaching I have spoken of. Then a second, and probably an even more important direction in which the action of the Science and Art Department might be developed is this, by the facilitation of practical instruction in the classes, and by gradually putting the screw on, so to speak, as the means of giving practical instruction are developed, gradually making it essential to all kinds of science teaching. Beyond this I doubt if it is essential—indeed, I doubt if it is desirable—that the State should go. I think that whatever else is needed should be done by the 'Owners and Trustees of Ancient Endowments' and by the employers of labour. With regard to those endowments, I should like to say that if you look at the educational endowments, which in this country are of enormous magnitude, you will find that a very large proportion of them, in the universities and elsewhere, were originally intended for poor scholars, men who had nothing to speak of in the way of wealth, but who had intellect, and whose intellect the founders of the endowments very wisely thought ought to be developed. Look at the actual state of things in our universities now. Those endowments are turned to the maintenance of what you may call a number of high schools; but, notwithstanding the endowments, the great majority of the persons who go to them find that they are extremely expensive. The colleges were founded by men who assumed that the persons who would take advantage of the scholarships and exhibitions might be utterly poor, and they were to be maintained there and educated for the glory of God and so forth. The practical result is that education in those places at this present time is only accessible to persons who are either moderately well off, or even wealthy. The State has permitted these endowments, which were originally intended for the poor, to be made into a kind of educational outdoor relief for the well-to-do. The State has permitted that, and is in the position of a sort of fraudulent trustee for the poor at the present time; and on that ground I think the State is bound to make good the wrong that it has done. I am not at all fond myself of needless State interference, or of State endowment, but I think that in this matter there is a great debt owing to the poorer classes, and that the State is bound to consider that. Then with respect to the action of the employers of labour, I think that the mere consideration of self-interest, if there were nothing better, should lead them to do their best to develope the system of scientific instruction; but I think there are many other very much higher considerations which should lead them to undertake and encourage it—the welfare of industry for example—and I know that in many instances employers of labour have taken this view of the case. I know, for example, that Sir William Armstrong has considered it his business and duty to develope a system of elementary scientific instruction at Elswick. They have one of the best elementary scientific schools extant in Elswick; and if employers, besides establishing such schools accessible to their work-people, would show that sort of practical interest in the matter which might be shown, if they would allow the more intelligent of the scholars some privileges in the way of extra time, or some kind of reward for doing well, which need not cost them very much, the encouragement to technical teaching would be very great. There are many honourable exceptions, but at present, if you take the great mass of the employers of labour, they do not care anything about the development of technical teaching. Very often there is a miserable sort of jealous feeling about the elevation of their workmen. Of all practical measures that could be taken for the advancement of technical education and scientific teaching, the most important would be that employers should show that they valued it, and that they would do something for the young people who in any way distinguished themselves.

3001. Is it not your opinion that employers might probably be more ready to institute and to support classes in connection with the Science and Art Department, if those classes were more special in their character than they are at present? Let us take for example the subject of metallurgy. Cannot you conceive that if that subject were divided so that in Cornwall the metallurgy of tin and copper, and in Staffordshire, or in the North Riding of Yorkshire, the metallurgy of iron formed the subject of examination, and if there were a similar division of subjects in regard to chemistry for instance, this might cause the employers to see that it would be more directly to the interest of the industry in which they were concerned that such classes should be promoted and encouraged?—I have no doubt whatever that employers would take the view you suggest, but at the same time I am perfectly certain that, in so doing, they would make a great mistake, and I say so on this ground: If you could, in the first place, keep your elementary scholars long enough to put them through a fair training in the principles of which the applications lie in the special directions of metallurgy; and, secondly, if you could secure that they should acquire a special proficiency in such subjects, I think that would be a most admirable thing to do; but I think the way it would work out under the present conditions would be this, that you would have special classes set up to grind young fellows, without any knowledge of principles, in that which would be no better than a rule of thumb learning of the present practice of their business; I do not believe that that would be of the smallest good. I believe, on the contrary, that it would do endless harm, because there would be a sort of pre-supposition that these young men really had the knowledge which would enable them to advance and improve their methods, whereas in reality the knowledge they possessed would be nothing more than a little of the ordinary technique of their business, varnished over with scientific phraseology.

3002. Might not that be guarded against by requiring that proved proficiency in the general subject should precede examination in the special subject?—If you provided that after one of these scholars had shown his acquaintance with general physical science he should go through a special examination in the metallurgy of steel or copper, that would be a different matter. What I am desirous to guard against is the encouragement of instruction in special subjects in cases where preliminary instruction in general physical science is ignored.

3003. In point of fact, is not it the case that you do not yourself guard against that in certain subjects? For instance, take the subject of steam, in which you do not require any previous training in heat and physics?—My reply to that, in the first place, is, that I did not organize the system; and, in the second place, as I ventured to say before, in looking at the system of the Department, you must take into consideration the circumstances under which it started. My own opinion is that you could not justify that course to which you refer. I should not, if I had been left to myself, have adopted it, but I take it that under the circumstances under which the whole system was started it was almost impossible to escape doing several things of that kind; the great practical industries of the country called for it. And let me remind the Commission of this, that in all these matters of teaching we have gone at a prodigious pace in the last quarter of a century. The whole system of scientific education has been absolutely revolutionised since the time that I first took an interest in these things. But as regards this question of steam, I should say that a man ought not to go in for an examination in steam till he had gone through an examination in physics and chemistry. [On referring to the regulations I find that a knowledge of physics, so far as heat is concerned, is required from examinees in steam. It is necessary therefore that my answer should be largely qualified.]

3004. In the same way, I presume, you would say that you would not allow a person to go in for an examination in copper and tin till he had passed an examination in chemistry and physics; but we may take it that you would say that when he had passed an examination of that sort he might be allowed to go in for an examination in tin and copper just as you would let him go in for an examination in steam only, after having gone through a preliminary teaching in physics?—Yes, certainly.

3005. (*Mr. Magnus.*) Have you heard that Sir William Armstrong has a class at Elswick in which the students who have passed through a general training in science receive special instruction in the metallurgy of iron and steel, this class being in connection with the City and Guilds of London Institute?—Yes. I should like to add, that while I quite agree with you that an examination devised upon that plan would be an admirable thing, I have my doubts whether that is the business of the State. I think it might be better done by employers and persons specially interested in technical education.

3006. (*Chairman.*) Would you, or would you not, like to see a more distinct line drawn between pure science on the one hand and its applications on the other, than is drawn at present by the Science and Art Department, if it could be shown that applied science might be provided for in some other efficient way?—If it could be shown that the branch of applied science could be provided for equally well in other ways, I quite think that it would be desirable to draw such a line.

3007. Will you now proceed to give the Commission your opinion with regard to the conduct of the examinations?—I stated at the outset that, in my opinion, the keystone of the whole system was in the examiners and the examinations. I remember telling Mr. Forster a number of years ago, when he spoke to me about the increasing expense of the system of the science and art examinations, that I thought that the question whether the system should be an extremely useful organization, or merely a gigantic job, depended on whether the examiners did their duty or not. I made that remark particularly because I felt quite convinced that unless the conduct of the examinations was kept in the hands of men of a high class, and unless the examinations were worked extremely conscientiously, and unless the examiners were supported against all clamour, the whole thing would become a nuisance, and, so far from doing good, would do harm. Perhaps the Commission would like to know the way in which we manage things, so that they may know what sort of precautions are taken to make the examinations fair. I can only speak of the conduct of my own examinations. For many years I conducted the examinations myself with assistants who were appointed yearly. Now I have the advantage of a colleague in Dr. Michael Foster, and our method is this: We have timely notice of the number of papers which are likely to be sent to us; in our own case they usually amount to about 7,000 or 8,000. We nominate a certain number of assistant examiners, and the assistant examiners whom we nominate are always young men of ability, often distinguished scholars of the universities, who have taken their degrees, and who are thoroughly competent men, with a knowledge of the methods of examination. Then we have a first meeting, which Dr. Foster and I and the assistant examiners attend; and we then explain to the assistant examiners the nature of the questions—sometimes we consult them beforehand about the questions,—telling them what we expect, and giving them, as far as we can, a sort of rough notion of what value should be attached to the answers, and the kind of blunders they will be likely to meet with. Then each takes away a batch of papers and is requested to send them up to the examiner-in-chief before a given date. Dr. Foster, of late years, has done the main work. A batch of something like 100 or 150 of these papers are read and marked by each assistant examiner. The examiner-in-chief then goes over these papers. Of course the habit of examination enables us to look at things pretty quickly, and we have a system of marking by which all errors are scored in a particular way, if fatal they have a red mark, if doubtful blue. The assistant examiner scores his papers with these marks and the examiner-in-chief then goes over them, paying particular attention to all the red or plucked answers, and he satisfies himself in this way that the assistant examiner is neither too lenient nor too severe. Very often we find them a good deal too severe—young examiners are always more severe than old examiners—and the examiner-in-chief forms his judgment of the way in which the work has been done. Then we have a second meeting for further discussion and consideration. Perhaps, as I say, one assistant examiner has a tendency to be too easy, and another to be too severe, and we talk over the thing; and if one has been too severe we point it out to him, and if another has been too easy we tell him that he ought to have been more severe. Then the rest of the examination is conducted on that foundation. Then there is a third meeting, at which the results are carefully compared with those of previous years. If any assistant examiner's percentages of passes are extremely low or extremely high, the thing is looked to to see if there is any mistake. If there is any suspicion in the mind of the assistant examiner that there has been copying or collusion, which is generally very easily betrayed, he is instructed to send the papers up at once and the case is forthwith reported. Another thing we are very careful about is this: If from seeing that they are all nearly word for word the same, the assistant examiner finds that a whole batch of papers have obviously been written off from memory, he is instructed to report that at once. Sometimes we have plucked a whole school, simply because we found that the boys had been stood up in rows and crammed like turkeys. Every now and then there is a terrible plucking in a particular school, and then of course there is a row. The teachers complain bitterly and say the examiners are unfair. In such cases, when there has been a great outcry, we have had the papers sent to us, and Dr. Foster or I, or both of us, have looked through them again and prepared a little précis for the advantage of the teachers, showing the style of answers. We have done that on several occasions and have caused the précis to be sent officially as our answers to the teachers. We have never heard from them any more. I may say this—I do not say it boastingly at all—that, during the 20 years that I have been an examiner, I have never known any case in which we have rejected a man, in which we have subsequently seen reasons to reverse our decision—if we had had to reconsider the case of some of those we had passed I am not at all sure that we might not have done so—but in the 20 years I have never seen any reasonable ground for reversing any of our decisions where we have plucked those whom we have examined. The upshot of that examination is roughly this: that taking one year with another we pluck on the average 40 per cent. whilst 8 or 9 per cent. of the papers are very good—in the advanced they are sometimes remarkably good, perhaps 5 per cent. The rest of the papers are very much like what you generally find elsewhere, good, bad, and indifferent.

Royal Commission on Technical Instruction, *B.P.P.*, 1884 (31).

10 SCIENCE AND PUBLIC HEALTH

10a Adulteration of Food

A TREATISE

ON

ADULTERATIONS OF FOOD,

AND

Culinary Poisons,

EXHIBITING

THE FRAUDULENT SOPHISTICATIONS

OF

BREAD, BEER, WINE, SPIRITUOUS LIQUORS, TEA, COFFEE,

Cream, Confectionery, Vinegar, Mustard, Pepper, Cheese, Olive Oil, Pickles,

AND OTHER ARTICLES EMPLOYED IN DOMESTIC ECONOMY.

AND

Methods of detecting them.

BY FREDRICK ACCUM,

Operative Chemist, Lecturer on Practical Chemistry, Mineralogy, and on Chemistry applied to the Arts and Manufactures; Member of the Royal Irish Academy; Fellow of the Linnæan Society; Member of the Royal Academy of Sciences, and of the Royal Society of Arts of Berlin, &c. &c.

London:

Printed by J. Mallett, 59, Wardour Street, Soho.

SOLD BY LONGMAN, HURST, REES, ORME, AND BROWN,

PATERNOSTER ROW.

1820

By courtesy of the British Museum

Preface

Old Compton Street, Soho

THIS Treatise, as its title expresses, is intended to exhibit easy methods of detecting the fraudulent adulterations of food, and of other articles, classed either among the necessaries or luxuries of the table; and to put the unwary on their guard against the use of such commodities as are contaminated with substances deleterious to health.

Every person is aware that bread, beer, wine, and other substances employed in domestic economy, are frequently met with in an adulterated state: and the late convictions of numerous individuals for counterfeiting and adulterating tea, coffee, bread, beer, pepper, and other articles of diet, are still fresh in the memory of the public.

To such perfection of ingenuity has the system of counterfeiting and adulterating various commodities of life arrived in this country, that spurious articles are every where to be found in the market, made up so skilfully, as to elude the discrimination of the most experienced judges.

But of all possible nefarious traffic and deception, practised by mercenary dealers, that of adulterating the articles intended for human food with ingredients deleterious to health, is the most criminal, and, in the mind of every honest man, must excite feelings of regret and disgust. Numerous facts are on record, of human food, contaminated with poisonous ingredients, having been vended to the public; and the annals of medicine record tragical events ensuing from the use of such food.

The eager and insatiable thirst for gain, is proof against prohibitions and penalties; and the possible sacrifice of a fellow-creature's life, is a secondary consideration among unprincipled dealers.

However invidious the office may appear, and however painful the duty may be, of exposing the names of individuals, who have been convicted of adulterating food; yet it was necessary, for the verification of my statement, that cases should be adduced in their support: and I have carefully avoided citing any, except those which are authenticated in Parliamentary documents and other public records.

To render this Treatise still more useful, I have also animadverted on certain material errors, sometimes unconsciously committed through accident or ignorance, in private families, during the preparation of various articles of food, and of delicacies for the table.

In stating the experimental proceedings necessary for the detection of the frauds which it has been my object to expose, I have confined myself to the task of pointing out such operations only as may be performed by persons unacquainted with chemical science; and it has been my purpose to express all necessary rules and instructions in the plainest language, divested of those recondite terms of science, which would be out of place in a work intended for general perusal.

The design of the Treatise will be fully answered, if the views here given should induce a single reader to pursue the object for which it is published; or if it should tend to impress on the mind of the Public the magnitude of an evil, which, in many cases, prevails to an extent so alarming, that we may exclaim, with the sons of the Prophet, 'There is death in the pot'.

For the abolition of such nefarious practices, it is the interest of all classes of the community to co-operate.

FREDRICK ACCUM

London, 1820

Poisonous Cheese

SEVERAL instances have come under my notice in which Gloucester cheese has been contaminated with red lead, and has produced serious consequences on being taken into the stomach. In one poisonous sample which it fell to my lot to investigate, the evil had been caused by the sophistication of the anotto, employed for colouring cheese. This substance was found to contain a portion of red lead; a method of sophistication which has lately been confirmed by the following fact, communicated to the public by Mr. J. W. Wright, of Cambridge*.

'As a striking example of the extent to which adulterated articles of food may be unconsciously diffused, and of the consequent difficulty of detecting the real fabricators of them, it may not be uninteresting to relate to your readers the various steps by which the fraud of a poisonous adulteration of cheese was traced to its source.'

'Your readers ought here to be told, that several instances are on record, that Gloucester and other cheeses have been found contaminated with red lead, and that this contamination has produced serious consequences. In the instance now alluded to, and probably in all other cases, the deleterious mixture had been caused ignorantly, by the adulteration of the anotto employed for colouring the cheese. This substance, in the instance I shall relate, was found to contain a portion of red lead; a species of adulteration which subsequent experiments have shewn to be by no means uncommon. Before I proceed further to trace this fraud to its source, I shall briefly relate the circumstance which gave rise to its detection.'

'A gentleman, who had occasion to reside for some time in a city in the West of England, was one night seized with a distressing but indiscribable pain in the region of the abdomen and of the stomach, accompanied with a feeling of tension, which occasioned much restlessness, anxiety, and repugnance to food. He began to apprehend the access of an inflammatory disorder; but in twenty-four hours the symptoms entirely subsided. In four days afterwards he experienced an attack precisely similar; and he then recollected, that having, on both occasions, arrived from the country late in the evening, he had ordered a plate of toasted Gloucester cheese, of which he had partaken heartily; a dish which, when at home, regularly served him for supper. He attributed his illness to the cheese. The circumstance was mentioned to the mistress of the inn, who expressed great surprise, as the cheese in question was not purchased from a country dealer, but from a highly respectable shop in London. He, therefore, ascribed the before-mentioned effects to some peculiarity in his constitution. A few days afterwards he partook of the same cheese; and he had scarcely retired to rest, when a most violent colic seized him, which lasted the whole night and part of the ensuing day. The cook was now directed henceforth not to serve up any toasted cheese, and he never again experienced these distressing symptoms. Whilst this matter was a subject of conversation in the house, a servant-maid mentioned that a kitten had been violently sick after having eaten the rind cut off from the cheese prepared for the gentleman's supper. The landlady, in consequence of this statement, ordered the cheese to be examined by a chemist in the vicinity, who returned for answer, that the cheese was contaminated with lead! So unexpected an answer arrested general attention, and more particularly as the suspected cheese had been served up for several other customers.'

'Application was therefore made by the London dealer to the farmer who

* Repository of Arts, vol. viii. No. 47, p. 262.

manufactured the cheese: he declared that he had bought the anotto of a mercantile traveller, who had supplied him and his neighbours for years with that commodity, without giving occasion to a single complaint. On subsequent inquiries, through a circuitous channel, unnecessary to be detailed here at length, on the part of the manufacturer of the cheese, it was found, that as the supplies of anotto had been defective and of inferior quality, recourse had been had to the expedient of colouring the commodity with vermilion. Even this admixture could not be considered deleterious. But on further application being made to the druggist who sold the article, the answer was, that the vermilion had been mixed with a portion of red lead; and the deception was held to be perfectly innocent, as frequently practised on the supposition, that the vermilion would be used only as a pigment for house-painting. Thus the druggist sold his vermilion in the regular way of trade, adulterated with red lead to increase his profit, without any suspicion of the use to which it would be applied; and the purchaser who adulterated the anotto, presuming that the vermilion was genuine, had no hesitation in heightening the colour of his spurious anotto with so harmless an adjunct. Thus, through the circuitous and diversified operations of commerce, a portion of deadly poison may find admission into the necessaries of life, in a way which can attach no criminality to the parties through whose hands it has successfully passed.'

This dangerous sophistication may be detected by macerating a portion of the suspected cheese in water impregnated with sulphuretted hydrogen, acidulated with muriatic acid; which will instantly cause the cheese to assume a brown or black colour, if the minutest portion of lead be present.

Food Poisoned by Leaden Vessels

VARIOUS kinds of food used in domestic economy, are liable to become impregnated with lead.

The glazing of the common cream-coloured earthen ware, which is composed of an oxid of lead, readily yields to the action of vinegar and saline compounds; and therefore jars and pots of this kind of stone ware, are wholly unfit to contain jellies of fruits, marmalade, and similar conserves. Pickles should in no case be deposited in cream-coloured glazed earthenware.

The custom which still prevails in some parts of this country of keeping milk in leaden vessels for the use of the dairy, is very improper.

'In Lancashire* the dairies are furnished with milk-pans made of lead: and when Mr. Parks expostulated with some individuals on the danger of this practice, he was told that *leaden* milk-pans throw up the cream much better than vessels of any other kind.'

'In some parts of the north of England it is customary for the inn-keepers to prepare mint-salad by bruising and grinding the vegetable in a large wooden bowl with a *ball of lead* of twelve or fourteen pounds weight. In this operation the mint is cut, and portions of the lead are ground off at every revolution of the ponderous instrument. In the same county, it is a common practice to have brewing-coppers constructed with the bottom of copper and the whole sides of lead.'

The baking of fruit tarts in cream-coloured earthenware, and the salting and preserving of meat in leaden pans, are no less objectionable. All kinds of food which contain free vegetable acids, or saline preparations, attack utensils

* Parks's Chemical Essays, vol v. p. 193.

covered with a glaze, in the composition of which lead enters as a component part. The leaden beds of presses for squeezing the fruit in cyder countries, have produced incalculable mischief. These consequences never follow, when the lead is combined with tin; because this metal, being more eager for oxidation, prevents the solution of the lead.

When we consider the various unsuspected means by which the poisons of lead and copper gain admittance into the human body, a very common but dangerous instance presents itself: namely, the practice of painting toys, made for the amusement of children, with poisonous substances, viz. red lead, verdigrise, &c. Children are apt to put every thing, especially what gives them pleasure, into their mouths; the painting of toys with colouring substances that are poisonous, ought therefore to be abolished; a practice which lies the more open to censure, as it is of no real utility.

F. Accum, *A Treatise on Adulterations of Food and Culinary Poisons*, London, 1820, p. ii.

10b Water and its Impurities

With this, the first number of The Lancet for the year 1851, we commence the publication of our Analytical Sanitary Commission; the new department, which we recently announced to the profession. We entertain no doubt that but one sentiment of satisfaction and congratulation will be experienced by our readers, when they shall have estimated the scope and purpose of our plan, from the specimen of its execution which is now before them.

We feel impressed with the conviction that we have undertaken a task, large and important enough to have engaged the attention of the Government of this country. But though the undertaking may be a vast one, incurring enormous labour and great expense, we have no fear of executing it in such a manner as will prove honourable to ourselves and to the profession, and beneficial to the public. We re-enter upon the labour, (which we actually commenced in 1831,) with a full sense of the responsibilities it involves, and a full determination to cope with them, come in what shape they may, satisfactorily and successfully.

As our Sanitary Commissioner remarks, we have our Boards of Health, and our Commissions of Sewers, which take ready cognizance of questions of air and light, effluvia and water. The whole force of the State is brought to bear upon the poisoning and deterioration of human blood, by miasms, by impure water, and by the absence of ventilation in our cities and dwellings. But with the exception of the appointment of officers to observe and report upon diseased meat or fish, the public authorities take no cognizance whatever of the deterioration and poisoning of our bodies by the slower, but equally sure modes of adulterations of food and drink. Questions of weight and measure are decided by the magistrates. The tradesman whose weights and balances are erroneous or unfair is summarily punishable; he must not measure an article unfairly, or fine or imprisonment will surely overtake him upon his detection; but it appears that he has hitherto been allowed to substitute one article for another with the most perfect impunity. As regards weight, the pound of 'coffee' must be correct to the penny-weight, but as regards quality, it may consist of one part coffee, and three parts chicory and roasted grain; and so on of a hundred other articles of consumption besides coffee, as we shall show in subsequent articles.

It cannot but strike every one as most extraordinary that this vitally important question should have remained in comparative neglect up to the present time. Organic chemists, PROUT, LIEBIG, DUNDAS THOMSON, and others, have devoted the aids and appliances of modern chemistry with the utmost success to the analyses of the ultimate elements of the solids and fluids which nourish the living body, but little or nothing has been done, or even attempted, in the systematic detection of adulterations of these great and universal agencies of health and disease.

Let not any tradesman, mechanic, or manufacturer, imagine for a moment that he can defeat our vigilance. We bring the microscope and the test-tube—those simple but mighty instruments of modern investigation—to bear with unerring truth upon things hidden and secret enough from the observation of the unaided senses. These instruments, assisted by the powers of chemistry and structural botany and anatomy, are our detective police, and we shall confess ourselves at fault if we find any difficulty throughout the whole intricacies of our subject, which they are unable to expose and make plain. No doubt the fabricators of the mixture sold as Coffee of Java, Ceylon, Berbice, and Costa Rica, thought the coffee-mill a screen for their malversations. The fine powder seemed to defy all inquiry into its composition; but the sunlight is not clearer than cell, granule, and fibre, when the microscopist brings his wonderful tube to bear upon their exploration. The burnt starch seems beyond identification, but the glass-rod of the magician-chemist, and a drop or two of iodine, instantly expose the cheat. A single glance of the practical microscopist, a single touch of the expert chemist, and the most intricate mystifications stand confessed and naked.

Of the immense service these researches will confer on the community we need not now speak. The coming season is one in which such an undertaking is especially necessary. During the presence of the concourse drawn to the metropolis by the Great Exhibition of the Industry of All Nations, the temptation to adulterate articles of food and drink will be immense. Against this evil our ANALYTICAL SANITARY COMMISSION will, we flatter ourselves, prove a powerful bar. We enter upon the task, then at an auspicious time, and though unaided, except in the support of the profession, we are determined to execute our self-imposed labour for the protection of the public, the advantage of the fair trader, and the ultimate exposure and punishment of the fraudulent one. That the duty shall be performed with the most strict and undeviating impartiality, we hope it is unnecessary to offer any pledge.

THE ANALYTICAL SANITARY COMMISSION

Records of the Results of Microscopical and Chemical Analyses of the Solids and Fluids consumed by all Classes of the Public

'Forewarned, forearmed'

WATER, AND ITS IMPURITIES

As the time is fast approaching when the great question of the Water Supply of London will again be brought under the notice of the public and of parliament, we propose to place before the reader, in a clear and concise manner, the whole of the facts connected with a subject of such vast social importance.

We shall discuss it under four heads or sections. Under the first head, the impurities of water will be considered; under the second, the different kinds of water will be treated of; under the third, the present sources of supply will be examined; and under the fourth, a detailed examination will be given of the proposal of the General Board of Health for the supply of water to the metropolis.

SECTION I. THE IMPURITIES OF WATER

The various contaminations to which water is liable may all be arranged under two heads, and are either *organic* or *inorganic.*

Organic Impurities

The organic impurities again naturally resolve themselves into the *dead* and the *living.*

DEAD ORGANIC IMPURITIES.—Inasmuch as the dead organic substances present in impure waters form the material or food upon which the living impurities are sustained, it is proper to allude first to them.

The dead organic contaminations again admit of still further division into those derived from the vegetable kingdom, or *dead vegetable impurities*, and those from the animal kingdom, or *dead animal impurities.*

Dead Vegetable Impurities.—The dead vegetable substances very frequently present in impure water are water-plants, leaves and other fragments of land plants, which find their way into it in a variety of ways, and at certain seasons, particularly at the period of the fall of the leaf.

These dead vegetable tissues are in a state of continual decay and change; deprived of life, they become subject to the action of the laws of chemical affinity, by which their structure is speedily destroyed, and their whole fabric resolved into substances of more simple forms and constitution, as *carbonic acid, carburetted hydrogen*, a little *ammonia*, sometimes *sulphuretted hydrogen*, as well as probably, in some cases, other bodies of more subtle nature, the characters of which have not hitherto been determined.

Dead Animal Impurities.—The dead animal organic impurities present are the bodies of infusoria, annelidæ, insects, fish, and animals, as well as the soluble proteine compounds, as albumen, gelatine, mucus, &c., derived from and dissolved out of these by the water.

These several animal substances and compounds are likewise in a state of constant change; their elements, obedient to the laws of chemical action, become rearranged, and form combinations more simple, as *ammonia, nitric acid, cyanogen, sulphuretted hydrogen, phosphuretted hydrogen*, water, and many other bodies, some of which, like certain of those just enumerated, are probably of a highly deleterious nature. These changes, decompositions, and re-combinations, are greatly facilitated and hastened by increased temperature, and therefore are effected more rapidly in summer.

LIVING ORGANIC IMPURITIES.—The living organic impurities are in direct dependence, as already remarked, upon the dead, from which they derive their nourishment and support, and like them, are either vegetable or animal.

Living Vegetable Impurities.—The living vegetable productions contained in impure water are algæ, which are the fresh-water types of the plants commonly known as sea-weeds and water-plants.

Living Animal Impurities.—The principal animal productions contained in water are, fish, zoophytes, worms, larvæ of insects, mollusca, and animalculæ or infusoria.

Circumstances which favour the Growth and determine the Kinds of Organic Life present in Water

Analysis has shown that *every vegetable* is composed of oxygen, hydrogen, and carbon, with a variable but usually a small quantity of nitrogen.

For the growth, therefore, of vegetables, it results as essential conditions that they should be supplied with material containing the whole of these substances.

Now as one vegetable resembles in its general constitution another, it follows that decaying vegetable matters contain the whole of the materials necessary to the life of a plant, and thus one generation of plants affords the food requisite for the growth and sustenance of a second.

But certain inorganic substances also favour greatly the growth of plants, as the phosphates and ammonia; the latter probably on account of the nitrogen it contains; and hence the value of animal manures, which contain these compounds largely, in promoting the growth of vegetation.

Again, analysis has shown that *every animal* contains precisely the same elements as the vegetable, with this important difference, however, that the proportion of nitrogen is much greater in the former.

It thus appears, that while vegetable substances in a state of decay are best adapted to the maintenance of vegetation, animal matter, in a similar state, is the best suited to sustain a large amount of animal life.

These particulars are capable of important practical application, inasmuch as they declare to us that every water which abounds in living animal forms, especially infusoria, must also contain dead animal matter in abundance.

The true interpretation to be placed on the presence of infusoria in water was not well understood until Dr. Hassall called particular attention to the condition of animalcular life in Thames water near to London, to which we shall hereafter have occasion again to refer.

The relation of dead to living organic matter is well shown by simply diffusing a little albumen through a wine-glass of distilled water: in a few days, and especially if the weather be warm, the infusion will be found literally to swarm with animalculæ.

But there are some other circumstances, independent of the amount of dead organic matter contained in water, which determine the kind of living productions present in it.

One of these, and a very important one, is the condition of the water as to alkalinity or acidity. If the water have a decided acid reaction, and contain a large amount of animal matter in solution, there will be an abundant development of fungi in the course of a few days; but if, on the other hand, it exhibits an alkaline reaction, then, instead of fungi, animalculæ will be developed.

These curious and important facts may be readily shown by experiment. Let two solutions of albumen in water be prepared, the one being rendered alkaline, the other acid; in the first, in the course of a short time, infusoria will appear, while in the second, in an equally short period, a scum of fungi will become visible to the naked eye.

Now fungi are usually regarded by the naturalist as vegetable productions, having, however, a resemblance to animals in the amount of nitrogen which they contain, and in their subsistence on decomposing animal matter.

But other differences may be observed between acid and alkaline animal solutions; while the animal matter contained in the alkaline fluids quickly undergoes putrefactive changes, and begins to smell disagreeably, that in the acid fluid scarcely alters at all, and emits no unpleasant odour.

From these facts it may be fairly concluded, that these two waters stand in very different relations to each other in a sanitary point of view, and from them important practical applications will in another part of this Report be deduced.

But there are occasionally found in nature, waters which, although largely impregnated with dead organic matter, yet contain neither animal nor vegetable productions in a living state; these are generally well waters, contaminated by sewage or the contents of cess-pools. The reason of the absence of life in these cases it is not difficult to divine; from the water in the deep and covered well, air, light, and warmth, are excluded, conditions essential to development.

The *kinds of vegetable and animal life* present in water vary much according to their nature and chemical composition.

In the *still waters* of ponds, cisterns, and reservoirs, which contain dead vegetable matter, in the spring and summer, which is the great season of the development of vegetation, we encounter living vegetable forms, especially confervæ.

In the same waters, holding in solution much animal matter, we meet with living animal productions, particularly infusoria and entomostraceæ.

The forms of life inhabiting the *running waters* of springs and rivers are usually very different from those which dwell in still waters; they belong to different species and genera, and are provided with an organization by which they are adapted to the very different life they are destined to lead.

The *quantity of animal and vegetable life* contained in water also varies according to its nature and composition.

Thus pure *distilled water* contains no organic matter, either dead or alive.

Rain water, caught at a distance from houses, towns, or other sources of contamination, contains none.

Uncontaminated well water contains no organic matter.

Neither does pure *spring water* contain any form or trace of organic matter.

In *River water* there is present usually a considerable but variable amount of organic matter; this amount varies according to certain well-determined circumstances: thus, at the source of the river, where the stream is comparatively pure and unpolluted, the quantity of organic matter, either dead or living, is but small.

As the stream flows along, however, it becomes exposed to numberless sources of contamination; it receives the contents of still smaller streams, and the impurities of innumerable ditches, the refuse of the dwellings of persons who live near its banks, the contents of drains, and very frequently the sewage of whole and populous towns, so that at last, when it reaches its outlet into the sea, great and manifold are the corruptions, and large is the amount of dead and living organic matter, pervading its waters.

Were it not that for the continually increasing impurities of river water there is a daily means of escape by the ocean, that description of water would be the most contaminated of all.

The *water of ponds* usually contains a large quantity of organic matter, because these also are exposed to many sources of contamination; the leaves of trees which grow near, fall each year into them, and successive crops of aquatic vegetation are developed and pass to decay in their waters.

The *water of reservoirs* likewise usually contains much organic matter; into these, water which is itself very impure is poured from time to time; on each occasion a portion of the contained organic matter subsides to the bottom, and thus at last a great accumulation of organic impurities occurs, sufficient in the course of a very short time to spoil and render unfit for use even the very best water.

Lastly, the *water of cisterns* generally contains much organic matter. A cistern may be compared to a reservoir; it is, in fact, a reservoir on a small scale, and in it also organic impurities increase from day to day, both by accumulation and growth.

It should be remembered that the actual quantity of organic matter present in a large body of water, or even in a river, cannot be estimated from an examination of samples of water, obtained by merely dipping a jar or bottle into it, because by far the greatest portion of the organic matter is contained in the mud or other deposit at the bottom of the reservoir or river. This fact has very generally been overlooked, and hence the proportion of organic matter given by chemists in many cases as present in certain waters is by far too low.

The effect of the presence of organic matter in the sediment of a reservoir or river on the entire body of the water contained in it is shown by placing a little of such sediment in distilled water in a bottle; in a short time, especially if the weather be warm, the water will begin to smell offensively.

But there are certain circumstances which are closely connected with the *kinds* as well as the *quantity* of organic life present in water, beyond those already noticed; these are, free exposure to air, light, and warmth, which all greatly favour development, and so assist in determining the kind and amount actually present.

For the purpose of showing clearly the influence of these agents, we instituted a considerable variety of experiments; as, however, the powerful operation of the causes referred to are well known and undisputed, it is unnecessary to detail their nature.

Now rivers, reservoirs, and other large bodies of water are exposed to the influences alluded to, whereby the water contained in them is greatly deteriorated.

We have dwelt thus long, and fully, on the *Organic Impurities* of water, because of their extreme and primary importance, for it is on these that the deleterious properties of impure water for the most part depend.

Until very recently, chemists did not in general attach sufficient importance to these organic contaminations, and in most of their analyses we find the different kinds of organic matter, vegetable and animal, living and dead, all lumped together, and included under the word 'Traces'.

Indeed, chemistry is but ill-adapted to investigate the nature of these organic matters; it gives but a very rough estimate only of their gross amount, and does not discriminate, as we have said, the animal from the vegetable, the dead from the living, and tells us nothing about the families, genera, and species, to which the numerous living productions contained in impure waters severally belong, or of their habits and modes of life &c.

This inquiry belongs to the naturalist, the physiologist, and the microscopist, and to Dr. Hassall is due the honour of having first applied the resources of these, extensively, and in a practical as well as a scientific manner, to an examination of the actual condition of water in general, and particularly the state of that now in use in the metropolis.*

Uses of Vegetables and Animals in Impure Water

In the existence in impure water of different kinds of organic productions, we recognise the fulfilment of wise and beneficial purposes.

If all the organic matter present in some waters were to be removed by the ordinary processes of putrefaction, decomposition, and the formation of offensive and deleterious gases and compounds, incalculable mischief would constantly ensue.

* Results of a Microscopical Examination of the Water supplied to the Metropolis, and the Suburban Districts.—THE LANCET, March, 1850.

A Microscopical Examination of the Water supplied to the Inhabitants of London and the Suburban Districts, Illustrated with coloured plates. Samuel Highley, Fleet-street.

Evidence before the General Board of Health; Report, Appendix III.

To obviate this, nature has ordained that some of the organic matter of impure water, in place of undergoing decomposition, should be imbibed by other and living forms, and these dying, that other generations should take their place, and fulfil a similar important office.

In the deep well, where the conditions necessary to animal life do not exist, the same end is accomplished by the formation, out of the nitrogen of the organic matter, of nitrates, to which we have already referred.

The purposes fulfilled by living vegetable and animal productions in water, are then of an eminently useful and preservative character; and this observation is true particularly of the vegetable productions contained in it, including both the lower forms, as the confervæ and other algæ, and the higher plants, as the flowering; for these not only remove the organic matter dissolved in water, by absorbing it into their own tissues for appropriation, but they still further purify water by the effect of their respiration.

Animals, during respiration, exhale a large amount of carbonic acid, vegetables absorb this, fix the carbon in their own tissues, and restore the oxygen to the air, if the plant be aerial, and to the water, if it be submerged and aquatic.

Inorganic Impurities

Water is a substance formed by the chemical combination of two gases, oxygen and hydrogen, in certain proportions: these alone being necessary to its constitution, all other inorganic matters contained in it are to be regarded as unnecessary, extraneous, and frequently injurious additions.

Under the head 'Inorganic Impurities,' therefore, we include, first, all those inorganic substances which are merely suspended in the water, as sand and grit; and, second, those substances which are in a state of actual solution, as the saline, earthy, and metallic salts, present in most waters, and the gases.

Inorganic Impurities in Suspension

The inorganic impurities which are merely suspended in the water are the least important of all, since they do not impart any deleterious properties; they consist chiefly of particles of a stony nature, which are therefore heavier than water, and are readily separated, either by subsidence or filtration.

The quantity of grit or sand in suspension in water varies with the condition of the water, and the state of the weather.

If the water be kept in continual agitation, as by the passage of boats, barges, and steam-vessels, it will always contain a very large quantity of earthy matter diffused through it, sufficient, indeed, to impart an opaque appearance and an earthy colour; the water of the Thames near to London is usually in the state described, and a sample of it is generally opaque and discoloured, by reason of the immense quantity of earth suspended in it.

Again, water kept in a state of motion by storms, wind, and rain, will be affected in a similar way, and will be laden with earthy matter.

The water just poured into a dirty reservoir or cistern remains in the state described for some hours after, in consequence of its stirring up the accumulated sediment contained in it.

Inorganic Impurities in Solution

The salts contained in water may be divided into the *earthy*, *alkaline*, and *metallic*, and it is to their presence that many of the sensible properties of water are due, as its softness, hardness, alkalinity, and acidity.

Softness is a property common to all the purer waters, as distilled, rain water, &c., and is possessed in greatest proportion by those waters which contain the least amount of earthy salts.

The alkalies or alkaline salts, however, possess the property of imparting softness, and hence their employment in soap and washing powders.

The degree of *hardness* of water depends upon the amount of earthy salts dissolved in it, that is, of sulphates and bicarbonates of lime and magnesia, but particularly of the bicarbonates.

The *alkalinity* of water is dependent upon the amount of alkalies present in solution, the principal of these being the carbonates of soda and potash.

The *acidity* arises of course either from the presence of acid salts, or some free acid, as the carbonic.

When we come to consider 'the different kinds of water,' we shall have again to refer to the sensible qualities we have thus briefly alluded to.

The two great metallic impurities or contaminations of water are those by lead and iron, each of which, but especially the first, we shall consider in detail.

On the Action of Water on Lead

Notwithstanding that lead has now been employed for ages in the storage and conveyance of water, the principles of its action on that substance have been, in general, but ill understood and defined; much uncertainty and contradiction have prevailed, even amongst scientific men, as to what waters act most on lead, and what least, some asserting that soft waters do so, and others, that hard waters affect lead most.

These principles will be found to be in strict accordance with the laws of chemical combination.

In speaking of lead in connexion with water, we commonly make use of the phrase, 'the action of water on lead:' this phrase, unless clearly explained, is liable to be associated in the mind with error.

Water, as we before observed, consists of oxygen and hydrogen: the only combination, therefore, which could by possibility result from the decomposition of water, and the union of its elements with lead, consists in the formation of the hydrated oxide of that metal—an almost insoluble compound.

But the waters in general use contain several gases and salts; these act on lead in a variety of ways, or, in other and more definite words, form various combinations with lead.

Lastly, water imbibes air and other gases, the oxygen of the former combining with the lead, and forming the oxide.

When, therefore, we speak of the action of water on lead, in general terms, water in the condition in which it is commonly met with is implied; but in our scientific inquiries we must carefully discriminate between the effects of the action of pure water on lead, and those effects which result from the presence of the salts and gases so commonly contained in water.

The effect of the action of pure water, then, on lead, supposing it to be really decomposed, is to give rise to the formation of a hydrated oxide, the same body likewise being formed as the result of the presence in water of free oxygen, derived from the atmosphere.

But in some cases lead does not only exist in water as a slightly soluble oxide, but is in a state of chemical solution, and we have now to inquire in what condition it may be supposed to be in such cases. The most common soluble salt of lead detected in water is the carbonate, which consists of carbonic acid in union with the oxide of lead.

Now an oxide is a body or base in combination with oxygen, which readily unites with any free acid with which it may come in contact: in a water, therefore, acid from the presence of free or uncombined carbonic acid, stored in a leaden vessel, and exposed to the air, all the chemical conditions necessary for the formation of a soluble and dangerous salt of lead exist; the cistern is the metal, the water itself, or the air contained in it, supplies the oxygen by which a portion of that metal is oxidized; and with the oxide thus formed the free acid also present readily combines.

EXPLANATION OF THE PLATES.

The samples of water from which the annexed engravings have been prepared were obtained either from the Thames itself, or from service-pipes, within the last few days.

By courtesy of the British Museum

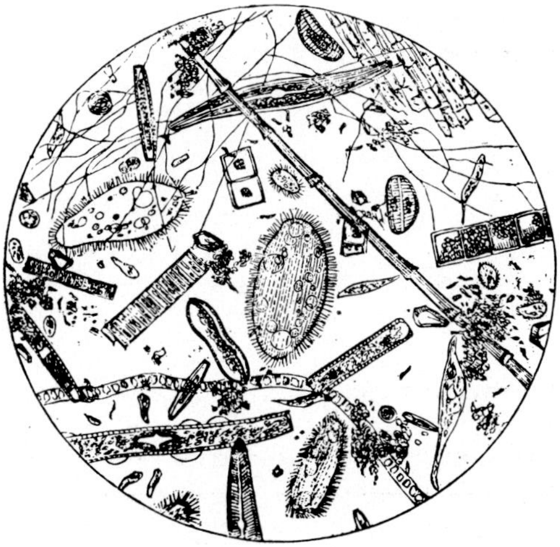

Fig. 1. Thames water at Richmond. This engraving exhibits the principal animal and vegetable productions contained in the water of the *Thames at Richmond*. Drawn with the camera lucida, and magnified 220 diameters.

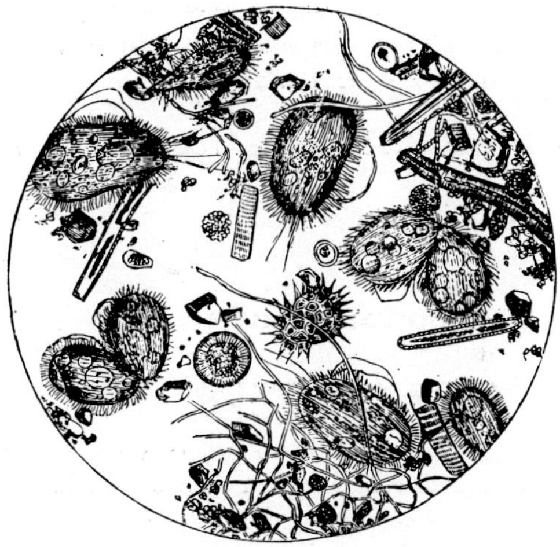

Fig. 2. Water from the Thames at Waterloo-bridge. This engraving shows the more remarkable animal and vegetable productions found in the water of the *Thames at Waterloo-bridge*. Drawn with the camera lucida, and magnified 220 diameters.

By courtesy of the British Museum

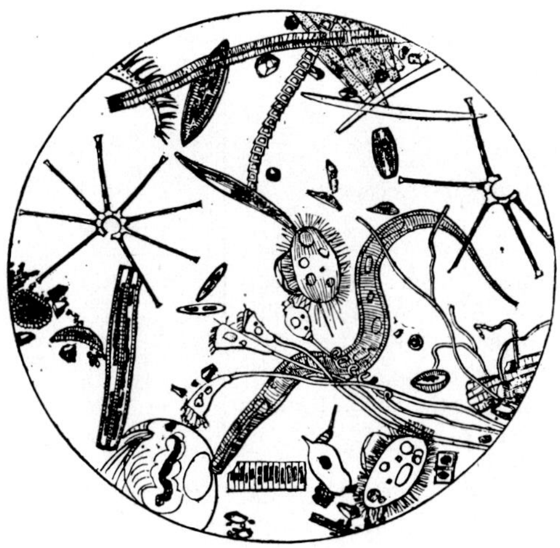

Fig. 3. Grand Junction Company. This engraving represents the chief animal and vegetable productions contained in the water as supplied by the *Grand Junction Company*.

Fig. 4. West Middlesex Company. The above engraving exhibits the principal animal and vegetable productions contained in the water supplied by the *West Middlesex Company*. Drawn with the camera lucida, and magnified 220 diameters.

Each drawing represents *only a very small portion* of the organic matters, living and dead, contained in several samples, of less than a quart each. On setting aside any water containing solid impurities, these of course subside, and form residues; but the living animal forms, mostly infusoria, possessing active powers of locomotion, diffuse themselves throughout the whole body of the water.

The forms exhibited by many of the lower orders of animals are often very singular, and we are not surprised that they should sometimes have been made the subject of jocose and harmless remarks. The important fact, however, should never be lost sight of, that the presence of these creatures, in considerable numbers, in water, shows that it contains a *large quantity of fluid organic matter*, and *is therefore in a state wholly unfit for human consumption*, more particularly as a beverage. Upon this point it would be simply a waste of words and argument to offer another observation.

'Water and its Impurities', *Lancet*, 1851, I, No. 1.

10c 'Food Additives'

CLASSIFIED LIST OF THE VARIOUS SUBSTANCES ASCERTAINED BY OURSELVES TO BE EMPLOYED FOR THE DIFFERENT PURPOSES OF ADULTERATION: VIZ. FOR BULK AND WEIGHT, FOR COLOUR, AND FOR SMELL, TASTE, AND OTHER PROPERTIES

	For Bulk and Weight	For Colour	For Taste, Smell, and other Properties
ARROW-ROOT -	Sago, Potato, and Tapioca Starches, and various mixtures and combinations of these with the inferior arrow-roots		
ANCHOVIES - -	Dutch, French, and Sicilian Fish	Bole Armenian, Venetian Red	
BREAD - -	Mashed Potatoes - - - -	- - - - - - -	Alum, Hards and Stuff
BUTTER - -	Water		
BOTTLED FRUITS AND VEGETABLES	- - - - - - -	Certain Salts of Copper, usually the Acetate	Salt
CINNAMON - -	Cassia, and most of the articles mentioned under Spices		
COLOURED CONFECTIONERY	East India Arrow-root, Wheat and Potato Flour, Hydrated Sulphate of Lime	Cochineal, Lake, Indigo, Prussian Blue, Antwerp Blue, Artificial Ultramarine, Carbonate of Copper or Verditer, Carbonate of Lead or White Lead, Red Lead, Vermilion; Chrome Yellow or Chromates of Lead, Lemon, Orange, and deep; Gamboge; the three Brunswick Greens, Emerald Green or Arsenite of Copper, Indian (*continued*)	

	For Bulk and Weight	For Colour	For Taste, Smell, and other Properties
COLOURED CONFECTIONERY (*continued*)		Red; brown ferruginous earths, chiefly Umber, Sienna, and Vandyke Brown, and various combinations of the above pigments	
COFFEE - -	Chicory, Roasted Wheat, Rye and Potato Flours, Roasted Beans, Mangel-wurzel, Acorns	Burnt Sugar, or Black Jack	
CHICORY - -	Roasted Wheat and Rye Flours, Burnt Beans and Acorns, Burnt Sugar, Sawdust, Mahogany Sawdust, Carrot, Mangel-wurzel	Ferruginous earths, as Venetian Red and Umber, Burnt Sugar and Black Jack	
COCOA AND CHOCOLATE	Maranta, East India, and Tacca or Tahiti Arrow-roots; Tous les Mois; the Flours of Wheat, Indian Corn, Sago, Potato, and Tapioca, and various Mixtures of these; Sugar, Chicory	Venetian Red, Red Ochre, and other ferruginous earths	
CAYENNE PEPPER	Ground Rice, Mustard Husk, Deal Sawdust, Salt	Red Lead, Vermilion or Bisulphuret of Mercury, Venetian Red, Turmeric	
CUSTARD AND EGG POWDERS	Wheat, Potato and Rice Flours	Chrome Yellow or Chromate of Lead, Turmeric	
CURRY POWDER -	Ground Rice, Potato-farina, Salt	Red Lead - - - -	Salt
FLOUR - -	- - - - - - - -	- - - - - - - -	Alum
GINGER - -	Wheat, Sago, and Potato Flours, Ground Rice, Mustard Husks	Turmeric Powder	
GIN - - -	Water, Sugar - - -	- - - - - - - -	Cayenne, Cassia or Cinnamon, Sugar, and flavouring of different kinds. For fining, Alum, Salt of Tartar
ISINGLASS - -	Gelatine		
LARD - -	Potato-flour, Water - -	- - - - - - - -	Salt, Carbonate of Soda, Caustic Lime
MUSTARD - -	Wheat-flour, Turmeric - -	Turmeric	
MILK - -	Water - - - - - -	Annatto	
MARMALADE -	Pulp of Apple or Turnip		
OATMEAL - -	Barley-flour, and the integuments of Barley called Rubble		
PORTER AND STOUT	Water - - - - - -	Sugar, Treacle - - -	Sugar, Treacle, Salt
PICKLES - -	- - - - - - - -	Salts of Copper, usually the Acetate of Copper	
POTTED MEATS AND FISH	Flour, probably Wheat-flour boiled	Bole Armenian, and sometimes Venetian Red	
PRESERVES - -	Salts of Copper, including the Acetate		

(*continued*)

	For Bulk and Weight	For Colour	For Taste, Smell, and other Properties
Pepper - -	Wheat and Pea Flour, Ground Rice, Ground Mustard Seeds, Linseed Meal, P. D., or Pepper Dust		
Snuff - -	- - - - - - -	The Chromates of Potash, Chromate of Lead, ferruginous earths, chiefly Umbers, Red and Yellow Ochre, Red Lead, or Oxide of Lead	The Chromates of Potash, Carbonate of Ammonia, Lime, Powdered Glass or Silex, Powdered Orris-root
Sugar - -	Wheat-flour in two cases only, Potato-flour, and Tapioca-starch		
Spices:			
Cloves - -	Powdered Clove-stalks in one case		
Cinnamon -	Cassia, Wheat-flour, Sago-meal, and mixtures of these; East India Arrow-root, Potato-flour		
Pimento -	Mustard Husk in one instance		
Mixed Spice -	Wheat, Sago, and Potato Flours, Ground Rice, Two Vegetable Substances, one of which resembled Linseed		
Sauces, as the Essences of Anchovies, Lobsters, and Shrimps, and Tomato Sauce	- - - - - - -	Red ferruginous earths, as Bole Armenian and Venetian Red	
Tea - -	Exhausted Tea Leaves; Leaves, other than those of Tea, British and Foreign, as, amongst the former, those of Sycamore, Horse-chesnut, and Plum; Lie Tea, Paddy Husk, Sand, Starch	Plumbago or Black Lead, Gum, Indigo, Prussian Blue, Turmeric, Chinese Yellow, China Clay, Soapstone or French Chalk	Sulphate of Iron, Catechu, Gum, Le Veno Beno, Chinese Botanical Powder
Tobacco - -	Water, Sugar, Treacle, and Salts	Sugar, Treacle - - -	Oil
Vinegar - -	Water - - - - - -	Burnt Sugar - - -	Sulphuric Acid

A. H. Hassall, *Food and its Adulterations*, London, 1855, p. vi.

10d Sanitary Conditions of the Poor

SUPPLEMENT, NO. 3. Account of a Personal Inspection of Bethnal Green and Whitechapel in May 1838, with a Supplement

Supplement, No. 3. Inspection of Bethnal Green and Whitechapel

On commencing an inspection of Bethnal Green, with a view to ascertain the causes of the great and constant prevalence in this district of fever and other diseases produced by malaria, I called on the medical officers of the Union (Mr. Taylor and Mr. Goodwin), who have the care of the out-door poor; I requested that they would accompany me over the district, and point out to me, first, those places in which fever is constantly prevalent, and secondly, those in which it is either altogether absent, or in which, when an individual case breaks out, it seldom or never spreads. I traversed a circle of from six to seven miles in extent. I wrote the account of the places I am about to notice on the spot. I entered many of the houses and examined their condition, as to cleanliness, ventilation, &c., as well as the state of the people themselves, who were at the time labouring under fever.

The district of Bethnal Green contains upwards of 70,000 inhabitants; in the greater part of it the streets are not close, nor are the houses crowded. On the contrary, large open spaces of ground intervene between them; but in one part the population is as densely crowded as in the closest and most thickly peopled parts of the city.

I notice the places about to be described in the order in which I visited them, and I give the account of the prevalence or absence of fever in the several localities as I received it from the lips of the medical officers who have the care of the poor. I record only the prevalence of fever, because the extent of the prevalence of this disease was readily ascertained, and because the extent of its prevalence might be taken as a sure indication of the presence and intensity of malaria.

Pitt-street.—A very narrow street; the houses have no sunk area; the ground floor is extremely damp; there is no drainage; in this street fever is often prevalent, and extends from house to house; it lately broke out in No. 21; extended to No. 20; spread to the houses on the opposite side, and proved very fatal.

Punderson's Gardens.—A long narrow street; the houses have no sunk area, and the ground floor is extremely damp. Along the centre of the street is an open, sunk gutter, in which filth of every kind is allowed to accumulate and putrefy. A mud bank on each side commonly keeps the contents of this gutter in their situation; but sometimes, and especially in wet weather, the gutter overflows; its contents are poured into the neighbouring houses, and the street is rendered nearly impassable. The privies are close upon the footpath of the street, being separated from it only by a parting of wood. The street is wholly without drainage of any kind. Fever constantly breaks out in it, and extends from house to house; it has lately been very prevalent here, and we have lately had several fatal cases from it in the London Fever Hospital.

Campden Gardens consist of a number of small ground-floor houses; each house contains only two rooms; that which is the largest is about seven feet by nine, and the smallest barely large enough to admit a small bed; the height is about seven feet. In winter the houses are exceedingly damp; the windows are very small; there is no drainage of any kind; it is close upon a marshy district, immediately to be described. In this street fever is very apt to break out, and to become more than commonly fatal. Often all the members of a family are attacked by it, and die one after another. During the present season, several entire families have been swept off by it; yet the houses stand in an open space,

Supplement, No. 3. Inspection of Bethnal Green and White-chapel

and have little gardens attached to them, so that if they were properly built, and well drained, they might be perfectly healthy.

Lamb's Fields.—An open area, of about 700 feet in length, and 300 feet in breadth; of this space about 300 feet are constantly covered by stagnant water, winter and summer. In the part thus submerged there is always a quantity of putrefying animal and vegetable matter, the odour of which at the present moment is most offensive. An open filthy ditch encircles this place, which at the western extremity is from 8 to 10 feet wide. Into this part of the ditch the privies of all the houses of a street called North-street open; these privies are completely uncovered, and the soil from them is allowed to accumulate in the open ditch. Nothing can be conceived more disgusting than the appearance of this ditch for an extent of from 300 to 400 feet, and the odour of the effluvia from it is at this moment most offensive.

Lamb's Fields is the fruitful source of fever to the houses which immediately surround it, and to the small streets which branch off from it. Particular houses were pointed out to me from which entire families have been swept away, and from several of the streets fever is never absent.

In several houses in *Collingwood-street* fever of the most severe and fatal character has been raging for several months. Part of the street called Duke-street is often completely under water. The street consists of about 40 houses. In 12 of them all the members of the families residing in them have been attacked with fever, one after another, and many have died.

Hare-street Fields.—An open space, close to the former, containing about 300 square feet, a large portion of which in rainy weather is completely inundated. It is surrounded on all sides but one with small houses, and several streets branch off from it. In all the houses forming the square, and in the neighbouring streets fever is constantly breaking out, and the character of the fever in this neighbourhood has lately been very malignant.

Mape's-street.—Running along the front of Mape's-street, and the back of Southampton-buildings, is a large open sewer, one branch of which also passes for a considerable extent along the backs of the houses in Teal-street. The privies of the houses, placed close to the street, pour their contents into this open sewer. Part of Mape's-street consists of houses of a good description, with gardens neatly cultivated; but all of them terminate at the margin of this open and filthy sewer.

Alfred and *Beckwith-rows* consist of a number of buildings, each of which is divided into two houses, one back, and the other front; each house is divided into two tenements, and each tenement is occupied by a different family. These habitations are surrounded by a broad open drain, in a filthy condition. Heaps of filth are accumulated in the spaces meant for gardens in front of the houses. The houses have common privies open, and in the most offensive condition. I entered several of the tenements. In one of them, on the ground floor, I found six persons occupying a very small room, two in bed, ill with fever. In the room above this were two more persons in one bed, ill with fever. In this same room a woman was carrying on the process of silk-winding. The window of the room is small, capable, if wide open, of ventilating the room but very imperfectly; yet this window is not only kept permanently closed, but is carefully and firmly pasted all round, so that not the slightest breath of air can enter. On remonstrating against this constant and total exclusion of the air, I was told by the woman at work that they are obliged to stop up the window, to prevent the drying of the silk, which is always weighed out to them when they receive it, and they are expected to return the same weight.

George Gardens.—A range of ground-floor houses, with neat gardens; the

Supplement, No. 3. Inspection of Bethnal Green and White-chapel

elevation is considerable; the soil dry, and the place unusually clean. Most of these houses are always tolerably healthy, but in a few of them fever is continually breaking out; these latter are smaller than the others; the ceilings are remarkably low, and the privies are close upon the windows of the dwelling-rooms.

North-street.—On the eastern side of this street, near Collingwood-street; the situation is close; the houses are in a most miserable condition, surrounded by vast collections of filth. Most of the houses are occupied by pig-dealers, and the filth produced by the pigs is seldom or never cleared away. The stench here is dreadful.

Manchester-place consists of a row of houses one story high. At the distance of five feet from the front of the houses is a dead wall, which rises as high as the middle of the first floor window, so that neither air nor light can get to the sitting or bed-rooms. Fever is frequently prevalent here.

Derbyshire-street.—Small ground-floor houses, each divided into two tenements containing two rooms, very similar to those in Campden Gardens. An open gutter runs close to the houses. Fever has been very prevalent in all these houses.

John's-passage and *Granby-row.*—Exactly similar to the above. In the middle of the street is a large gutter always full of filth, the overflowing of which sometimes renders the place impassable, and the noxious matter is with difficulty kept from flowing into the houses. The street, for the extent of 300 or 400 yards, consists of hovels of the last degree of wretchedness.

St. John-street.—A close and densely-populated street, in which malignant fever has been remarkably prevalent, and has stalked from house to house. In one room which I examined, eight feet by ten, and nine feet high, six people live by day, and sleep at night; the closeness and smell almost intolerable. Adjoining this house is a cow-yard in a filthy state.

In *Fleet-street* and the streets and alleys passing off from it the population is dense, the houses are high, and different families inhabit each flat; the ventilation is imperfect, and the houses dirty. Fever has lately raged here, and been very destructive.

Shackwell-street.—A close, narrow, densely-populated street, where fever has been extremely prevalent. I went over one of the houses, in every room of which fever has lately existed. The ground floor consists of one room; behind this there is a room perfectly dark, about four feet wide and ten feet long, at one extremity of which is the privy. There is no outlet of any kind. Parallel with the privy, in the adjoining room, is the bed. The room above is occupied by four persons, three of whom have been attacked with fever. In this room there is only one bed, and that is on the floor.

I asked a young man, 18 years of age, whom I found in the room, how they managed at night; he told me that some slept in the bed, and the others upon the chairs. Five persons live in the room above.

Turk-street.—Some time ago, at No. 29, in one room, at one time, all the inmates, consisting of six persons, the father, the mother, and four children, were all ill of fever together.

Mount-court.—A small court, in which fever has prevailed to a dreadful degree. The houses consist of two rooms: one a small dark room on the ground floor, without any outlet of any kind; and the other a still smaller bed-room above. At No. 2 seven persons have been attacked in succession; in No. 3, six; in the adjoining house three; and in the next, one.

Collingwood-court, Mount-street.—In this court, out of seven houses, fever has prevailed in five; in No. 6, six people have been attacked.

Supplement, No. 3. Inspection of Bethnal Green and Whitechapel

On the north side of Bethnal Green Road the houses are of a better description, and the district is better drained. In this part of the district there is always much less fever.

Hackney Road Division.—Gibralter Walk, consists chiefly of houses of one story; along the centre of the street is an open gutter in which all sort of animal and vegetable matters accumulate and putrefy, and the odour of which is most offensive. In the upper part of this street, especially, fever has been extremely prevalent.

Virginia-row.—In the centre of this street there is a similar gutter, into which potatoe parings, the refuse of animal and vegetable matters of all kinds, the dirty water from the washing of clothes and of the houses are all poured, and there they stagnate and putrefy. In a direct line from Virginia-row to Shoreditch, a mile in extent, all the lanes, courts, and alleys in the neighbourhood pour their contents into the centre of the main street, where they stagnate and putrefy.

Read-street and Vincent-street.—The condition of these streets is exactly similar to that of the preceding. Families live in the cellars and kitchens of the undrained houses, dark, and extremely damp. In some or other of these houses, fever is always prevalent. 'My assistance here,' said the medical officer who was attending me, 'is always required: I am never without cases of fever here.'

Even in Church-street, Bethnal Green, the main thoroughfare, there is no drain, the water runs off as it can; and now and then the parish authorities send round a mud cart, to gather up what becomes so thick as to block up the way.

At the lower end of Orange-street the mass of putrefying matter on each side flows into a stagnant ditch in Wellington-row. In the houses adjacent to this ditch fever is often extremely prevalent.

At the end of Wellington-row, and at right angles with it, a ditch from eight to ten feet broad, extends nearly to the Hackney Road. In the greater part of its course gardens neatly cultivated extend from adjacent houses to its edge. The stench arising from this ditch at this moment is intolerable. The poor people inhabiting the neighbouring houses, while cultivating their little gardens with so much care as a recreation, and in the hope of promoting their health, little think that at every moment they are inhaling a deadly poison.

From a common sewer, which comes from Hackney, and which is covered until it reaches Goldsmith's-place, in the Hackney Road, commences a broad uncovered ditch, which empties itself, at about the distance of a quarter of a mile, into a pond called Wellington Pond. This is a piece of stagnant water, which covers at least three acres of ground. It is black and filthy, more especially that part of it into which the above-mentioned ditch directly opens. This pond is surrounded by houses, and the inhabitants complain bitterly of the stench which often arises from it. Several of the residents, of whom I inquired respecting its condition in different states of the weather, assured me, that in the summer especially, when they are suffering greatly from the heat, they dare not open the windows of their houses, on account of the intolerable stench from this pond.

Appendix to the Fourth Report of the Commissioners under the Poor Law Amendment Act, *B.P.P.*, 1838 (28).

10e The Cholera Report

LONDON:

SATURDAY, OCTOBER 22, 1853

THE splendid Report on Cholera drawn up by Mr. FARR, which exhibits the most valuable analysis of the facts relating to this dire disease that we possess, opens with the following passage:—

'If a foreign army had landed on the coast of England, seized all the sea-ports, sent detachments over the surrounding districts, ravaged the population through summer, after harvest destroyed more than a thousand lives a day for several days in succession, and, in the year it held possession of the country, slain fifty-three thousand two hundred and ninety-three men, women, and children, the task of registering the dead would be inexpressibly painful; and the pain is not greatly diminished by the circumstance that in the calamity to be described the minister of destruction was a Pestilence that spread over the face of the island, and found in so many cities quick poisonous matters ready at hand to destroy the inhabitants.'

Well, indeed, might the Registrar mourn over his task! But what should we say of the astounding folly, of the blank stupidity of the Government and of the nation, if they were tamely and in dumb terror to suffer that foreign army to keep its hold on our sea-ports, and to extend its ravages through the land, unopposed and unmolested? Such a total prostration of energy and spirit as this implies, Englishmen are not likely to witness. But may not the most torpid, the most bigoted, the most ignorant perceive in the other part of the picture, in the invasion by Pestilence, a foe as implacable and as deadly, and experience a reproach as foul and as disgraceful?

In the case of the foreign army landing on our coasts—nay, even in the event of a suspicion of such an outrage, every patriotic resentment, every native energy would be roused; no appliances would be spared; no sums of money would be grudged to destroy the enemy, or to drive him from our shores. And yet could the foreign army inflict more misery upon individuals, or cause greater national calamities than the Pestilence? Ought the invasion of the one to be regarded as a ground of deeper humiliation than that of the other? Is there not an equal dereliction of duty, an equal moral turpitude in passively submitting to the devastation of the Pestilence, when it is in our power to combat it successfully? Does any one still doubt that the cholera may be fought and conquered? If there be such an one, let him turn to the records of its march; let him observe the localities where it takes up its quarters; let him study the conditions of its existence.

What is to be done to arrest its deadly course, to expel the cholera from our land, to deprive it of so much as a footing on our soil, and if we cannot extirpate utterly this scourge of humanity, at any rate to confine it within its indigenous limits, the festering mud of the Ganges? It will be our present object to point out the course which in our opinion an enlightened government is imperatively called upon to pursue.

The Registrar-General's report, the reports of the Board of Health, the common experience of the medical profession and the public, plainly establish certain important propositions. The cholera first seizes on the sea-ports in populous places built on alluvial soil, or the delta of rivers; it is mainly destructive in low-lying situations, and loses much of its virulence as higher locality is gained; it runs riot and revels where the drainage is bad, the water supply deficient and of bad quality, the air contaminated and befouled by the exhalations of a crowded population and of unwholesome manufactures; it seizes with unrelenting avidity upon the feeble in body, the depressed in mind, the half-starved, the half-naked, and the badly-housed. The cholera has its favourite habitats and its chosen victims. These truths are forcibly exemplified by the fact, that the epidemic of 1848–9 followed the same track as it did in 1831–2. It is true that it was much more fatal in some places and less fatal in others at the two periods.

So true are the facts we have mentioned, that Mr. FARR has pointed out on the map of England the head-quarters of the pestilence, which he has graphically named the Cholera-fields. There is the Bristol Cholera-field, extending along the Severn upwards from its mouth; the Liverpool Cholera-field, embracing the surrounding great manufacturing populations; the Hull Cholera-field, stretching over the basin formed by the Ouse, the Trent, and the Humber; there is the Tynemouth Cholera-field, whose centre is Newcastle, and where the pestilence has lately raged with a virulence and destructiveness far greater than in 1832; and, lastly, omitting many of lesser note, there is the great London Cholera-field, extending along the polluted and deadly banks of the Thames, and which every indication points to as the next scene of contest with the enemy.

But when we have learned all this,—when we have proved it to demonstration, what is the real extent of our acquisition,—what its available application? Is there anything new or specially useful in the discovery that bad air, bad water, confined habitations, or alluvial soils are unhealthy? Do not these and cognate conditions favour typhus, scarlatina, measles, puerperal fever, and every form of zymotic disease? Could we not, *à priori*, have pronounced that they would be equally favourable to cholera?

What is the inevitable conclusion? Irresistibly this—that our inquiry must go deeper; we must extend our observations beyond the mere *accidental circumstances* which lend virulence to the pest, and which accompany its career, but do not constitute it or create it. Mr. FARR has recorded and analyzed a mass of facts of inestimable value indeed, but which are far from exhausting the subject. When he has detailed the ages at which the victims perished, and the leading circumstances amid which they fell, he has recorded nothing but *results*. When Mr. FARR and the Board of Health, and others, have shown the conditions under which cholera gains strength and deals destruction, they have simply shown that cholera is subject to similar laws as are typhus and other forms of epidemic disease; they have succeeded in illustrating some of the laws of its development and progress. But we contend that they have done little more: they have left the cardinal point of the inquiry untouched; they reveal nothing of the mysterious essence of the pestilence. The question, What is cholera? is left unsolved. Concerning this, the fundamental point, all is darkness and confusion, vague theory, and a vain speculation. Is it a fungus, an insect, a miasm, an electrical disturbance, a deficiency of ozone, a morbid offscouring from the intestinal canal? We know nothing; we are at sea, in a whirlpool of conjecture. The consequence of this ignorance is, that in spite of what we know—in spite of empirical warnings to improve our sanitary condition, there is little reasonable hope in anything that will or can be done that the cholera will not again be as terrible and as fatal as ever it has been before. True, much may be usefully done in pruning down the rank luxuriance of abuse—in diminishing the abominations of filth and slothfulness. But diminish all this as much as possible, how can we hope to destroy the fundamental peculiarities upon which the insalubrity of alluvial soils depends? Can we subvert the geological foundations of the globe? We may render the favourite habitats less conducive to the development of cholera, and this is much; but it is impossible to anticipate such complete success as to render them altogether unfit for its sojourn. Our greatest success is but one of degree; it cannot be absolute. Yet why should we despair of crushing the enemy in embryo? What forbids us from making the attempt to track the full-grown monster back to its origin? We say again that this is the only rational course to explore.

Every analogy leads to the conclusion that the essential cause of cholera consists in a morbid poison, which under certain congenial conditions becomes developed into activity, and ferments in the blood. Science has detected the poison of small-pox; it has also discovered the antidote. Should we despair of similar success in the case of cholera? We believe not. Well-directed systematic investigations, on a scale commensurate with the importance and the difficulty of the undertaking would assuredly lead to the detection of that pestilential germ, that morbific something, which has hitherto eluded imperfect research, and which has consequently been able to develope itself to the destruction of mankind.

Shall we point out the insufficiency, the hopelessness—a hopelessness which weighs upon every one—of our present means of studying this subject? The most efficient public machinery we possess is undoubtedly the General Register Office; but, as we have seen, that office can but chronicle results: it cannot investigate causation. The Epidemiological Society contains much of the requisite science and good-will, but the means at its disposal are totally inadequate. The Board of Health has claims of greater pretension; and some of its proceedings have not been deficient in energy or ability. But this Board, whether deservedly or not we cannot now stop to inquire, has lost the confidence of Parliament, of the country, and of the profession. It also has overlooked the main point of the inquiry. That Board is absolutely self-disqualified for the purpose; and the fatal reason is this:—*It is already wedded to a theory*. It scoffs at the notion of infection. The question is prejudged by its members. Everything they witness is made to tend to support their dogma. They must miss the solution. The hidden truth can only be discovered by the humble and the unbiassed. Special machinery of vast capabilities is required. We unhesitatingly and broadly declare,—and confident are we that our proposal will receive the unanimous sanction of the members of the medical profession in this country,—that it is the imperative duty of the responsible advisers of the Crown to appoint forthwith a

GOVERNMENT CHOLERA COMMISSION

A reliance upon private resources, or the efforts of voluntary societies, at a moment like this, with the pestilence hourly counting its victims, is a dalliance with death, and indicates a want of foresight, an absence of judgment, and an indifference to the sufferings of the least fortunate of our fellow-creatures, really disgraceful to the character of the nation. Of necessity such a Commission as we have named would be specially charged with the task of determining not only the laws of cholera, but its nature and its essence. When this has been done, as by patience, by skill, by adequate appliances, it assuredly may, then the resources of modern science will not be slow in discovering not a mere palliative remedy, such as are our sanitary nostrums, but a real prophylactic, as certainly, as absolutely protective as is vaccination against the small-pox.

Is not such an object worthy of the most energetic efforts? Shall we stand by in passive despair, and neglect this great and obvious duty, out of doubt of success? Shall we, out of ignorant parsimony, haggle about the cost of the inquiry? Shall we be guilty of the folly and the iniquity of yielding to a contemptible spirit of penny-wisdom, and show ourselves profligate of human life? What would be the cost? We cannot say: but this is certain, that at *any cost* the thing must be done. Commissions, far from economical, and of very doubtful utility are common enough. Commissions are daily appointed to find out what every one knows. Did it need, for example, a commission to discover that bribery and corruption had long flourished as moral plagues in a score or two of parliamentary boroughs? And shall we hesitate, on account of the expense, to institute a commission properly organized and invested with sufficient funds and power to scrutinize successfully the nature of cholera, and to devise the means of destroying, in the germ, that raging pestilence which threatens to become endemic in these islands. It has been said, with truth, that one of the main causes of the British supremacy in commerce is the greater salubrity of this country. Shall we, by the most culpable and idiotic parsimony and supineness, foster the growth of this horrible plague amongst us, and think to save our pockets at the risk of encountering a fatal blow to our national prosperity?

It may be said that the working of such a Commission as we demand would cost a large sum of money,—that the expenditure would amount to £50,000 or £100,000; but when the objects contemplated are considered—namely, a saving of the lives of hundreds of thousands of human beings, if the expense were to amount to ONE MILLION of pounds sterling, the objection to such expenditure could only be made by an idiot, a lunatic, or a monster in human form. The inexpressible importance of this subject will induce us next week to enter into some details respecting the appointment of the Commission we suggest.

'Cholera Report', *Lancet*, 1853, **2**, 393.

10f The Main Drainage of London

By courtesy of the Institute of Civil Engineers

SECTION OF OVERFLOW CHAMBER AT JUNCTION OF NORTHERN HIGH & MIDDLE LEVEL SEWERS.

By courtesy of the Institute of Civil Engineers

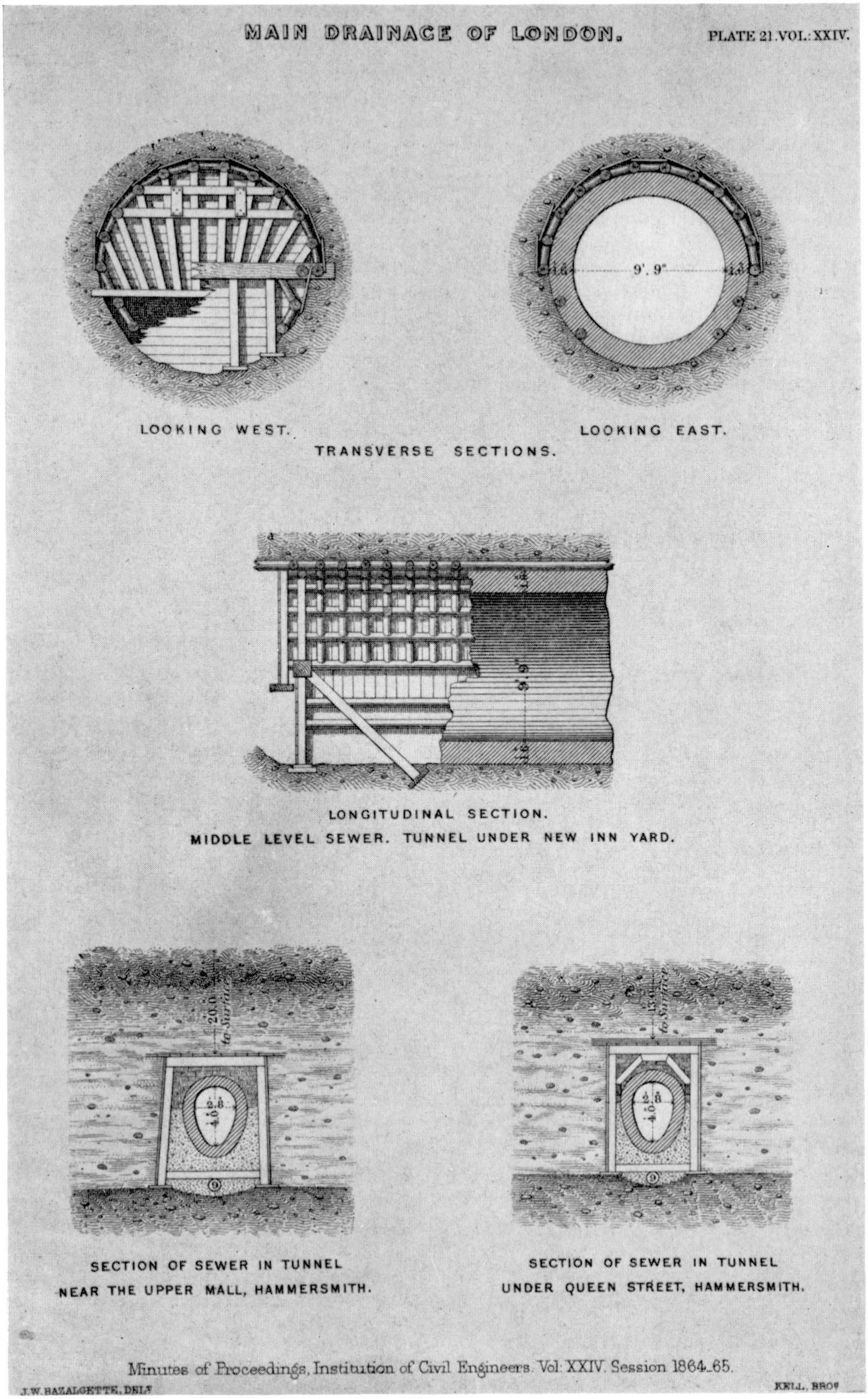

By courtesy of the Institute of Civil Engineers

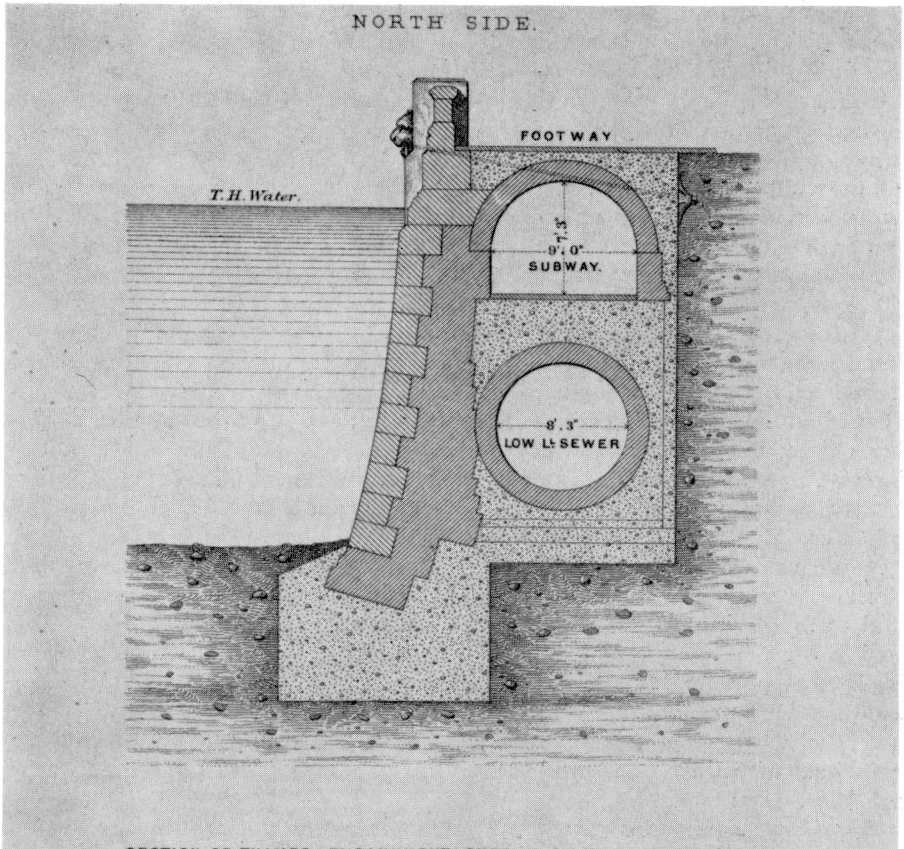

SECTION OF THAMES EMBANKMENT, SUBWAY & LOW LEVEL SEWER.

J. W. BAZALGETTE, 'On the Main Drainage of London, and the Interception of the Sewage from the River Thames', *Proceedings of the Institute of Civil Engineers*, 1864–5, **24**, plates 16, 17, 21.

10g Eighth Report of the Medical Officer of the Privy Council

MEDICAL OFFICER'S REPORT
Scientific researches

. . . 4. It remains to be mentioned under the present head that, in aid of the more immediately practical objects of the department, my Lords, early in 1865, ordered certain *scientific researches* to be begun. Inquiries already made and reported on had set before the Legislature and the public much information, collected according to the current lights of medical science, with regard to the distribution and causes of all our most destructive diseases. The relations of typhoid fever and other diarrhœal infections to the presence and decomposition of excremental impurity in air and water,—the relation of the most fatal afflictions of artisan life to mere want of common ventilation, or of dust-withdrawing ventilation, in places of industry,—the relation, similarly, of miners' lung-diseases to the insufficient ventilation of mines,—the relation of our highest infantine

MEDICAL OFFICER'S REPORT

death-rates to that neglect of infants, too often an almost murderous neglect, which follows the extensive industrial employment of women, whether in factories or in gang-agriculture,—the relations of unconquered smallpox to ill-performed vaccination, and of this to demonstrable faults of local arrangement, —the relations of contagion to the spread of disease in different circumstances of life, and of overcrowding to the power of contagion,—these very important ætiological relations had been studied, in their broadest bearings, in a large number and variety of local inspections throughout England, as well as in elaborate mortuary statistics; and the results of the study had been laid before Parliament, with other sanitary information, in the successive annual reports of the department. My Lords had now to determine what was to be done in the way of further like proceedings. It was submitted to them that by renewed inquiries, of exactly the same kind as heretofore, they might extend, but perhaps not importantly strengthen for practical purposes, the evidence already elicited on the great subjects which had been in hand; and that inquiries, tending chiefly to produce repetitions of evidence, would be but of secondary interest. Also it was submitted that, in the establishment of new principles for the prevention of disease, great steps of progress were now not likely to be made otherwise than with improved methods of ætiological observation;—that scientific researches must first have created a far more intimate knowledge than is yet current as to the nature of the morbid processes which are to be prevented, and as to the physical and chemical conditions of their development;—that such researches, on the scale and system which were to be desired, were not likely to be undertaken here by private investigators;—that, if their Lordships' inquiries were to continue to give fruit of national interest and importance, collateral work, aiming at such knowledge as I have just spoken of, must be combined with their Lordships' former line of inquiry, and serve hereafter in some sort as its pioneer.*

My Lords having instructed me to act in accordance with these views, investigations of the kind in question became in 1865 part of the work of the department. Hitherto they have had relation only to the chemical processes of disease, and here, under their Lordships' authority, I have had the advantage of Dr. Thudichum's assistance. At present I have only to name these investigations as of the category of work in progress. On future occasions I shall hope to particularise the proceedings, so far as their very technical subject-matter will permit, and of course from time to time to give their practical results. It may be, however, that these will not so much admit of popular appreciation as they will conduce to urgent professional requirements. The work is not of the kind which can promise either immediate returns, or such returns as are most popularly striking. But that medicine in its most practical aspects, whether preventive or curative, needs constant aid and guidance from purely scientific researches, is a fact on which it would be superfluous to insist. Of the convictions which in this respect are universal among enlightened persons, perhaps no stronger illustration could be cited than the urgency, almost the clamour, with which such researches were recently called for in relation to the steppe-murrain

* In relation to one very important disease, opinions like the above had, ten years previously, been pressed upon the attention of Government by the Medical Council, which the President of the then General Board of Health had called together to advise him on the then prevailing epidemic of Asiatic Cholera. See Report of Medical Council, p. 8, and Report of the Council's Committee for Scientific Inquiries, pp. 65–6, among papers of General Board of Health laid before Parliament with reference to the Cholera-Epidemic of 1853–4. But cholera is only one of innumerable illustrations which might be adduced of our present practical stand-still for want of deeper scientific insight.

now in England, and the scale on which the Royal Cattle-Plague Commission ordered them. Also in another point of view I may refer to those researches for an illustration. Undertaken as they were in the medical interests of horned cattle, they yet elucidated some processes of disease in a way which is of interest to human pathology, and, through it, to the preventive medicine of mankind.

B.P.P., 1866 (33).

10h Ninth Report of the Medical Officer of the Privy Council

MEDICAL OFFICER'S REPORT
Scientific investigations of cholera

. . . One sort of proceeding which my Lords took in relation to cholera during the year remains yet to be mentioned. I had submitted to their Lordships that many pathological and medical questions, as to the nature and habits of the disease, and the principles on which it ought to be treated, were still matters of scientific uncertainty; and that, in my opinion, it was very greatly to be desired, in the interests of mankind, that, during the presence of the disease among us, methodical attempts should be made, by the researches and observations of skilled persons, to narrow, if possible, the limits of those large uncertainties. My Lords were pleased to accept this view, and afterwards to sanction a scheme, which at their desire I had submitted to them, for the organization of some such work under their auspices.

The proposed branches of inquiry were as follows:—1) examination of the degrees of success attained by different methods of *treatment* of cholera, especially as practised in the London hospitals; 2) study of the successive *chemical changes* undergone by the body in cholera, and of the relations subsisting between those changes and the symptoms which the patient presents during life; 3) similar study, chiefly microscopical, of the successive *anatomical changes* of the affected body; 4) verification of alleged experimental proofs of the *communicability of cholera*; 5) collection of facts, in the line of study opened by Professor v. Pettenkofer, of Munich, as to the non-coincidence of local epidemics of cholera with such conditions of the local ground-water as are indicated by a full state of surface-wells. For the first of these objects, four physicians of leading metropolitan hospitals—Dr. Wilks of Guy's, Dr. Martin of St. Bartholomew's, Dr. Hughlings Jackson of the London, and Dr. Bristowe of St. Thomas's, consented to act together as a committee of investigation for the Privy Council; and it was intended that, if the epidemic extended westward in the metropolis, other representative physicians should be added to their number. The second object was referred to Dr. Thudichum, as a development of the work on which he was already engaged for the department. The third, from unavoidable circumstances, remained in abeyance. The fourth was undertaken by Dr. Sanderson. And the fifth (likely in part to join on to meteorological studies) was put into the hands of Mr. Glaisher, of the Royal Observatory, Greenwich.

The results of these several studies are in part given in my appendix, but in other part are still incomplete. Happily for our endangered population, opportunities for scientific investigation of the disease neither became so general in London, nor continued for so long a time, as to enable any great programme to be fulfilled.

Towards the end of the year, Dr. Wilks, on the part of the Treatment Committee, informed me that the information before the committee was not enough

By courtesy of the British Museum

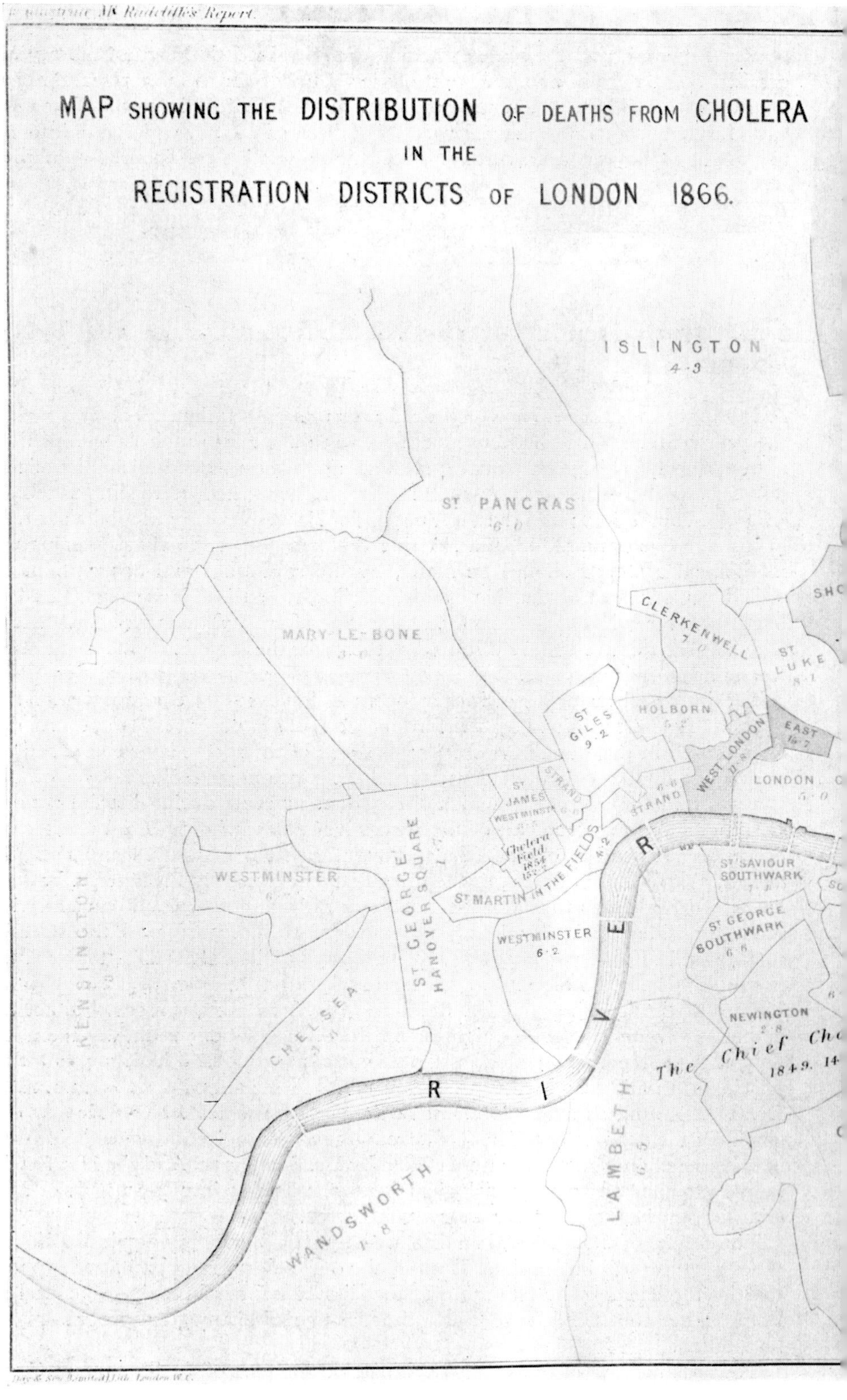

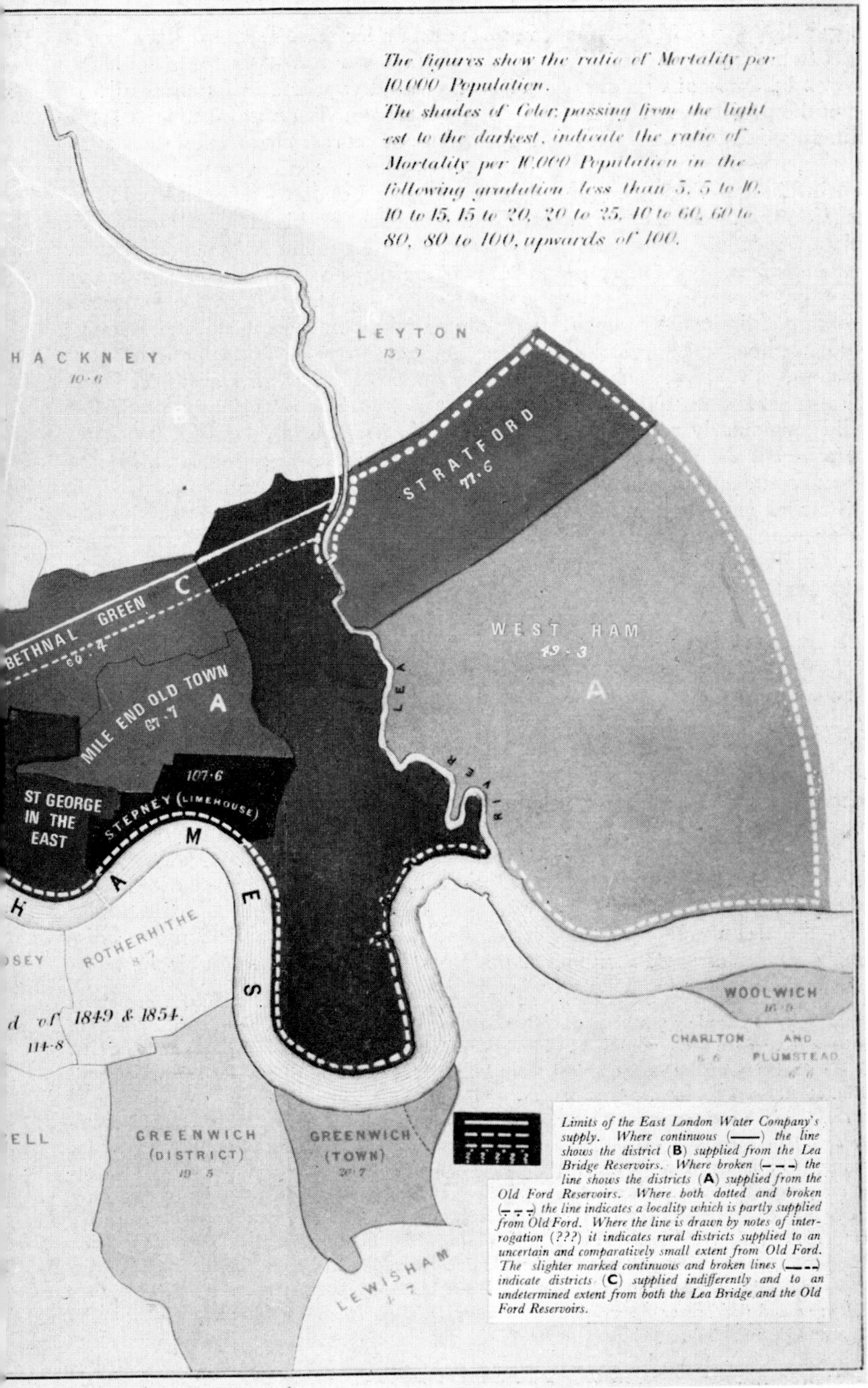
The figures show the ratio of Mortality per 10,000 Population.
The shades of Color passing from the lightest to the darkest, indicate the ratio of Mortality per 10,000 Population in the following gradation less than 5, 5 to 10, 10 to 15, 15 to 20, 20 to 25, 40 to 60, 60 to 80, 80 to 100, upwards of 100.
HACKNEY
10·6
LEYTON
STRATFORD
77·6
WEST HAM
49·3
A
BETHNAL GREEN
C
MILE END OLD TOWN
67·7
A
107·6
ST GEORGE IN THE EAST
STEPNEY (LIMEHOUSE)
THAMES
RIVER LEA
ROTHERHITHE
WOOLWICH
CHARLTON AND PLUMSTEAD
d of 1849 & 1854.
114·8
GREENWICH (DISTRICT)
19·5
GREENWICH (TOWN)
LEWISHAM
Limits of the East London Water Company's supply. Where continuous (——) the line shows the district (B) supplied from the Lea Bridge Reservoirs. Where broken (– – –) the line shows the districts (A) supplied from the Old Ford Reservoirs. Where both dotted and broken (- - -) the line indicates a locality which is partly supplied from Old Ford. Where the line is drawn by notes of interrogation (???) it indicates rural districts supplied to an uncertain and comparatively small extent from Old Ford. The slighter marked continuous and broken lines (— - -) indicate districts (C) supplied indifferently and to an undetermined extent from both the Lea Bridge and the Old Ford Reservoirs.

MEDICAL OFFICER'S REPORT

to form the basis of a satisfactory report; and under these circumstances I conveyed to him their Lordships' request that the committee—since the epidemic as yet might perhaps be only suspended—would continue in function till the end of the present year. At the same time, as it seemed important to have some testimony to the quality of this epidemic (so far as yet seen) in relation to medical practice, I had their Lordships' authority to request from Dr. Sutton, who had been in charge of Miss Sellon's temporary East London Cholera Hospital, a report on the medical experiences of that most serviceable charity; and I now subjoin (App. No. 8) the report which Dr. Sutton has in consequence written. In this context, too, I may permit myself the pleasure of referring to various interesting papers in relation to the East London epidemic which are contained as an appendix in the recent (third) volume of 'Clinical Lectures and Reports by the Medical and Surgical Staff of the London Hospital,' and which probably represent, as well as if they had been contributed from a wider area, the latest relations of cholera to medical treatment. Among them, in this point of view, I would particularly refer to the account which Mr. Little gives of a few cases where he tried with much thoroughness the practice of injecting, and, in case of need, re-injecting, saline fluid into the blood of persons in collapse.

The chemical investigation of cholera was pursued most laboriously by Dr. Thudichum at St. Thomas's Hospital; and I subjoin (App. No. 10) the report which he has made of the very interesting results hitherto obtained by him in that study.

That our last year's arrangements for investigation could not include the anatomy, and especially the microscopy, of the body in cholera, was matter for much regret;* the more so, as that line of study in Germany has led some of our fellow-labourers there to new and striking observations, which, under more favourable circumstances, if they had not been simultaneously made here, might at least have been verified and perhaps extended. For it is again alleged, and now as the discovery (or rather re-discovery†) of two independent German observers, Ph. Dr. Thomé of Cologne, and Professor Klob of Vienna, that the stomach and intestines of cholera patients contain within them microscopical fungic bodies, innumerable and vehemently multiplying, whereof swarms are shed, with prolific and infective power, in each characteristic evacuation of the sick.‡

The fourth branch of the inquiry was conducted by Dr. Sanderson to a satisfactory conclusion; and I subjoin (App. No. 9) his report on the subject.

* I am glad, however, to notice that some independent study of the affected mucous membrane was made by Dr. Beale, with his eminently skilful use of the microscope, and that the results of this work have since been published by him as contributions in the Medical Times and Gazette.

† See section 5 of Boehm's classical work 'die kranke Schleimhaut in der asiat. Cholera,' Berlin, 1838. See also, below, App. No. 11.

‡ Dr. Thomé's observations are contained in a paper, published (with date of 'end of November, 1866') in Virchow's Archiv, for last February. Dr. Klob's, dated 'December, 1866,' are published as a pamphlet with the title 'Pathologisch-anatomische Studien über das Wesen des Cholera-processes.' Professor Parkes gives me the interesting information that he recognizes in the descriptions and plates of these writers the 'corpuscles' or 'granules' which attracted much of his attention in 1849, and are often spoken of by him (with no suspicion of their being of vegetable nature) in his paper, then published in the London Journal of Medicine, 'on the intestinal discharges in Cholera;' also that he had noticed these bodies again in 1865 and 1866, when they vividly recalled the former observations to his mind; and that he noticed like bodies once in the sediment of a first cholera urine.

B.P.P., 1867 (37).

10i Lister on 'The Germ Theory'

*A Contribution to the Germ Theory of Putrefaction and Other Fermentative Changes, and to the Natural History of Torulae and Bacteria**

PART I

ALTHOUGH the subject of the following communication has of late years attracted a great deal of attention among the public generally, it may, nevertheless, be well for me to preface my statements by a few elementary remarks.

It is well known that organic substances, when left exposed under ordinary circumstances, undergo alterations in their qualities. For example, an infusion of malt experiences the alcoholic fermentation; a basin of paste prepared from wheaten flour becomes mouldy; or, again, a piece of meat putrefies when so treated. The microscope shows that each of these changes is attended by the development of minute organisms. In the fermenting sweet-wort the yeast which falls to the bottom of the containing vessel is found to consist of budding cells, constituting the yeast plant, *Torula Cerevisiae*, represented in Plate 22, Fig. 2.† In the mouldy paste the blue crust which is the most frequent appearance, owes its colour to the spores of a species of filamentous fungus, *Penicillium Glaucum*, the commonest of all moulds, of which Fig. 1 in Plate 22 represents a pencil of fructifying threads; and the putrid flesh will be probably found teeming with bodies which, in the most typical form, consist of two little rods, connected endways as by a joint, such as are seen at *a*, Fig. 3, Plate 22, characterized by astonishing powers of locomotion, and, from their rod-like form, termed *Bacteria*.

The Germ Theory supposes that the organisms are the causes of the changes; that the germs of these minute living things, diffusible in proportion to their minuteness, are omnipresent in the world around us, and are sure to gain access to any exposed organic substance; and, having thus reached it, develop if it prove a favourable *nidus*, and by their growth determine the chemical changes; and further, that these organisms, minute though they appear to us, form no exception to the general law of living beings, that they originate from similar beings by parentage.

Of those who oppose this theory, some attribute the changes to the oxygen of the air; others, while convinced of the insufficiency of the oxygen theory, hold the doctrine of so-called chemical ferments, and ascribe the alterations we are considering to organic principles destitute of vitality, the organisms being regarded as accidental accompaniments; while others, admitting perhaps the fermentative agency of the organisms, believe that they do not necessarily spring from parents like themselves, but may arise, *de novo*, from the inorganic world by spontaneous generation.

The philosophical investigations of Pasteur long since made me a convert to the Germ Theory, and it was on the basis of that theory that I founded the antiseptic treatment of wounds in surgery. The results of the treatment pursued

* This communication was originally made orally to the Royal Society on the 7th April, 1873. In preparing it for the press I have introduced various details which I was unable to enter upon at the time. I have also added facts ascertained at subsequent periods; but the dates of the observations being always mentioned, there will be no difficulty in distinguishing between those made before and after the delivery of the original address.

† In the present state of uncertainty regarding the true affinities of the yeast plant, it seems justifiable to retain for it the old name *Torula Cerevisiae*, a practice which has the advantage of enabling us to apply to similar budding cells the generic name Torula, and the adjective toruloid.

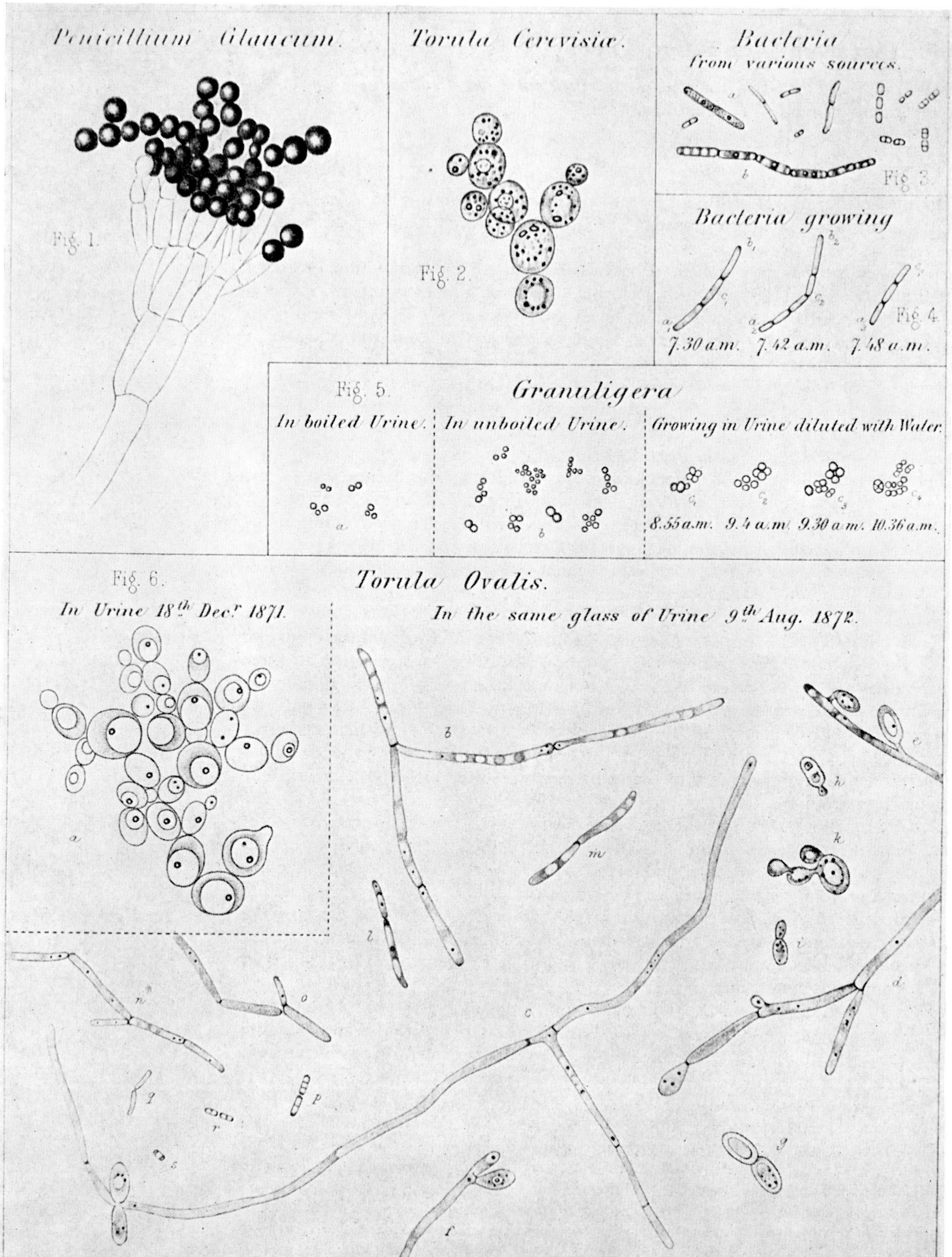
Penicillium Glaucum.
Fig. 1.
Torula Cerevisiæ.
Fig. 2.
Bacteria
from various sources.
Fig. 3.
Bacteria growing
Fig. 4.
7.30 a.m.
7.42 a.m.
7.48 a.m.
Fig. 5.
Granuligera
In boiled Urine.
In unboiled Urine.
Growing in Urine diluted with Water.
8.55 a.m.
9.4 a.m.
9.30 a.m.
10.36 a.m.
Fig. 6.
Torula Ovalis.
In Urine 18th Decr 1871.
In the same glass of Urine 9th Aug. 1872.
0 1 2 3 4 5
Scale in Ten-thousandths of an Inch.

constantly on this guiding principle have convinced me more and more of the truth of the theory upon which it was based; and if I were to put together the facts which I have had presented to me in surgical practice, proceeding on the antiseptic system, I should be able to present an array of evidence in favour of the Germ Theory as good and convincing as experiments performed in a laboratory.

J. LISTER, 'A Contribution to the Germ Theory of Putrefaction and Other Fermentative Changes, and to the Natural History of Torulæ and Bacteria', *Transactions of the Royal Society of Edinburgh*, 1876, **27**, 313.

PLATE 22

By courtesy of the Royal Society of Medicine

Fig. 1, a pencil of fructifying threads of the common blue mould *Penicillium Glaucum.*

Fig. 2, a group of cells of the yeast plant, *Torula Cerevisiae.*

Fig. 3, Bacteria from various sources. The pair below the letter *a* are examples of a granular appearance of the protoplasm, and the presence of a distinct nucleus in each segment. *b* a Leptothrix filament, some of the segments being nucleated.

Fig. 4 illustrates the ordinary mode of growth of Bacteria, viz. by increase of the segments lengthwise and transverse segmentation. When first sketched, at 7.30 a.m., the object consisted of three segments, a_1, c_1, b_1. During the few minutes that elapsed between the completion of this sketch, and that at 7.42 a.m., the uppermost segment b_1 is seen to have increased in length to the size shown at b_2, and the two lower ones are not only longer, but each presents a transverse line of segmentation, while the middle segment is bent at this new place of division, c_2. Three minutes later the three lowest of the five segments of which the object now consisted separated from the other two, and in the sketch taken at 7.48 they are shown thus detached, the lower two obviously increased in length. Two minutes later one of these three was found to have separated and moved off, and the remaining pair were observed to swim away as an ordinary double bacterium.

Fig. 5 represents a minute organism, consisting of granules grouped in a different manner from that which commonly prevails among Bacteria. The difference of arrangement is explained by difference in the mode of growth, as is illustrated by the sketches c_1, c_2, c_3, and c_4, which represent the same granules in process of fissiparous generation. It will be observed that the granules, instead of increasing like ordinary Bacteria in one direction only, swell up in all dimensions and afterwards undergo segmentation, either into pairs or into fours, . . .

Fig. 6 represents a form of Torula which resulted from the mingling of a drop of rain with fresh uncontaminated urine. Appearing in the first instance as an unmixed Torula (*a*), it changed in course of time to a delicate filamentous fungus (*b* and *c*) bearing buds, some of which were more or less toruloid in aspect and habit of growth, while others were morphologically identical with Bacteria, as described in detail in the text.

10j 'Antisepsis in Surgery'

On the Antiseptic Principle in the Practice of Surgery. By* Joseph Lister, *F.R.S., Professor of Surgery in the University of Glasgow*

In the course of an extended investigation into the nature of inflammation, and the healthy and morbid conditions of the blood in relation to it, I arrived several years ago at the conclusion that the essential cause of suppuration in wounds is decomposition, brought about by the influence of the atmosphere upon blood or serum retained within them, and, in the case of contused wounds, upon portions of tissue destroyed by the violence of the injury.

To prevent the occurrence of suppuration with all its attendant risks was an object manifestly desirable, but till lately apparently unattainable, since it seemed hopeless to attempt to exclude the oxygen, which was universally regarded as the agent by which putrefaction was effected. But when it had been shown by the researches of Pasteur that the septic property of the atmosphere depended not on the oxygen, or any gaseous constituent, but on minute organisms suspended in it, which owed their energy to their vitality, it occurred to me that decomposition in the injured part might be avoided without excluding the air, by applying as a dressing some material capable of destroying the life of the floating particles. Upon this principle I have based a practice of which I will now attempt to give a short account.

The material which I have employed is carbolic or phenic acid, a volatile organic compound, which appears to exercise a peculiarly destructive influence upon low forms of life, and hence is the most powerful antiseptic with which we are at present acquainted.

The first class of cases to which I applied it was that of compound fractures, in which the effects of decomposition in the injured part were especially striking and pernicious. The results have been such as to establish conclusively the great principle that *all the local inflammatory mischief and general febrile disturbance which follow severe injuries are due to the irritating and poisoning influence of decomposing blood or sloughs.* For these evils are entirely avoided by the antiseptic treatment, so that limbs which otherwise would be unhesitatingly condemned to amputation may be retained, with confidence of the best results.

In conducting the treatment, the first object must be the destruction of any septic germs which may have been introduced into the wound, either at the moment of the accident or during the time which has since elapsed. This is done by introducing the acid of full strength into all accessible recesses of the wound by means of a piece of rag held in dressing forceps and dipped in the liquid.† This I did not venture to do in the earlier cases; but experience has shown that the compound which carbolic acid forms with the blood, and also any portions of tissue killed by its caustic action, including even parts of the bone, are disposed of by absorption and organisation, provided they are afterwards kept from decomposing. We are thus enabled to employ the antiseptic treatment efficiently at a period after the occurrence of the injury at which it would otherwise probably fail. Thus I have now under my care, in the Glasgow Infirmary, a boy who was admitted with compound fracture of the leg as late as eight hours

* Read in the Surgical Section before the annual meeting of the British Medical Association in Dublin, on August 9th, 1867.

† The addition of a few drops of water to a considerable quantity of the crystallised acid, induces it to assume permanently the liquid form.

and a half after the accident, in whom, nevertheless, all local and constitutional disturbance was avoided by means of carbolic acid, and the bones were soundly united five weeks after his admission.

The next object to be kept in view is to guard effectually against the spreading of decomposition into the wound along the stream of blood and serum which oozes out during the first few days after the accident, when the acid originally applied has been washed out or dissipated by absorption and evaporation. This part of the treatment has been greatly improved during the last few weeks. The method which I have hitherto published (see the *Lancet* for March 16th, 23rd, 30th, and April 27th of the present year) consisted in the application of a piece of lint dipped in the acid, overlapping the sound skin to some extent and covered with a tin cap, which was daily raised in order to touch the surface of the lint with the antiseptic. This method certainly succeeded well with wounds of moderate size; and indeed I may say that in all the many cases of this kind which have been so treated by myself or my house-surgeons, not a single failure has occurred. When, however, the wound is very large, the flow of blood and serum is so profuse, especially during the first twenty-four hours, that the antiseptic application cannot prevent the spread of decomposition into the interior unless it overlaps the sound skin for a very considerable distance, and this was inadmissible by the method described above, on account of the extensive sloughing of the surface of the cutis which it would involve. This difficulty has, however, been overcome by employing a paste composed of common whiting (carbonate of lime), mixed with a solution of one part of carbolic acid in four parts of boiled linseed oil, so as to form a firm putty. This application contains the acid in too dilute a form to excoriate the skin, which it may be made to cover to any extent that may be thought desirable, while its substance serves as a reservoir of the antiseptic material. So long as any discharge continues, the paste should be changed daily, and, in order to prevent the chance of mischief occurring during the process, a piece of rag dipped in the solution of carbolic acid in oil is put on next the skin, and maintained there permanently, care being taken to avoid raising it along with the putty. This rag is always kept in an antiseptic condition from contact with the paste above it, and destroys any germs that may fall upon it during the short time that should alone be allowed to pass in the changing of the dressing. The putty should be in a layer about a quarter of an inch thick, and may be advantageously applied rolled out between two pieces of thin calico, which maintain it in the form of a continuous sheet, which may be wrapped in a moment round the whole circumference of a limb if this be thought desirable, while the putty is prevented by the calico from sticking to the rag which is next the skin. When all discharge has ceased, the use of the paste is discontinued, but the original rag is left adhering to the skin till healing by scabbing is supposed to be complete. I have at present in the hospital a man with severe compound fracture of both bones of the left leg, caused by direct violence, who, after the cessation of the sanious discharge under the use of the paste, without a drop of pus appearing, has been treated for the last two weeks exactly as if the fracture were a simple one. During this time the rag, adhering by means of a crust of inspissated blood collected beneath it, has continued perfectly dry, and it will be left untouched till the usual period for removing the splints in a simple fracture, when we may fairly expect to find a sound cicatrix beneath it. . . .

There is, however, one point more that I cannot but advert to, viz., the influence of this mode of treatment upon the general healthiness of an hospital. Previously to its introduction the two large wards in which most of my cases of accident and of operation are treated were among the unhealthiest in the whole surgical division of the Glasgow Royal Infirmary, in consequence apparently of those wards being unfavourably placed with reference to the supply of fresh air; and I have felt ashamed when recording the results of my practice, to have so often to allude to hospital gangrene or pyæmia. It was interesting, though melancholy, to observe that whenever all or nearly all the beds contained cases with open sores, these grievous complications were pretty sure to show themselves; so that I came to welcome simple fractures, though in themselves of little interest either for myself or the students, because their presence diminished the proportion of open sores among the patients. But since the antiseptic treatment has been brought into full operation, and wounds and abscesses no longer poison the atmosphere with putrid exhalations, my wards, though in other respects under precisely the same circumstances as before, have completely changed their character; so that during the last nine months not a single instance of pyæmia, hospital gangrene or erysipelas has occurred in them.

As there appears to be no doubt regarding the cause of this change, the importance of the fact can hardly be exaggerated.

J. Lister, 'On the Antiseptic Principle in the Practice of Surgery', *British Medical Journal*, 1867, **2**, 246.

10k 'Virulent Diseases'

*On Virulent Diseases, and especially on the Disease commonly called Chicken Cholera.** *By* M. Pasteur

Virulent diseases may be ranked among the greatest of the evils that afflict living beings. To prove this, we have merely to name measles, scarlatina, variola, syphilis, glanders, the carbuncular disease, yellow fever, typhus, and the cattle plague. This list is far from being complete; the pathology of the most important diseases may find a place here.

When the ideas of Liebig on the nature of ferments were in vogue, each virus was considered as a substance undergoing an internal change, which could be communicated to living organisms, turning the constituents of these into a virus of the same nature. Liebig was well aware that the first apparition of the ferments, their multiplication and their power of decomposition, present the greatest analogies with the phenomena of life, but, in the introduction to his 'Organic Chemistry,' he tells us that these analogies may be considered as deceitful illusions.

All the experiments which I have communicated to this Academy for the last twenty-three years have demonstrated, either directly or indirectly, the inaccuracy of the opinions of Liebig. A single method has guided me in the study of microscopic organisms. This method has been essentially the cultivation of these minute beings in a pure state; that is, by eliminating the heterogeneous substances, living or dead, which accompany them. By the use of this method, the most difficult questions are often solved in the easiest and most decisive manner. I will here recall one of the first applications which I made of this method (1857–1858).

* Translated from the *Comptes Rendus de l'Academie des Sciences*, of February 9, 1880, page 239, by P. Casamajor. From the *Journal of the American Chemical Society*, vol. ii.

Ferments, according to Liebig, are the nitrogenous substances of organisms, such as fibrine, albumen, casein, &c., in a state of decomposition, resulting from contact with air. There was no fermentation known in which these nitrogenous substances were not present and active. One character of fermentations, as well as of diseases, was that they were spontaneous in their origin and development. In order to show that the hypothesis of the learned German chemist was, to use his own words, 'but a deceitful illusion,' I made up artificial mixtures whose only constituents were as follows:—Water, the mineral constituents essential to life, fermentable substances, and the germs of the ferments which act on these substances. With these mixtures, fermentation took place with a regularity and a purity, if I may use the word, which are never found in the spontaneous fermentation of nature. As every albuminoid substance had been excluded from these mixtures, the ferment appeared as a living being, which borrowed from the fermentable substance all the carbon of its successive generations, and, from the mineral constituents, the nitrogen, phosphorus, potassium, magnesium—elements, the assimilation of which is an indispensable condition to the formation of all living beings, be they great or small.

After these experiments, not only was the theory of Liebig left without any foundation, but the phenomena of fermentation presented themselves as simple phenomena of nutrition, taking place in exceptional conditions, the most extraordinary of which is the possible absence of any contact with air.

Human, as well as veterinary, medicine made use of the light which shone from these new results. Many investigators made experiments to discover if every virus or contagion was not an animated being. Dr. Davaine, in 1863, endeavoured to show the functions of the *bacteridia* of carbuncular disease, which he had discovered in 1850. In 1868, Dr. Chauveau tried to show that virulence was due to the solid particles previously noticed in every virus. Dr. Klebs, in 1872, attributed traumatic virus to microscopic organisms. In 1872, Dr. Kock obtained, by artificial cultivation, the germs of *bacteridia*, which were similar in every respect to those which I had pointed out in *vibrios* (1865–70), and the causes of several other diseases were ascribed to microscopic organisms. To-day those who are most opposed to the theory of germs are wavering. Still the greatest obscurity prevails on the most important points.

In the great majority of virulent diseases, the virus has not as yet been isolated, and still less has it been shown, by artificial cultivation, that it is a living organism, and everything contributes to make us regard these 'unknown quantities' of pathology as mysterious morbific causes. The study of the diseases which they cause presents many extraordinary circumstances, among which the most remarkable is their non-recurrence. Human imagination can hardly venture to present a hypothetical explanation having any experimental foundation. Is it not still more surprising to find that vaccine, a virulent but mild disease, is a preventive, not only of vaccine itself, but of a more serious disease—the small pox? These facts were known from the remotest antiquity. Variolisation and vaccination have been practised in India from immemorial times, and when Jenner demonstrated the efficacy of vaccination, the common people of the locality in which he practised medicine knew that cow-pox was a preservative from variola.

Vaccination appears as an isolated fact, but the non-recurrence of virulent diseases appears to be general. The organism does not go twice through measles, scarlatina, typhus, the plague, variola, syphilis, &c.; at least it may be said that the immunity persists for a certain time.

Although in the presence of such mysteries, it behoves us to be humble, I dare to hope that the Academy will find that the facts which I am about to have the honour of presenting before it throw unexpected light on the problems raised by the study of virulent diseases.

There occurs sometimes in poultry yards a fearful disease, commonly called chicken cholera. The victim overtaken by it loses its strength, and stumbles about with drooping wings. Its feathers stand on end, and give it the appearance of a ball; it seems overcome by drowsiness; if we open its eyes it seems to awake from profound sleep, and soon its eyelids close again. Generally death comes after a dumb agony, without the victim even moving from the position it has occupied during the last stages of the disease. In rare cases, it beats its wings for a few seconds. The internal disorders are of the most serious nature. This disease is caused by a microscopic organism which, according to Tundel's Dictionary, was first suspected by M. Moritz, a veterinary surgeon in upper Alsatia; which was drawn more accurately, in 1878, by M. Peroncito, a veterinary surgeon in Turin; and which was found again, in 1879, by M. Toussaint, professor at the Veterinary School of Alfort, who demonstrated, by cultivation in neutralised urine, that this organism was the cause of the virulence in the blood.

In the study of microscopic parasites, the first, the most useful, condition to fulfil, is to obtain a liquid in which the infectious organism may be cultivated with ease, and without any admixture of other organisms of different species. Neutralised urine, which I have used with so much success to show that the product obtained by the cultivation of the *bacteridia* of Davaine, is identical with the virus of carbuncular disease (1877, Pasteur and Jaubert), does not fulfil the double end in view. But a liquid marvellously adapted to the life of the germ of chicken cholera, is a broth made from chicken's muscles, neutralised with potassa, and made sterile by a temperature superior to 100°C. (110° to 115°). The ease with which the microscopic organism multiplies in this liquid seems prodigious. In a few hours, the most limpid broth becomes turbid, and is filled with an infinite multitude of small articulations of extreme tenuity, slightly thinner in the middle, and which at first sight have the appearance of isolated dots. These small articulations have no motion of their own, and they certainly belong to a very different group from that of *vibrios*. I imagine that they will be classified some day with other forms of virus, now unknown, when we cultivate these, as I hope we are on the eve of doing.

The cultivation of this microscopic organism presents some very interesting peculiarities.

In my former researches, one of the liquids which I used with the greatest success, was a decoction of beer-yeast in water, after filtering it to obtain it perfectly limpid, and after rendering it sterile by a temperature superior to 100°C. The most various microscopic organisms thrive on the food presented by this liquid, particularly after being neutralised. For instance, the *bacteridia* of carbuncular disease multiplies surprisingly in a few hours. It is a strange thing that this liquid is entirely unsuited to the life of the organism of chicken cholera, which dies in it in less than forty-eight hours. Is not this entirely analogous to what happens when a microscopic organism is entirely innocuous towards an animal on which it has been inoculated? It remains inoffensive because it does not develop in the body of the animal, and it does not reach the organs essential to life.

The sterility of the decoction of yeast, with respect to the microscopic organism of chicken cholera, affords us an excellent criterion for the purity of the cultivation of this organism in chicken broth. If the cultivation be pure, upon transferring it to a decoction of yeast, no development takes place, and the yeast solution remains limpid. If, however, other organisms are present, they are developed, and the solution becomes turbid. I will, in the next place, call your attention to a still more extraordinary peculiarity of the cultivation of the germ of chicken cholera. The inoculation of this organism on guinea pigs is not so surely fatal as in the case of chickens. In guinea pigs, particularly in the older animals, the only thing that can be observed is a local lesion, at the point of inoculation, which ends in an abscess, of greater or lesser volume. This abscess opens spontaneously and heals, and meanwhile the guinea pig eats his food as usual, and seems to possess all the characteristics of health. These abscesses last sometimes for several weeks before discharging, being surrounded by a membrane full of creamy pus, in which the microscopic organism exists in infinite numbers, side by side with the globules of pus. It is the life of the inoculated organisms which causes the abscess, which is as a closed vessel, from which we may obtain the organism without endangering the life of the animal. The microscopic organism remains, mixed with pus, in a great state of purity, without losing its vitality. This may be proved by inoculating on chickens a small portion of the contents of the abscess. From the effect of these inoculations, the chickens very soon die, while the guinea pig, which has furnished the virus, is entirely cured after a short time. This is an instance of the localised evolution of a microscopic organism, which causes the formation of pus and of a closed abscess, without, at the same time, causing internal disturbances or the death of the animal on which it exists. It is, however, always able to cause the death of other species on which it may be inoculated, and even the death of the animal on which it exists in a closed abscess, if through some fortuitous circumstances, it should pass into the blood or into the viscera. Chickens and rabbits, living in company with guinea pigs, affected with abscesses of this kind, might, all at once, sicken and die, without any great change being observable in the health of the guinea pigs. This could easily happen if the abscesses of the guinea pigs discharged a small portion of their contents on the food of the chickens and rabbits.

An observer who witnessed these facts, and was ignorant of all the points, might well be astonished to see chickens and rabbits die in great numbers, without any apparent cause, and he would be apt to believe in some spontaneous disease. Certainly, he would not suppose that the guinea pigs were the cause of all the trouble, when he saw them all in good health, and particularly if he knew that the guinea pigs themselves often suffer from the same disease. Many of the mysteries in the history of contagions will some day be solved in easier ways than the one I have just mentioned. We may reject theories which are in contradiction with known facts, but we must not reject them solely because some of their applications elude our grasp. The combinations of nature are both simpler and more varied than those of human imagination.

I may easily convince you of the truth of these statements, if I add that, if a few drops from a cultivation of our microscopic organism, be placed on bread or meat given to chickens, they are sufficient to propagate the evil to their intestines, in which the little organism propagates with such remarkable rapidity, that the excrements of chickens so poisoned cause the death of those on whom they are inoculated. These facts enable us to understand the manner in which this fearful disease develops in poultry yards. Evidently, the excrements of the diseased chickens have most to do with the contagion. Nothing would be easier than to prevent the spread of the disease by isolating the chickens for only a few days; by washing the poultry yard with plenty of water, and particularly with water containing a little sulphuric acid, which kills the germ of this disease. The excrements should be carried off to a distance. After a few days, the chickens that are still alive could be brought together again with perfect safety, because this disease is so rapidly fatal that in a short time all the diseased animals would be dead.

If the cultivation of the infectious organism in chicken broth is repeated many times over, by passing from one cultivation to the next by sowing an infinitely small quantity, such as may be gathered on the point of a needle, the virulence of the germ is not weakened by the process. This is analogous to the ease with which it multiplies in the bodies of the *Gallinaceæ*. This virulence is so great, that the inoculation of a minute fraction of a drop will cause death in two or three days, and most generally in less than twenty-four hours.

Having established these preliminaries, I now come to the most important portion of this communication.

By operating certain changes in the process of cultivation, the virulence of the infectious germ may be much lessened. This is the vital part of the subject. I beg the Academy's permission to withhold a description of the processes by means of which I determine this diminution of virulence. My object is to insure independence in my studies.

The diminution of virulence is seen in cultivations by a slower development of the infectious organism, but, in reality, the two varieties of virus are identical. In the first or very infectious state, the inoculated germ may kill twenty times in twenty. In the milder state, it may twenty times in twenty give rise to illness, but not to death. These facts have an importance which is easily understood, as they allow us to form an opinion, in this particular disease, of the problem of its recidivation or non-recidivation. If we take forty chickens, and inoculate twenty of them with the very virulent virus, these twenty will die. If we inoculate the other twenty with attenuated virus, these will all be ill, but they will not die. We let the twenty chickens be entirely cured, and then if we inoculate them with the very infectious virus, they will not die. The conclusion from this is evident. The disease is its own preventive. It has the character of virulent disease, which do not recidivate.

Let us not be astonished at the singularity of this result. All things are not here as new as they appear at first. In one important particular, however, there is a novelty which will be pointed out. Before the time of Jenner, who himself practised this method, as I have already mentioned, there was a practice of inoculating variola to preserve from variola. In our day, sheep are inoculated with murrain to preserve them from murrain, and cattle are inoculated with peripneumonia to preserve from this fearful disease. Chicken cholera shows us an immunity of the same kind. It is an interesting fact, but it does not show any theoretical novelty. There is, however, an important novelty in the preceding observations, a novelty which gives food for reflection on the nature of virus. It consists in this, that we have here a disease whose virulent cause is a microscopic parasite, which may be cultivated outside of the animal economy. The virus of variola, the virus of vaccine, those of glanders, syphilis, the plague, &c., are unknown in their nature.

This new virus is a living organism, and the disease to which it gives rise has one thing in common with virulent diseases, properly so-called, a quality heretofore unknown in virulent diseases, caused by microscopic parasites: it is that it does not recidivate.

The existence of this disease is a connecting link between virulent diseases caused by a living virus, and other diseases, in the virus of which life has never been recognised.

I would not have it believed that the facts present the constancy and mathematical regularity which I have mentioned. To believe this would be to ignore the great variability in the constitution of animals, taken at haphazard from among domestic animals, and also the variability in the manifestations of life in general. The very virulent virus of chicken cholera does not always kill twenty times in twenty. Sometimes this virus only kills eighteen times in twenty, but generally twenty times in twenty. We may also remark that the virus, when reduced in virulence, does not save life twenty times in twenty. Sometimes this happens only eighteen times in twenty, and even sixteen in twenty. Neither is it an *absolute* preservative by one inoculation. We may more surely prevent recidivation by two than by one inoculation.

If we compare the results above stated with what is known of vaccine and its relations to variola, we may see that the less vigorous organism which does not cause death, is analogous to a vaccine, relatively to the one that kills, for it gives rise to a disease which may be called mild, as it does not cause death, and, at the same time, it preserves from the disease in its most deadly shape. What other condition must this organism fulfil to be a true vaccine like that of cow-pox? This condition is that it should be a definite variety, and that we should not be obliged to prepare it *de novo*, when we wish to use it. We find here the same difficulty which presented itself to Jenner. After he had demonstrated that inoculated cow-pox is a preservative against variola, he thought that it was necessary to start from the cow-pox of a cow. Jenner soon discovered, however, that he could get along without cows, and make vaccine pass from one arm to another. We may try to do the same by causing our germ to pass from one cultivation to another. Under these circumstances, will the germ become actively virulent or will it remain moderately so? Although this may appear very astonishing, I can say that the last supposition is the correct one. The virulence of the germ, in the small number of cultivations which I have attempted, has not increased, and everything seems to point to the existence of a true vaccine. I may even add that one or two trials favour the idea that the attenuated virus keeps its character of mildness after passing through the bodies of Guinea pigs. Will the same thing happen after repeated cultivations and repeated inoculations? Only by experiments can such a question be answered.

At any rate, we now know of a disease caused by a microscopic parasite, which may be obtained in such a condition that it does not recidivate, as other diseases caused by similar parasites. Moreover, we have a variety of its virus, which behaves towards it as vaccine towards variola.

The Academy may allow me a digression worthy of attention. From what has been said, we can easily obtain chickens affected with the disease called *chicken cholera*, in which death is not a necessary consequence of the disease. We may then witness as many cases of cure as we may wish. Now, I do not believe that experimental surgery has ever met with more curious phenomena than those which are present when the animal returns to health, after inoculations have been made in the large pectoral muscles. The germ of the disease multiplies in the substance of the muscle as it would in a vessel. At the same time the muscle swells, hardens, and becomes bleached at the surface and below the surface. It becomes filled with globules of pus, but does not suppurate. Its histological elements are easily torn, because the parasitical germ is scattered through them in numerous pockets, and it feeds on a portion of their substance. I will, later on, exhibit coloured figures showing the disorders caused by the parasitical germ in case of cure. The parasite is gradually arrested in its development and disappears, while, at the same time, the portion of muscle which has been attacked unites, hardens, and lodges itself in a cavity whose surface resembles that of a healthy granulating wound. The portion which has suffered from the disease finally forms a sequestrum, and is so well isolated in the cavity that holds it that it may be felt by the finger under the skin, and, by the least incision, it may be seized with forceps and extracted. The small wound left in the skin heals immediately, and the cavity is gradually filled by the renewed elements of the muscle. I will now place some of these demonstrations before the Academy.

I have now to close by an explanation relating to the non-recidivation of the disease which occupies our attention. Let us take a chicken thoroughly vaccinated by one or more previous inoculations of the enfeebled virus. What will happen if we inoculate the same chicken again? The local lesion will be insignificant, while the first inoculations, and, in particular, the very first, had been the cause of such marked change in the muscle, that a large sequestrum can be easily felt by the touch. The cause of the difference in the effects of these inoculations is to be found entirely in a greater relative facility of the development of the germ of the disease at the first inoculations, and, in the last inoculation, in the development being either entirely wanting or very feeble and promptly stopped. The consequence of this seems evident, and it is that the muscle, which has been seriously diseased, has become, even after it has been cured, unfit for the cultivation of the germ of the disease, as if this germ, by a preceding cultivation, had suppressed some principle which life does not bring back, and whose absence prevents the development of the microscopic organism. I have no doubt that this explanation, to which we are led by palpable facts in this case, will be found to be generally applicable to all virulent diseases.

It must appear superfluous to point out the principal consequences of the facts which I have had the honour to present before this Academy. There are, however, two of these which may be mentioned. One is, that we may hope to obtain artificial cultivations of every virus, and the other is, the idea of obtaining vaccines of the virulent diseases which afflict humanity, and which are the greatest plague of agriculture in the breeding of domestic animals.

M. PASTEUR, 'On Virulent Diseases', *Chem. News*, 1880, **42**, 4.

101 'Atmospheric Pollution?'

THE ALKALI WORKS REGULATION ACT

THE Alkali Works Regulation Act has now been in force twelve months. We say in force; but it is, we believe, a fact that the machinery for setting it in practical operation was not completed until some months after the Act legally took effect. There has been time enough, however, to draw some conclusions as to the practical results of the operation of the Act. Our readers will remember that the object of the Legislature was to compel manufacturers to condense at least 95 per cent. of the hydrochloric acid evolved in the production of soda from common salt. An opinion seems to be prevalent that the object was to prohibit the escape of noxious vapours from all chemical works; but, in fact, no such general enactment exists. The operation of the present Act is strictly confined to alkali works. Sulphurous fumes from copper works are allowed to escape as before; and the most careless management of chemical works still remains without Government interference. In a late action against a copper work in Lancashire, it was argued that the Act applied universally; but in the face of the precise words of the Act, it is needless to say the argument broke down. We have only, then, to see what has been the effect of the law, so far as regards alkali works. Now, it is a curious fact that in the course of our inquiries we have found that it is precisely in places, like St. Helens, most complained of (for an account of its condition see CHEMICAL NEWS, vol. vi., p. 202) that the Act has been most thoroughly effective. Indeed, we were informed that it has been so to an extent never dreamt of by the Legislature. It will be remembered that in their evidence before Lord Derby's Commission some of the principal members of the alkali trade insisted that perfect condensation was impossible; and that at least 5 per cent. of the acid must be allowed to escape. Lord Derby yielded to this representation, and 95 per cent. was the amount fixed upon to be condensed. Upon trial, however, it would seem that perfect condensation is by no means an impossibility. It is found, for example, that a cubic foot of the escaping gases from some manufactories may be passed through a solution of nitrate of silver without affording the least trace of a precipitate, whilst in one case the merest trace was found after the gas has passed through the solution without interruption for three days and nights. This carefulness, we are happy to learn, is spreading rapidly through the trade, and it is found that the manufacturers are not at all satisfied with keeping within the Act. They perceive, in fact, what was evident enough, that if 95 per cent. of the acid could be condensed, there could be no difficulty in condensing the whole, and they are doing their best to effect this, and so remove every possible cause of complaint. Nothing less was to be expected from gentlemen who gave evidence before Lord Derby's Commission, sometimes against themselves, with a fairness and straightforwardness which could not be too highly praised.

We have said that there is no difficulty in the way of complete condensation, nor is there when close furnaces are used. With open furnaces, however, it would seem that there must be some escape, but in the best of these the escape does not much exceed 1 per cent. At St. Helens we understand there are no open furnaces, so that this town has been so far favoured. And yet the wilderness around that town is not likely soon to blossom as the rose. It is still what was described by a witness before the Commission, 'one scene of desolation. You may look round for a mile, and not see a tree with any foliage on whatever.' The fact is, that only one 'noxious vapour' has been suppressed. Others no less deadly, but almost as easily prevented, are still allowed to escape as freely as before, and while these are at large the country round St. Helens can never be like what it was fifty years ago. We have made these last remarks chiefly to call attention to the fact that the Alkali Act has a very narrow territory, and no influence beyond it. But the satisfactory results of its operation will no doubt encourage the Legislature to proceed further in the same direction; and much as we deprecate, in a general way, Government interference with private enterprise, we are bound to express our belief that an extension of this enactment may be productive of much good.

'The Alkali Works Regulation Act', *Chem. News*, 1865, **11**, 7 January.

10m 'Panorama of the Blackwall Railway'

By courtesy of the British Museum

PANORAMA OF THE BLACKWALL RAILWAY.

Enjoyment of the scenery constitutes more than half the pleasure of travelling; and in order that the voyager along the Blackwall Railway may know what to expect, we have had a panorama prepared by a first-rate artist of the scene that will greet the eye and sink into the heart of the passenger from Fenchurch Street.

Talk of the "feast of reason and the flow of soul," what are they compared with "the feast of chimneys and the flow of smoke," on the line alluded to? There is something elevating in the idea of running for miles on a level with the weathercock and the cowl, while we mark the fierce cat revelling uncontrolled along the sky-piercing parapet. The chamois hopping and skipping over rocky peaks, or the goat frisking about in the caves of the Colosseum, are both objects of grandeur; but the unfettered cat, roaming at ease through the lofty gutters of Limehouse and Poplar, is a sight which none but the passenger on the Blackwall Railway is privileged to contemplate.

But we are growing sentimental, and the tear-drop is trickling down the cedar pen-holder, till, mingling with the murky ink, it becomes darkened with that it mixes with; like the gushing dew-drop, which—but we are losing sight of the Panorama of the Blackwall Railway, to which we entreat the serious attention of all lovers of the beautiful.

Punch, 1845, **9**, 124.

11 MINERAL RESOURCES AND THE GROWTH OF TECHNOLOGY

11a Kingzett on the Alkali Trade

ENGLAND is open to similar charges of neglect of its best sons, and even now, in these days of boasted civilisation, no fact is more striking than that science is not sufficiently recognised by the Government of this country. It is and must be admitted on all sides that scientific research necessarily precedes the application of science to industry, and yet, even now, the pay of the chemist is little or no better than the recompense of an unskilled labourer. True it is that the patent laws afford some protection to those who are in the position to patent, but many even of our best chemists are not possessed of the means, while for the most part chemical inventions and discoveries do not admit of protection in this way. We cannot imagine, for instance, Scheele patenting the mode of preparation and uses of chlorine, and yet his discovery has led to an industry, viz., the manufacture of bleaching powder, of which substance this country alone produces annually more than 100,000 tons.

We can only hope that the time is near at hand when we shall have a Government sufficiently enlightened to appreciate that it is to its best interests to foster and encourage science in all its multiplicities, for by so doing alone, can England maintain that supremacy over all other nations which it now possesses in its varied industries.

With these preliminary observations we will now return to our more immediate subject. The alkali trade as now worked in Great Britain is planned out on the adjoining scheme, and in the following chapters it will be our endeavour to give a complete sketch of the history, products, and processes involved in each branch, together with some remarks on the uses of the manufactured products. In doing this we shall pay especial attention to those recent improvements which may be said to have almost revolutionised the whole trade, by effecting the utilisation of what were previously waste products, or by their avoidance. While, therefore, the work will stand alone as the only one of its kind, it is hoped that it will serve as a basis for further thought and research by chemists, who, as a body, may be safely said to be unacquainted with those chemical processes which form the foundation stones of the alkali trade; and if, in assisting to destroy and overthrow the use of imperfect empiric, or 'rule-of-thumb' methods often largely employed in alkali works, it indicates a more scientific method, it will accomplish some good.

Products of the Trade:—Iron, copper, silver, gold, sulphuric and hydrochloric acids, bleaching powder, potassic chlorate, calcic chloride, sulphate of soda, caustic soda, sodic carbonate and bicarbonate, sulphur, sulphate of copper, hyposulphite of sodium, phosphate of sodium, and Epsom salts.

Waste Products:—Part of the hydrochloric acid as chloride of calcium, all the lime or nearly so, part of the sulphur.

Regenerated Products:—Nitre (partly), manganese and part of the sulphur.

The introduction of recent improvements in the alkali trade was attended with much opposition from manufacturers themselves.

Scheme of the Alkali Trade as now worked.

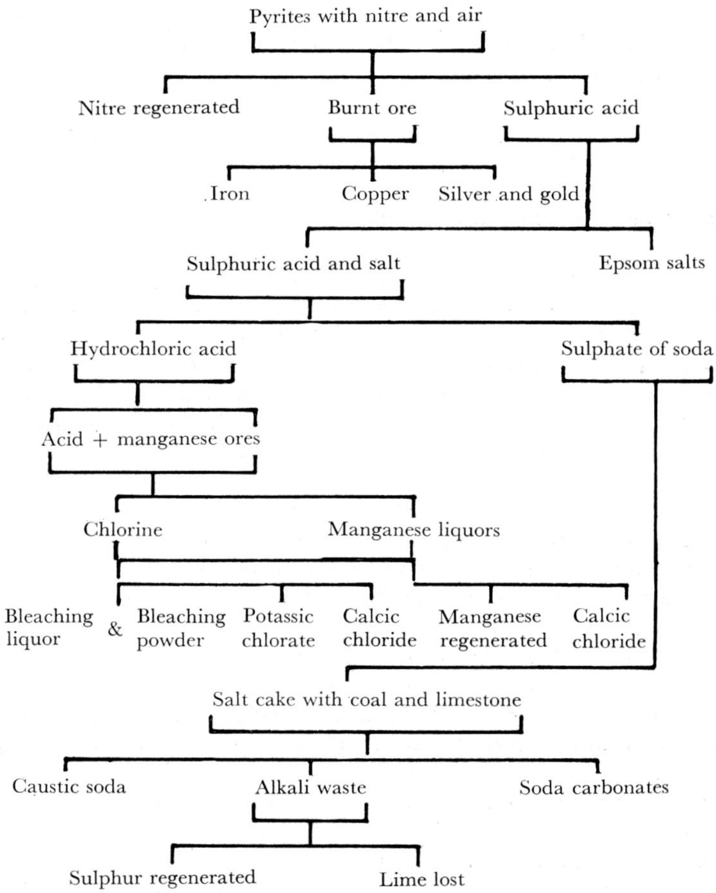

However strange at first sight it may seem, that manufacturers should so stand in their own light, it is not so inexplicable if we bear in mind that they were unscientific, and, moreover, it must be admitted that there is good reason why they should be chary of adopting new processes until the strongest evidence is forthcoming that the new processes give better results than the old ones; for while, however good the improvements may be, there is always the greater possibility at such times than at others, that the improvements should themselves be improved upon; and every change of plant means, for the time at least, the sacrifice of so much capital.

It is gratifying, however, to find that many manufacturers are now well-known scientific men; to the truth of this assertion the names of Deacon, Pattinson, Young, Scheurer Kestner and others testify most fully. But it is to be deplored, perhaps, that so much scientific ignorance should exist as is known to exist among many managers and the second-rate chemists often employed at alkali works. These men often boast loudly of their rule-of-thumb methods of working, when those methods are known to be unreliable and therefore worse than useless, inasmuch as they mislead. Of course, when a process becomes familiar to those working it, it is often possible to adopt methods of testing certain products in an abbreviated way, and so save time and trouble, but there can be no doubt that many of these methods at present largely used are *not* reliable, and their use should therefore be discontinued. We refer, for instance, to the estimation of nitrous compounds in oil of vitriol by simple dilution with water; the determination of the temperature of vat liquor by the hand alone, etc. In short, when it is recognised that the chemist's duty in an alkali works consists in the examination and analysis of materials consumed, and products manufactured, and so forth, it becomes evident that his chemical ability should be such as to enable him to accomplish these things in a reliable manner; and also sufficient to enable him to trace any loss in a manufacturing process to its source, or to detect any tampering of the materials by the workmen.

Thus, suppose the production of 'caustic' in a soda works to be below the average, then by the analysis of the successive intermediate products, as the salt cake produced in the pots, the vat liquors and so on, he should be able, generally, to discover the source of the loss. In the case under consideration, this might arise from the employment of 'salt cake' (sodic sulphate) below average strength, or from the 'balls' (crude sodic carbonate) being 'burned' or 'green,' or from imperfect washing of the lime mud from the causticiser, etc.

To any one acquainted with the status of chemical science in alkali works, the foregoing remarks will not be matter for surprise. Thus, quite recently, the author of a paper on caustic soda, alluded to laboratory work as a 'kind of out-at-elbows trade' and actually stated his opinion that 'a manager can be too much of a chemist.' To this we must strongly object, for it is absurd to suppose that because a manager may be likewise a good chemist, he will be misguided by his knowledge in attempting more than reasonable 'maximum productions'; on the other hand, his chemical knowledge will serve him as a rudder serves a boat, namely, as a guide by which to obtain a certain result, without loss of time or labour.

Chemical industry is just emerging from ignorance and prejudice, and any attempt to return to that state should therefore be opposed. For, as Parkes tells us in his 'Chemical Essays,' it is not so many years ago 'that Mr. Steer, a very large *aqua fortis* maker at Birstall, near Leeds, was for long accustomed to lay his glass retorts in a shallow rivulet which ran through his fields, where, in a few weeks, the indurated salt was dissolved and washed away in the water.' In short, industrial chemistry has yet much to attain, as is shown by its history, by the existence of the Alkali Acts, and by the mountains of alkali waste that stagnate the atmosphere for miles round the districts where alkali works exist. But, at the same time, these things demonstrate to us what chemical science, aided by engineering, has done for the alkali trade of Great Britain. We cannot do better than conclude this chapter with some observations due to Dr. Lunge, who, in a recent address to the Tyne Chemical Society, stated that one of the largest works in Germany employs six chemists, at from 300*l*. to 400*l*. per annum, and in addition retains the services of a chemist of reputation, *exclusively for theoretical work in the laboratory*, at a salary of nearly 2,000*l*. per annum. This wise conduct gives rise to happy results, as is exemplified (to those who are conversant with the development of the soda trade in Germany) by the following figures:—

From 1867 to 1872, the German production of sulphuric acid increased from 57,825 tons to 84,264 tons; sulphate of sodium from 35,767 to 51,618 tons; and calcined soda ash from 26,250 to 36,227 tons.

C. T. Kingzett, *The History, Products and Processes of the Alkali Trade*, London, 1877, p. 5.

11b 'The Origin of the Ammonia-soda Process'

On the Origin of the Ammonia-soda Process, By Ludwig Mond

ALL writers on this subject from Brande in 1841* to Lunge† in 1880, fix the date of this invention at the 30th of June, 1838, when Messrs. Harrison Grey Dyar and John Hemming, both chemists of London, filed the petition for their English patent. Only of recent years an attempt has been made by several French savants‡ to put the date of the invention sixteen years later, in order to claim it as a French invention and to give the credit of it to two of their countrymen, Messrs. Schloesing and Rolland, who took out a patent relating to the subject in 1858, preceded by a patent taken out by Mr. Schloesing alone in 1854.

It would seem superfluous to contradict an error so palpable and so easily refuted by reference to documents accessible to everyone, had not this statement been made by three scientific men of the highest standing, whose name is sufficient to obtain acceptance for it by numbers who have not themselves inquired into the facts.

In justice to Messrs. Schloesing and Rolland, I must state that these gentlemen in their various publications have fully acknowledged that the process they endeavoured to improve had been invented long ago, and that the various chemical operations constituting the process had been described so exactly in a patent taken out in France by Delaunay as to leave to future inventors very little to do.

On referring to Delaunay's patent, dated 27th of May, 1839, this will be found an exact translation of the English specification of Messrs. Dyar and Hemming, of October, 1838, whence it is clear that Delaunay was simply the agent of Messrs. Dyar and Hemming. Inquiring further into this French specification we come, however, upon a remarkable and important fact which has hitherto been left unnoticed. This patent is followed on the 18th of May, 1840, by an additional patent, a facility enjoyed under the French patent law almost free of expense, which has evidently in this instance led the inventors to protect in France most important improvements found out shortly after taking out their original patent, while the high cost of a second English patent has prevented their doing the same in England.

As is well known, the original process of Messrs. Dyar and Hemming consisted in mixing a solution of common salt with solid carbonate of ammonia, separating the bicarbonate of soda formed from the residual liquor, boiling this liquor to dryness, and heating the dry product with chalk in order to recover the carbonate of ammonia employed. It is evident that this series of operations would involve so serious a loss of ammonia as to make the economical application of the process impossible, and if Messrs. Dyar and Hemming had left their process in this very incomplete form, their claim to be considered the inventors of the process now known as the Ammonia-soda Process might well be called in question. The additional French patent referred to above, however, puts the matter in an entirely different light. I give an exact translation of their specification:—

* Brande: Manual of Chemistry, 5th edition, pp. 661.

† Lunge: Manufacture of Sulphuric Acid and Alkali, viii. p. 1.

‡ Frémy: Presidential Address to the French Association for the Advancement of Science, 1878. Bouley: Président de l'Académie des Sciences; *Comptes Rendus,* T.C. No. 14. Scheurer Kestner: Conférence à la Société d'Encouragement sur Nicolas Leblanc, 28 Mars, 1885.

18th May, 1840. *Additional Patents for Improvements.* The operations can be reduced to the following:

Preparations of the Carbonate of Soda

1st. To a concentrated solution of chloride of sodium is added an equivalent quantity of sesquicarbonate of ammonia in order to decompose it.

2nd. A current of carbonate-acid gas is passed through the mixture to saturate the ammonia, and to make the decomposition as perfect as possible.

3rd. After a sufficient time the saline deposit is collected on filters.

4th. The deposit is drained and washed either with water or with a concentrated solution of bicarbonate of soda.

5th. The saline deposit is submitted to pressing, or put immediately in a reverberatory furnace or in a retort to drive off the water, the excess of carbonic acid, and the ammoniacal salts which it may contain.

Extraction of the Ammonia from the Residues of the Operation

1st. The liquid collected in the third operation contains ammoniacal salts; it is run into a distilling vessel.

2nd. To this liquid is added a sufficient quantity of lime to decompose the entire quantity of the ammoniacal salts, heat is applied and thus ammoniacal gas is obtained.

The ammonia gas can be employed in several different ways.

It can be simply dissolved in water.

It can be passed into a concentrated solution of chloride of sodium.

When the gas has been dissolved in water chloride of sodium is added in sufficient quantity relative to the alkalimetric value of the liquor, and then a current of carbonic acid gas is passed through to effect the decomposition of the chloride of sodium.

If the ammonia gas is added to the solution of chloride of sodium, one can at the same time or successively pass a current of carbonic gas through it.

The order of the preceding operations shows that the sesquicarbonate of ammonia is employed only once, and that after it has been decomposed, the ammonia it contains is continually used over.

In all cases, and whatever be the mode of proceeding, the chloride of sodium is decomposed under the influence of ammonia and carbonic acid.

In this operation the ammonia must be considered as an intermediate agent which takes hold of the carbonic acid to give it off again to the chloride of sodium, which it can then decompose, and produce carbonate of soda through the intervention of the elements of water.

The ammonia plays here a part similar to that of the binoxide of nitrogen in the preparation of sulphuric acid, and to that of the acetate of lead in the manufacture of white lead by the process of Thénard.

In order to make this process economical, the principal object should be the preservation of the ammonia, the employment of which ought to be continued till it is, so to say, used up.

As to the carbonic acid likewise employed in this process, it can be produced economically by the combustion of coal, or by the calcination of limestone as practised in other manufactures.

Here we have described in detail all the operations as they are this day carried out in the manufacture of the immense quantities of ammonia-soda now produced, and we may well agree with Messrs. Schloesing and Rolland that, as far as the chemistry of the ammonia-soda process is concerned, Messrs. Dyar and Hemming left very little to do to future inventors. Still there is the fact that from 1840, when the chemistry of the process was fully developed, until 1865, this remarkably simple and beautiful process found no application, although many very eminent engineers, chemists, and manufacturers spent years of their time and large sums of money in endeavouring to realise it on an industrial scale. All these attempts, including the one made by Messrs. Schloesing and Rolland between 1855 and 1858, ended in failure. If it could be shown that these gentlemen, by the apparatus they have patented, or by the experiments they have made to work the process on a large scale, had advanced the ultimate success of the process, they would, although not being the inventors of the process, deserve great credit, and no objection could be raised to the award of the diploma of honour which they have obtained at the Vienna International Exhibition. A close examination of the facts unfortunately does not show that they have in this direction done more than their predecessors and contemporaries.

The fact that bicarbonate of soda is formed by mixing carbonate of ammonia with a strong solution of salt has been known, although it was never published, long before the date of Messrs. Dyar and Hemming's patent. A. Vogel* tells us that he found this fact mentioned in his father's note-book as far back as 1822. Dr. Angus Smith has told me that his friend Mr. John Thom has shown him this reaction, performing it in the palm of his hand, several years before 1838. I owe to Mr. Watson Smith fuller particulars of Mr. Thom's early work in connection with this subject, contained in several letters from Mr. Thom himself, which are corroborated by a letter from Mr. William Henderson, of Glasgow. These letters leave no doubt that Mr. Thom actually manufactured soda by the action of carbonate of ammonia upon salt in 1836 at the works of Messrs. Turnbull & Ramsay at Camlachie, where he was employed as chemist. Mr. Thom mixed the two salts intimately with a small quantity of water, and put the resulting magma in bags into a press in order to separate the liquid from the bicarbonate formed. The bicarbonate was used for making soda-crystals, of which about two hundred-weights per day were produced. The liquid was evaporated, mixed with chalk, and heated in iron retorts to recover carbonate of ammonia. After working the process for about a year, Mr. Thom left the works at Camlachie, and did not follow up the subject any further. Here we have in all probability the first attempt to turn the reaction between salt and carbonate of ammonia to practical use. But the operations were still so crude, the separation of the bicarbonate of soda from the chloride of ammonium so unsatisfactory, the reaction between salt and sesquicarbonate of ammonia without the aid of carbonic acid so imperfect, and the process used for recovering the carbonate of ammonia so impracticable, that however highly we may think of Mr. Thom's merits as the discoverer of the principal reaction, and as the first who attempted to manufacture soda by its aid, yet he cannot be considered to have invented a chemical process industrially applicable for this purpose. The merit of having invented the series of reactions now forming the ammonia-soda process remains thus undiminished with Messrs. Dyar and Hemming, who, as far as can be inferred from Mr. Thom's letters, have worked quite independently of him and without knowledge of what he had been doing.

According to information, for which I am indebted to Mr. Richard Muspratt, Messrs. Dyar and Hemming built a small works to demonstrate their process in Whitechapel very shortly after taking out their patent. These works were regularly visited by Dr. Sheridan Muspratt and Mr. James Young of paraffin celebrity, then assistant to Professor Graham in London, in order to study the process. Shortly afterwards, probably in 1840, Mr. James Muspratt engaged

* *Chem. Centralblatt*, 1874, p. 98.

Mr. Young to erect plant for the process on a considerable scale, at his large chemical works at Newton. This plant was worked under Mr. Young's direction for about two years, when the process was given up because it was found more expensive than the Le Blanc process, principally on account of the heavy loss of ammonia. Mr. Carrol, at the time foreman of the works at Newton, has told me that about £8000 were spent in this experiment.

It is well known that various other attempts to work the process were made about this time on the Continent, amongst others, by Kunheim in Berlin, and Seybel in Vienna. A few years later a small plant was erected near Leeds by Mr. Bowker, and kept at work for several years, the product being regularly sold in the district, but this also was subsequently abandoned. The first attempt to construct a more perfect plant for carrying out the process was made by Mr. William Gossage (patent of 21st February, 1854). In the same year, and following each other in close succession, patents were taken out by Turck (26th May), Schloesing (21st June), Henry Deacon (through his agent, Henry Johnson, 8th July). I am informed by Mr. Holbrook Gaskell, who entered into partnership with Mr. Deacon in 1854, that Mr. Deacon leased land for the purpose of erecting ammonia-soda works in 1853, and that in the following year he was already engaged in the manufacture of ammonia-soda, which he carried on for about two years at the rate of several tons per day. After a sum of from five to six thousand pounds had been expended upon it, the firm decided to abandon the ammonia-soda process, and commenced the erection of the large Le Blanc Soda Works, which have since become so well known, and which have proved such a fruitful field for the display of Mr. Deacon's great powers as a manufacturer and as an inventor.

Judging from Mr. Deacon's patent, and from what he has told me himself many years ago, the plant which he used for making ammonia-soda was very similar to that described by Messrs. Schloesing and Rolland in their patent of 1858, except that Mr. Deacon went a considerable step further than these gentlemen by using carbonic acid under pressure. He used horizontal cylinders with agitators for absorbing the carbonic acid, and a still of the Coffey type for recovering the ammonia.

Messrs. Schloesing and Rolland commenced in 1855 to erect works at Puteaux, near Paris, which were kept going for two years, and during this time produced 316 tons of soda—86 tons during the first 14 months, and 230 tons during the last ten. At the end of the two years the experiment was given up, like all the preceding ones had been, although the soda manufactured was sold at a price varying from 65 to 75 francs per 100 kilogrammes, equal to £26 to £30 per ton, or about five times its present value. Messrs. Schloesing and Rolland assert, in a very lengthy paper published in 1868 (*Ann. de Chimie und Physique*, T. xiv.), and this assertion has been repeated by various writers, that their want of success has been due, not to the insufficiency of their apparatus, but solely to the high tax levied on salt at the time in France. They state that they had to use 180 kilogrammes of salt per 100 kilogrammes of soda, where the Le Blanc makers used 150 kilogrammes, and that the tax they had to pay on this extra quantity of salt made competition impossible. This duty amounted to 10 francs per 100 kilogrammes of salt. Messrs. Schloesing and Rolland would thus have to pay 3 francs per 100 kilogrammes of soda more than their competitors, an amount very much below the difference in cost of production between the two processes at the present time. The total tax they had to pay on salt would have amounted to 18 francs per 100 kilogrammes of soda manufactured. The selling price of soda at the time being 65 to 70 francs per 100 kilogrammes, Messrs. Schloesing and Rolland ought to have realised enormous profits in spite of the tax if they had at all understood the manufacture. The selling price to-day is below 18 francs per 100 kilogrammes, and this leaves a very handsome profit. Moreover, it would have been easy to reduce the quantity of salt consumed per ton of soda in the ammonia-soda process to a less quantity than is used in the Le Blanc process, by recovering the undecomposed salt from the residual liquor of the distillation. We can, therefore, not accept the assertion that but for the salt duties Messrs. Schloesing and Rolland could have manufactured at a profit; their own figures prove plainly that their experiment was no more successful than their predecessors; that it was not carried out on any larger scale, nor persevered with for a longer period of time.

After 1858 no further serious attempts were made to overcome the engineering difficulties of the ammonia-soda process, until Mr. Ernest Solvay undertook this task in 1863. It is now a matter of history how completely Mr. Solvay succeeded, after five years of incessant work, under trying circumstances, in inventing and working out new apparatus for carrying out on a large scale all the various operations of Messrs. Dyar and Hemming's beautiful but hitherto barren invention; how quickly the ammonia-soda manufacture has developed in his hands, and what enormous proportions it has attained in a comparatively short time. While recognising Messrs. Dyar and Hemming as the inventors of the chemical operations which constitute the ammonia-soda process, we have to give to Mr. Solvay the honour of being the inventor of the apparatus, which alone has made this process of value to the public, and of having thus become the founder of the ammonia-soda industry.

LUDWIG MOND, 'On the Origin of the Ammonia-soda Process', *J. Soc. Chem. Ind.*, 1885, **4**, 527.

11C 'Stassfurt Salt Deposits' (*see* p. 208)

G. C. NAPIER HAKE, 'Stassfurt Salts and their Mode of Treatment', *J. Soc. Chem. Ind.*, 1883, **2**, 146.

FIG. I.

TRANSVERSE SECTION OF THE MINES: V. D. HEYDT & LUDWIG II.

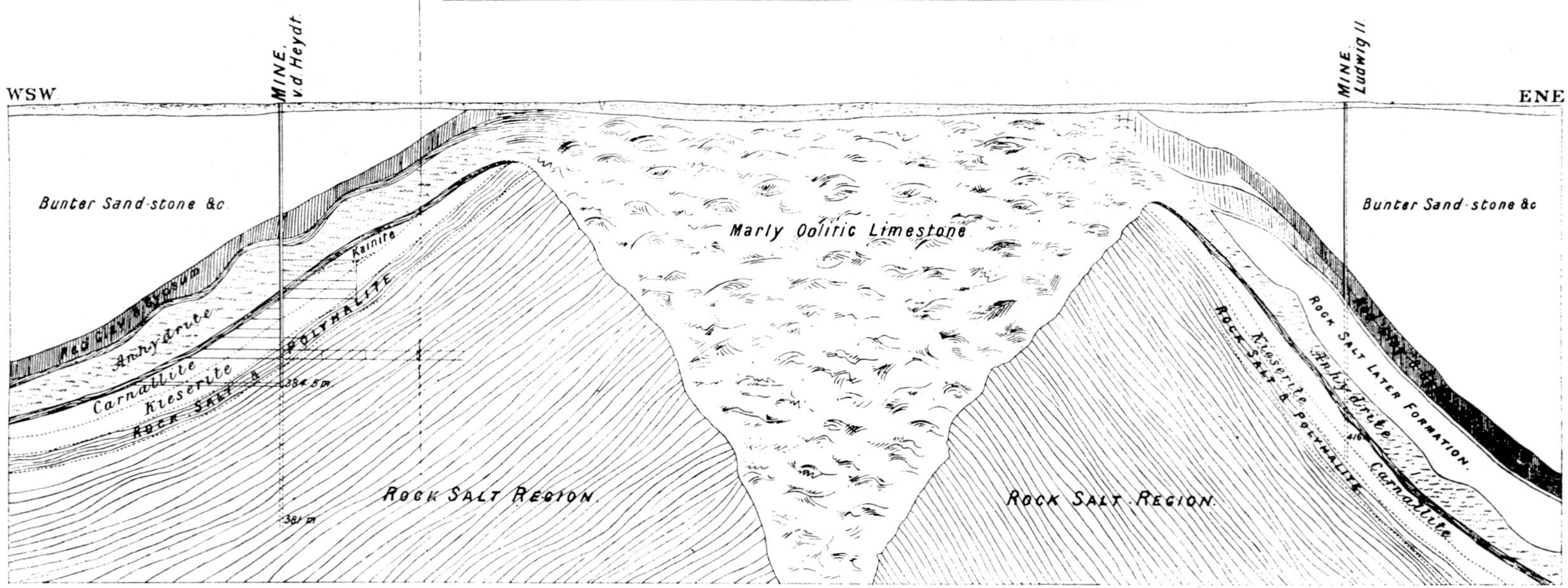

SCALE 1/5000

By courtesy of the
Chemical Society

12 PETROLEUM

12a 'Drake's Oil-Well'

By courtesy of Northwestern University Press

H. F. Williamson and A. R. Daum, *American Petroleum Industry, 1859–1899*, Evanston, U.S.A., 1964, p. 80.

12b 'Early Lubricating-oil Refining'

By courtesy of Northwestern University Press

H. F. Williamson and A. R. Daum, *op. cit.*, p. 243.

12c Early Drilling Operations, U.S.A.

By courtesy of the American Petroleum Institute and Northwestern University Press

H. F. Williamson and A. R. Daum, *op. cit.*, p. 95.

12d 'Medicinal Claims for Petroleum'

By courtesy of Northwestern University Press

PETROLEUM, OR ROCK OIL.

A NATURAL REMEDY!

PROCURED FROM A WELL IN ALLEGHENY COUNTY, PA.

Four hundred feet below the Earth's Surface!

PUT UP AND SOLD BY

SAMUEL M. KIER,

CANAL BASIN, SEVENTH STREET, PITTSBURGH, PA.

The healthful balm from Nature's secret spring,
The bloom of health, and life, to man will bring;
As from her depths the magic liquid flows,
To calm our sufferings, and assuage our woes.

CAUTION.—As many persons are now going about and vending an article of a spurious character, calling it Petroleum, or Rock Oil, we would caution the public against all preparations bearing that name not having the name of S. M. KIER written on the label of the bottle.

PETROLEUM.—It is necessary, upon the introduction of a new medicine to the notice of the public, that something should be said in relation to its powers in healing disease, and the manner in which it acts. Man's organization is a complicated one; and to understand the functions of each organ, requires the study of years. But to understand that certain remedies produce certain impressions upon these organs, may be learned by experience in a short time. It is by observation in watching the effects of various medicines, that we are enabled to increase the number of curative agents; and when we have discovered a new medicine and attested its merits, it is our duty to bring it before the public, so that the benefits to be derived from it may be more generally diffused, but have no right to hold back a remedy whose powers are calculated to remove pain and to alleviate human suffering and disease. THE PETROLEUM HAS BEEN FULLY TESTED! About one year ago, it was placed before the public as A REMEDY OF WONDERFUL EFFICACY. Every one not acquainted with its virtues, doubted its healing properties. The cry of humbug was raised against it. It had some friends,—those that were cured through its wonderful agency. These spoke out in its favor. The lame, through its instrumentality, were made to walk—the blind, to see. Those who had suffered for years under the torturing pains of RHEUMATISM, GOUT and NEURALGIA, were restored to health and usefulness. Several who were blind have been made to see, the evidence of which will be placed before you. If you still have doubts, go and ask those who have been cured! Some of them live in our midst, and can answer for themselves. In writing about a medicine, we are aware that we should write TRUTH—that we should make no statements that cannot be proved. We have the witnesses—crowds of them, who will testify in terms stronger than we can write them to the efficacy of this Remedy, who will testify that the PETROLEUM has done for them what no medicine ever could before—cases that were pronounced hopeless, and beyond the reach of remedial means—cases abandoned by Physicians of unquestioned celebrity, have been made to exclaim, "THIS IS THE MOST WONDERFUL REMEDY EVER DISCOVERED!" We will lay before you the certificates of some of the most remarkable cases; to give them all, would require more space than would be allowed by this circular. Since the introduction of the Petroleum, about one year ago, many Physicians have been convinced of its efficacy, and now recommend it in their practice; and we have no doubt that in another year it will stand at the head of the list of valuable Remedies. If the Physicians do not recommend it, the people will have it of themselves—for its transcendent power to heal, *will and must* become known and appreciated—when the voices of the cured speak out, when the cures themselves stand out in bold relief, and when he who for years has suffered with the tortures and pangs of an immedicable lesion, that has been shortening his days, and hastening him "to the narrow house appointed for all the living," when he speaks out in its praise, who will doubt it? THE PETROLEUM IS A NATURAL REMEDY—it is put up as it flows from the bosom of the earth, without anything being added to or taken from it.

[4]

Rheumatism yields to the Power of Petroleum.

I feel myself under a debt of gratitude to the proprietor of the Petroleum—the use of which great medicine has entirely cured my wife of a violent attack of the Rheumatism. She had labored under an attack of the disease for about two months, suffering the most intense pain; the greater part of the time confined to her bed, and unable to do anything. The pain in the limbs was very great, attended with a great deal of swelling. The Petroleum had a happy effect; for the first two or three applications the swelling diminished, and the pain left her. I continued to apply the medicine daily for about two weeks, which entirely relieved her, and she is now as well as ever she was in her life. I feel confident that the Petroleum is one of the greatest medicines in the world for Rheumatic pains and swellings. (Signed) PETER PUHL, *Perry Street, Allegheny City.*

Read the following certificate. It is from Mr. Harvy Mann, the manufacturer of Mann's celebrated axes:

"There is nothing like it for Burns."

BELLEFONTE, February 21st, 1850.

I got three dozen of Mr. Kier's Petroleum of Mr. Howell, and I am disposed to think there is nothing equal to it for burns; having burnt my hand, and expecting nothing less than a long, tedious sore, but found no soreness at all from its only leaving a red scale, without any soreness when pressed upon, which quite surprised me. I therefore have great faith in its efficacy, so farat least, and will introduce it in this section. Yours, respectfully, &c.

HARVEY MANN.

MR. S. M. KIER:—Allow me to express to you my heart-felt thanks for the great benefit I have received from an article called Petroleum or Rock Oil, of which you are sole proprietor. I had occasion to use it about the 1st of January, in a violent attack of Rheumatism, which was very painful, flying about from place to place, accompanied with much swelling, so as to keep me in constant torture. I used the Petroleum externally, a few applications of which removed all pain, and every symptom of the disease. I am now entirely well, and would take this occasion to recommend the Petroleum to all who may be suffering under the agonizing pains of Rheumatism or kindred diseases.

Pittsburgh, February 12, 1850. ZIBEON WILBAR.

A Bad Case of Inflammatory Rheumatism.

The following extracts are from a letter received from a young gentleman of Middleburg, Summit County, Ohio. They are attested by Charles Belden and Dr. Elijah Curtis, of that place:

MR. KIER: *Dear Sir,*—Gratitude compels me to express my thanks for the discovery of your invaluable medicine I was afflicted with that awful and disheartening disease, the Inflammatory Rheumatism, on the 21st of September last, whilst I was in Cleveland; and was brought home from that place to Middleburg, where I lay on my back for over two months. For three weeks of this time, I was in the most intense suffering and agony. I lay with my limbs in one position, and could not stir a muscle, nor have my bed changed. My screams could be heard all over the neighborhood, so intense was my suffering. When every other remedy proved valueless, I had recourse to your invaluable Petroleum, of which I had made but three applications, until I commeneed to get better. I used altogether but two bottles and a half, and am now well. I can get any number of persons to testify to the above facts, if necessary.

Attest. CHARLES BELDEN, (Signed) FREDERICK H. BLECKER.
DR. ELIJAH CURTIS.

CHRONIC COUGH CURED.

Public attention is most respectfully invited to the plain, unvarnished statement of John Watt, who was cured of an old cough by the use of the Petroleum:

"This may certify that I have been cured of an old Chronic Cough, by the use of four bottles of Petroleum. The cough attacked me a year ago last December, and I had lost all hopes of getting well, as I had taken the advice of several physicians without any benefit. I was benefitted almost instantly by the Petroleum. I coughed up, during the use of the Petroleum, a hard substance resembling bone. I make these statements without any solicitation from any one to do so, and solely for the purpose that others who are suffering may be benefitted. You are at liberty to publish this certificate. I am an old citizen of Pittsburgh, having resided here 33 years. My residence, at this time, is in Second Street. JOHN WATT.

Pittsburgh February 21st 1851."

H. F. WILLIAMSON and A. R. DAUM, *op. cit.*, p. 21.

12e By-products from Petroleum: an Early Patent

By courtesy of the Patent Office and Petroleum Times

A.D. 1853 . . . . N° 1748.

Treating Naphtha or Tar to Procure therefrom Lubricators, Fuel, and Solvents.

LETTERS PATENT to Warren De la Rue, of Bunhill Row, in the County of Middlesex, for the Invention of "**Means of Treating and Preparing certain Tar or Naphtha, and Applying Products thereof.**"

Sealed the 6th October 1853, and dated the 25th July 1853.

PROVISIONAL SPECIFICATION left by the said Warren De la Rue at the Office of the Commissioners of Patents, with his Petition, on the 25th July 1853.

I, Warren De la Rue, of Bunhill Row, in the County of Middlesex, do hereby declare the nature of the said Invention for "**Means of Treating and Preparing certain Tar or Naptha, and Applying Products thereof**" to be as follows:—

The object of this Invention is the treatment of that description of mineral substance known as Burmese naptha or Rangoon tar in order to procure therefrom fluid and solid substances, and applying the same as lubricators, as fuel for lighting purposes, and as solvents.

In order to effect this the mineral substance is submitted to a process of distillation, which is effected, in the first instance, at as low a temperature as possible, and with the aid of steam of 212° Fahrenheit passed through the tar, which is kept heated to the temperature of the

2 **A.D. 1853.—N° 1748.**

De la Rue's Impts. in Treating & Preparing certain Tar or Naphtha, &c.

steam by means of a steam jacket; the oils are passed over with the steam, and condensed with it in a refrigerator, when they are easily separated.

When all the products obtainable at this temperature (212°) are run off, then steam of higher temperature is passed through the tar, which is then heated by an open fire or otherwise to a higher temperature; and, lastly, the distillation is completed at a temperature very little short of a red heat; a small carbonaceous residue is ultimately left in the retort or still. Some of the products may be used without further distillation, but the several hydrocarbons are separated one from the other more completely by repeated distillations, with or without the use of steam, and treatment with mineral acids and alkalies.

Paraffine, one of the products, has to be so treated to render it pure.

Most of the products are of value for mixing with the common kinds of oil, such as seal and whale oil, for the production of lubricators.

SPECIFICATION in pursuance of the conditions of the Letters Patent, filed by the said Warren De la Rue in the Great Seal Patent Office, on the 18th January 1854.

TO ALL TO WHOM THESE PRESENTS SHALL COME, I, Warren De la Rue, of Bunhill Row, in the County of Middlesex, send greeting.

WHEREAS Her most Excellent Majesty Queen Victoria, by Her Letters Patent, bearing date the Twenty-fifth day of July, in the year of our Lord One thousand eight hundred and fifty-three, in the seventeenth year of Her reign, did, for Herself, Her heirs and successors, give and grant unto me, the said Warren De la Rue, Her special licence that I, the said Warren De la Rue, my executors, administrators, and assigns, or such others as I, the said Warren De la Rue, my executors, administrators, and assigns, should at any time agree with, and no others, from time to time and at all times thereafter during the term therein expressed, should and lawfully might make, use, exercise, and vend, within the United Kingdom of Great

Petroleum Times, 17 June 1949, p. 487.

12f Candle-making, 1849

By courtesy of Petroleum Times

Petroleum Times, 17 June 1949, p. 447.

12g Petrol Advertising, 1900

By courtesy of Petroleum Times

Petroleum Times, 17 June 1949, p. 474.

12h 'The Russian Petroleum Industry'

The Russian Petroleum Industry. By Boverton Redwood, *F.I.C., F.C.S.*

SHORTLY after my return from a visit to Baku and the principal distributing centres of the petroleum trade in Russia, last autumn, it was suggested to me that some account of the Caucasian petroleum trade would be of interest to the members of the Society of Chemical Industry. Much has been already published on the subject; but it appeared to me, upon consideration, that the whole ground had not been covered, and that the members of this society might justifiably be addressed on the technical and commercial aspects of the question.

I propose to divide my remarks under three heads—viz., (I.) Production (as the obtaining of the crude petroleum is termed), (II.) Refining, and (III.) Distribution. Under the second head I shall deal with the properties and uses of the various products, and shall bring the third section to a conclusion with a very few words in reference to the routes to Baku, in the hope that some of those present to-night may be induced to make the trip. The majority of the statistics I shall give are from notes taken during my journey; but some are quoted from the valuable official reports of Mr. Peacock, H. M. Vice-Consul at Batoum, and from Mr. Charles Marvin's publications. I am indebted to Mr. Ludwig Nobel, as well as to Mr. Crusell, Mr. Ryden and many other officials of the Nobel Company, particularly Dr. Lagerwall, the courteous Secretary of the Company, for replies to many questions, and for affording me special opportunities of making myself acquainted with the industry. I have also to thank Mr. Burkhardt of Batoum, Mr. D. N. Sacker, agent in London for Mr. S. E. Palaschkowsky and the Batoum Naphtha Industry and Commerce Company, and Mr. E. G. von Glehn, representative of the Ragosine Company, for much valuable information.

I. PRODUCTION

Anyone whose experience of oil-producing territory has been gained in the United States, cannot fail to be impressed, on first visiting the neighbourhood of Baku, with the very limited extent of the area within which, practically the whole of the crude oil is being collected. The oil-field from which the Baku refiners are supplied, is not in fact more than $3\frac{1}{2}$ miles square; and on this small tract of land there are flowing wells or fountains which apparently could supply the whole world with lamp oil and lubricating oil. It must not, however, be assumed that the oil-bearing lands of Russia are of small area. On the contrary, it has been estimated that petroleum deposits underlie no less than 14,000 square miles, and admitting that this estimate is not based upon very reliable data, there is, I believe, no reason to doubt that something like 1000 or 1200 square miles of the Apsheron peninsula may be fairly regarded as more or less productive oil territory. No systematised attempts have yet been made to explore this large field, and strictly define its limits, for the reason that more than sufficient oil is obtained within the narrow limits I have referred to. The locality of the existing wells is termed the Balakhani or Balakhani-Saboontchi district, and it is distant about eight miles from the town of Baku, at an elevation of something less than 200 feet above sea level. A few miles away, at Surakhani, is another group of wells on the site of the prehistoric Sacred Fires; but these wells are of minor importance. Petroleum has also been obtained in quantity at Kouban, about 60 miles inland from Novorossisk, on the shores of the Black Sea; and evidence of the extension of the oil-field in a north-westerly direction has been furnished by the result of trial borings in the Crimea. In the neighbourhood of Tiflis, and in a large number of other places throughout the Apsheron peninsula, there are also petroleum pits and natural outflows of petroleum, some of which have been locally utilised. While, as I have said, the petroleum field appears to extend into the Crimea (though it is possible that there may be a succession of deposits rather than an actual belt), there is evidence of its continuous extension from Baku across the Caspian Sea to the Trans-Caspian territory. Near the Baku coast the surface of the sea is frequently covered with oil, which has floated to the surface; and in still weather the petroleum gas may be ignited. The island of Tcheleken, immediately opposite, and distant about 200 miles, contains apparently immense deposits of ozokerit. I hoped to have been able to visit this interesting place, but the Russian Government has, for political reasons, prohibited Englishmen from landing there; and in Russia the rigid passport-registration system makes every movement of the traveller known to the authorities. On the mainland, in the Trans-Caspian territory, there is stated to be abundance of petroleum; but except to supply fuel for the railway this source is not utilised.

The existence of the petroleum deposits of Baku has probably been known for 2500 years, for it is believed that the Persian fire-worshippers frequented Surakhani from the time of Zoroaster, who lived not later than 600 B.C. Gibbon states that in A.D. 624, the Emperor Heraclius wintered at the mouth of the river Kura, 70 miles south of Baku, and that his 'soldiers extinguished the fire and destroyed the temples of the magi.' Marco Polo, who wrote in the thirteenth century, is supposed to have referred to the petroleum of Baku when he stated that there was in this neighbourhood an abundant spring of oil not good for food, but good to burn, or to anoint camels that had the mange; and he added that people came from great distances to fetch it. In Jonas Hanway's time (the middle of the eighteenth century), Indians were worshipping at the Surakhani temple, as many as 50 pilgrims congregating there at one time. Hanway described the eternal fire as a flame in colour and gentleness not unlike a lamp that burns with spirit, only more pure, and sometimes rising to a height of eight feet, when the wind blows. I have on the table a photograph of this temple as I recently saw it. In the centre of a walled quadrangle stands a solidly built pillared and roofed rostrum of white stone, surmounted by a small trident. Around the boundary walls, with entrances from the quadrangle, are stone cells, which bear evidence of having been used as dwellings. Within these cells, and in the courtyard, there are many fissures in the ground, and upon applying a lighted match to some of these, I had no difficulty in igniting the gas, which burned with a flame of so little luminosity that in the bright daylight it was difficult to trace its outline. In some places channels of stone had been constructed to convey the gas. At the time of my visit there were neither officiating priests nor devotees, but a bell that had been used to summon the faithful, still remained suspended. Returning to the historical aspect of the subject, as to which I have only a very few more words to say, I may mention that Tcheleken island was referred to by Hanway, under the name of Naphtonia or Cherriken, as being peopled, when he visited it in 1743, by about 36 families, who had 24 large boats, owned several wells of naphtha, and carried on a brisk trade in selling the oil to Persians, Turcomans, etc. Colonel Yule states that the quantity of naphtha collected from the springs about Baku was, in 1819, estimated at 241,000 poods (more than one million gallons), the greater part of which went to Persia. Crude petroleum is still burned as a source of light in some parts of Russia, in earthenware lamps of primitive form, one of which I have on the table.

Very little is known concerning the geology of the Southern Russian oil-fields; and experience gained in petroleum-well boring in the United States, does not appear to be of much value in the Apsheron peninsula. Instead of the oil being found, as it is in America, in well-defined strata of sand-rock, it appears to exist in the Caucasus for the most part in separate cells or pockets. Professor Abich says the rocks can be assigned to the miocene formation, of which there are traces of two layers, above these being tertiary clays and sands; but the whole region has evidently been affected by volcanic action, and the strata have been broken up. Not far from Baku is an extinct mud volcano, from the sides of which petroleum exudes. The limestone found in the neighbourhood, much of which is so soft that it is squared for building with axes, is full of shells of molluscs; and this particular stone and petroleum seem to go together in the Caucasus. I remember, when visiting the Bradford oil-field in the United States, having a somewhat similar stone (a specimen of which is on the table) pointed out to me as a surface indication of petroleum. I do not propose to say much on the present occasion in reference to the various theories put forward as to the origin of petroleum. It is well known that while some are disposed to consider petroleum as a product of the destructive distillation of coal, many more look upon it as derived from the remains of marine animals. In addition, we have Professor Mendelejeff's ingenious theory that petroleum results from the action of water at a high temperature, and under great pressure, upon compounds of iron, manganese and carbon; and I would refer those who are interested in the question to this distinguished chemist's essay, entitled—'L'Origine du Petrole,' published in the *Revue Scientifique*, of November 3, 1877.

The peninsula of Apsheron projects in an easterly direction from the mainland of Trans-Caucasia, and upon the southern shore of the bay thus formed lies the town of Baku. The northern side of this peninsula is tolerably fertile; but the neighbourhood of Baku presents the appearance of a desert of sand and rock, here and there stained black by the exudation of petroleum, and with occasional deposits of salt glittering in the sun. The only vegetation which thrives in this district is the prickly camel-thorn; and it may be imagined that a more dreary prospect is nowhere presented to the eye. The oriental-looking Tartar villages, with solid flat-roofed stone buildings, surmounted by the characteristic domed chimneys, form inconspicuous features in the landscape, hardly distinguishable at a distance from the surrounding desert; and at intervals along the so-called roads water wells of primitive form are met with. Strings of camels, bearing skins of petroleum to Daghestan villages; tall sheepskin-hatted Tartars riding on small donkeys, and gaudily-painted Persian 'arbas,' or two-wheeled carts, to which I must again refer, add to the thoroughly Eastern character of the scene.

The contrast between this silent desert and the bustle and confusion of the oil wells is most marked. Within the area of the present producing territory derricks are crowded together, a network of pipes like so many gigantic serpents covers the oil-sodden ground, miniature mountains of sand thrown up by the spouting wells abound, and all the remaining space is occupied by lakes of crude oil, as well as storage tanks for its reception before it is pumped to the refineries.

I had seen wells in America that were considered remarkable in regard to the quantity of oil yielded, and had read accounts of the productiveness of the Russian wells, but I was, I must confess, wholly unprepared for the evidence of abundant supply afforded by the flowing wells, or fountains, as they are locally termed, which have been struck in the Balakhani district. These wells are invariably capped to prevent waste and damage to neighbouring wells, as I shall presently explain. One of the fountains was opened in my presence, and immediately with a roar a mighty column of oil more than a foot in diameter shot up to a height exceeding 100 feet, and continued to flow without diminution of violence as long as the valve remained open, forming a lake of oil to leeward of the well. The derrick over the well was 70 feet in height, and the oil column occasionally rose to at least double this height, while stones were distinctly seen to be projected to an even greater altitude. An American who was with me said he could compare the spectacle with nothing but the falls of Niagara, and there was certainly the same awe-inspiring effect in watching nature disgorging her treasure that one experiences when standing on the brink of the stupendous Falls. The photograph exhibited (see Fig. 1) represents a Caucasian oil fountain. The well I saw spouting (Nobel's No. 18) yields at the rate of 1,125,000 gallons per 24 hours whenever opened, and is by no means the most productive that has been struck, the Droojba well and Nobel's No. 9 well having for a time yielded about double that quantity per 24 hours. In the case of these two wells the pressure of the oil amounted to about 200lb. on the square inch, and the oil column was from 200 to 300 feet in height. In the instance of Nobel's No. 9 well no great proportion of the oil was lost before the well was capped, but the Droojba fountain was for no less than four months quite uncontrollable, and it is estimated that from 68,000,000 to 136,000,000 gallons were ejected during this period. This quantity of American crude oil was worth in the United States at the time from £700,000 to £1,400,000 sterling, but in consequence of the impossibility of storing it, the oil spouted by the Droojba well was practically unsaleable. Not only was the greater part of the oil lost, but great damage was done to surrounding well property, engine houses being buried by the sand thrown up. It may afford a means of realising the enormous productiveness of these wells, if I state that the well known as Armstrong No. 2 struck last autumn in Butler county, Pennsylvania, flowed for a time at the rate of 260,000 gallons per 24 hours, and is considered one of the most prolific wells ever known in the oil regions of the United States. The yield of this well, it will be noted, is only about one-eighth of that of the Droojba well or of Nobel's No. 9 well.

FIG. 1.
By courtesy of Bristol Central Library

The first step in the operation of boring a well is the erection of a derrick. I have on the lecture table a model made for me in the United States which illustrates the method of construction, derricks of the same form being used in Russia. It will be seen that the structure consists of a timber framework, pyramidal in form and strengthened with diagonal braces. The sides of the derrick are frequently, I may say usually, enclosed with boarding to a considerable distance from the ground, to keep out wind and rain. The height of the structure is commonly about 70 feet, and it is about 20 feet square at the base. Immediately outside the derrick stands what is termed in America the Sampson-post, a massive pillar of wood some 18 inches square, which supports the 'walking beam.' The walking beam is connected to the crank of the band-wheel by what is called a pitman, and a rocking movement is thus imparted to it by the steam engine in the adjoining shed. There are two methods of boring in use in the Caucasus, termed respectively the rod system and the rope system, the latter being the more expeditious, and admitting of from 7 feet to 8 feet being bored per working day of 10 hours. In the rod system of boring the bit is screwed into an iron rod which may be as much as 40 feet in length, and a vertical movement is imparted to the apparatus by the walking beam already alluded to. As the hole increases in depth, additional lengths of iron rod are added, in the same way that the chimney sweeper manipulates his machine. At frequent intervals, however, it becomes necessary to remove the detritus from the bore, and when the well has been sunk to a considerable depth (say 200 to 300 feet), the time occupied in unscrewing each rod as the drill is drawn up, and reconnecting the rods after removing the detritus, is considerable. In the rope system, the set of drilling tools, consisting of temper screw, rope socket, auger-stem, the jars, and bit, are suspended by a stout cable of manilla. The aggregate length of this set of tools is over 50 feet, and the weight about a ton. The bits used in America are about 5½ inches in diameter, and are faced with about 40 pounds of steel, but in Baku the wells are of greater diameter and the bits correspondingly larger. As the boring of the well progresses it is cased with iron tubing, which is lowered in successive lengths. Tubing as much as 16½ inches in diameter is occasionally used in Russia, but the usual size with which the casing of the well is commenced is 15 inches diameter, the tubing being added in successive lengths of 7 feet. As the depth of the well increases and the larger tubing can no longer be driven down, smaller tubing is used, and the lower lengths are sometimes less than 8 inches in diameter. I have already stated that from time to time the detritus has to be removed. For this purpose the drilling tools are drawn up through the medium of the strong windlass (technically known as the bull-wheel) fixed to the side of the derrick, and actuated by the same steam engine that moves the walking beam, and then a few bucketsful of water having been poured into the well, the 'sand pump' is lowered by means of a small windlass. This sand pump consists of a tube or cylinder of metal some 6 feet or more in length, having at the lower end a valve opening inwards. The detritus having entered the tube through the opening of the valve, the tube is withdrawn with its contents, and this operation is repeated as many times as are necessary. The cost of sinking a well is about £2000, more or less, according to circumstances, the irregular character of the strata rendering the operation more difficult and costly than it is in the United States. As soon as oil is struck, the great anxiety is to affix a cap to the upper end of the well tubing before the flow becomes unmanageable. A difficulty arises from the fact that the oil contains a great deal of sand, the violent flow of this sand-laden oil quickly cutting through a considerable thickness of iron. I saw in Balakhani several caps which had thus been rendered useless. The caps used are provided with a valve so that the flow may be controlled and regulated. In many cases it is found necessary to artificially strengthen the ground surrounding the well, in order to prevent the oil from making its way to the surface outside the tube when the flow is checked, and in the case of some wells the closing of the valve is attended with the risk of the cap being blown off by the immense pressure of the oil. It has more than once happened that on the oil being struck the outflow has taken place with such violence that the drilling tools, weighing as I have said a ton, have been projected from the bore. The employment of nitro-glycerine for increasing the production of a well, so common in the United States, has not, I believe, hitherto been resorted to in the Caucasus.

It is stated that there are over 400 wells in the neighbourhood of Baku (as compared with more than 20,000 in the oil-fields of the United States), but of these I was informed that only about 100 were at the time of my visit producing. Rather more than half the producing wells are owned by the firm of Nobel Brothers. Of the 100 wells only some 20 are flowing wells, the remainder requiring to be pumped; but from what I have stated as to the productiveness of the fountains, it will be readily believed that if the oil were allowed to flow, these 20 wells would for a time yield far more than sufficient raw material for the manufacture of the 1,800,000 gallons of petroleum burning oil which the world is now daily consuming.

The oil containing too much sand to admit of ordinary pumping, it is lifted to the surface by a contrivance similar to that which I have described as being employed for removing the detritus during the sinking of the well. The valved tube thus used holds about 45 gallons, and it is stated that from 18,000 to 22,000 gallons can thus be raised from each well per working day of 10 hours.

The average depth of the wells last year was 450 feet, having been in 1873, 154 feet; in 1880, 273 feet; and in 1882, 350 feet. It is estimated that the average level of the oil is lowered to the extent of 56 feet for every 500 million gallons extracted.

The No. 18 well (Nobel Brothers), which I saw spouting, has a depth of 721 feet. The boring was at first stopped at 500 feet, but the flow having ceased, the well was deepened. This well was struck three years ago. The depth of the famous Droojba well is 574 feet. I may add for comparison that 1500 feet or more is now by no means an uncommon depth for petroleum wells in the United States.

Soon after the annexation of Baku by Russia in 1801, the attention of the Russian Government was turned to the natural wealth of the district, and the production of petroleum was at first farmed out as a monopoly to Meerzoeff. This continued till 1872, and from then till 1877 there was an excise duty. During the period extending from 1821 to 1872 the yearly revenue from the petroleum monopoly was never far short of, and often exceeded, 100,000 roubles, and in the latter year it had reached the sum of 136,000 roubles. The output increased from 3565 tons in 1840 to 24,800 tons in 1872, but in two previous years—viz., 1869 and 1870—the output exceeded 27,000 tons. In 1872 there were only two drilled wells, in addition to 415 pit wells, and the price of the crude oil was 45 copecks per pood (about £3, 10s. per ton). The following tabular statement shows the growth of the industry from the time when the monopoly was abolished:—

PRODUCTION AND PRICE OF CRUDE PETROLEUM

	Tons	Price per Ton s. d.
1873	64,000	7 9
1874	78,000	6 3
1875	94,000	15 6
1876	194,000	7 9
1877	242,000	12 6
1878	320,000	8 8
1879	370,000	6 3
1880	420,000	3 8
1881	490,000	2 6
1882	680,000	2 6
1883	800,000	2/6 to 0 3½
1884	1,130,000	

The production of crude petroleum in the United States for 1884 has amounted to about 3 million tons, and the price has ranged from about 16/- to 32/- per ton.

The specific gravity of the crude oil is usually between ·870 and ·875; some wells have, however, yielded oil as heavy as ·885, and others oil as light as ·850. The specific gravity of the oil first obtained from Nobel's No. 18 well was ·860, but upon the well being deepened the specific gravity fell to ·854. The specific gravity of the crude oil now found in Pennsylvania ranges from about ·800 to about ·820. I have placed on the table samples of Russian and United States crude oil for comparison.

The crude oil is in the first instance conducted from the well through wooden channels to ponds or lakes, where it deposits the sand held in suspension, and it is then pumped into storage tanks prior to being piped to the refineries.

Formerly—that is, up to 1875—the transport of the oil to the refineries took place exclusively in the Persian carts or 'arbas' I have already mentioned. These strange-looking vehicles are entirely constructed of wood, the body of the cart simply resting on the wooden axle, and the latter revolving with the wheels, which are often 8½ to 9 feet in diameter. The felloes of the wheels, being unprotected by tyres, are made of great depth to allow for the rapid wear to which they are subjected in traversing the rough roads, and the wheel becomes in use of smaller and smaller diameter until its dissolution, which generally seems to occur unexpectedly at the most inconvenient time. In these carts two casks were conveyed, one in the body of the vehicle, and the other slung beneath the axle. Thousands of these carts were in use prior to the era of pipe-lines: it is stated that as much as £100,000 per annum was paid for thus transporting the crude oil.

The first pipe-line from the wells to the refineries was laid by Messrs. Nobel Brothers, in the face of the most determined opposition on the part of the proprietors of the 'arbas,' who recognised that their business was threatened with extinction; and it is evidence of the lawlessness prevailing in the Caucasus that Messrs. Nobel were compelled to erect a line of watch-houses, wherein they stationed armed men to guard their pipe. Pipe-lines have some time since been accepted by the native carters as a necessary evil, but there is still much jealousy among the well-owners and refiners, and as the Tartars have no scruples about taking life, it is not safe for the visitor to traverse the district around Baku after dark without an armed escort, while even in daylight it is well to carry a revolver.

At the present time there are seven pipe-lines for the conveyance of crude oil, the two largest, which are of the respective diameters of 6in. and 5in., being the property of Messrs. Nobel Brothers. A certain amount of crude oil is also transported to the refineries by rail. More than 700 millions of gallons per annum could thus be delivered at the refineries, the Nobel Company's two pipes alone having a carrying capacity of 176 million gallons per year. Several telephone wires connect the wells with the refineries.

B. Redwood, 'The Russian Petroleum Industry', *J. Soc. Chem. Ind.*, 1885, **4**, 70.

13 CHALLENGE OF NEW MATERIALS

13a 'Parkesine'

On The Properties of Parkesine, and its Application to the Arts and Manufactures. By Alex. Parkes, Esq., *of Birmingham*

In introducing to this meeting the subject of Parkesine, the author wishes to explain the reasons that led him to the production of this substance.

For more than twenty years the author entertained the idea that a new material might be introduced into the arts and manufactures, and in fact was much required; he succeeded in producing a substance partaking in a large degree of the properties of ivory, tortoise-shell, horn, hard wood, india rubber, gutta percha, &c., and which will, he believes, to a considerable extent, replace such materials, being capable of being worked with the same facility as metals and wood. This material was first introduced, under the name of parkesine (so called after its inventor), in the Exhibition of 1862, in its rough state, and manufactured into a variety of articles in general use; it then excited the greatest attention, and received a prize medal, Class IV., 1112.

Parkesine is made from pyroxyline and oil, alone or in combination with other substances; the various degrees of hardness or flexibility are obtained in the easiest and most expeditious manner by varying the proportions of pyroxyline, oil, and other ingredients.

The pyroxyline used as the base in the manufacture can be made from any vegetable fibre, or fibre-producing grasses, starch, &c., but preferably of waste from cotton and flax mills, old rags, paper makers' half-stuff, or any fibrous waste material capable of being reduced into a soluble condition by the action of acids. To subdue the inflammable nature of this compound, the inventor has introduced several substances such as iodide of cadmium, tungstate of soda, gelatine, chloride of zinc, several carbonates, sulphates, phosphates, and other substances.

The oils employed are some of the vegetable and some of the animal kingdom; they may be used alone or combined, either in their normal condition, or changed by a solidifying agent, chloride of sulphur being preferred, which has the remarkable property of completely solidifying the oils almost instantaneously; but the chemical combination can be modified according to the per-centages of the chloride of sulphur employed, which may be varied to meet the exigencies of commerce. These solidified oils, although unchanged by ordinary re-agents, are readily soluble in the author's solvents of pyroxyline, by which means the two ingredients are combined to form one of the descriptions of parkesine.

The inventor, after much research, labour, and investigation, observed that the solid residue left on the evaporation of the solvent of photographic collodion produced a hard, horny, elastic, and waterproof substance. This led him to employ, in all his experiments, pyroxyline, xylodin, or some collateral matter, as his base for future operations. By the word pyroxyline the author wishes to be understood a less explosive preparation than the more highly converted compound 'gun cotton,' and his constant aim has been to apply to peaceful industrial purposes a material hitherto only used for military, blasting, and photographic purposes. The solutions of collodion known at the time of his first patent, in October, 1855, were practically unsuited to carry out the manufacture in solid masses and other large forms. This necessitated a new series of experiments, to discover a more economical mode of production, and he found that, by improving the manufacture of pyroxyline, and using different solvents, considerable success was attained; as an illustration to show the impracticability of using collodion for the manufacture of solid articles, we have here a bottle of ordinary collodion, which is submitted to your notice. This is a solution of pyroxyline in the well-known solvents, ether and alcohol, and when you are told that, in one pint of these mixed solvents there is only one-third of an ounce of solid material, and the whole of this pint of solvents must be evaporated to obtain this small quantity of hard substance, you will readily conceive that the cost of production would be much too high for large commercial purposes. The cost of one pound of the mixed solvents would be 2s. 6d., the pyroxyline being 10s. per lb., and as one pound of pyroxyline will require 48 pounds of solvent to make photographic collodion, there will be a total cost of 130s. per lb. for the solid material which could be obtained from the evaporation of such solvents. This small specimen of pyroxyline now exhibited is the exact weight of that contained in the bottle, whereas, by employing the same quantity of the author's patented solvents, this large bulk of unconverted cotton, which, for the sake of safety, represents a similar bulk of pyroxyline, can be dissolved sufficiently for the manufacture of parkesine.

The author commenced his investigations in face of the above-named difficulties, and endeavoured, by more economical methods of manufacturing the pyroxyline or other similar compounds, and by the use of improved and less costly solvents, to produce a new and cheap material; this, after many years of labour and thousands of experimental trials, he has succeeded in doing, and at the present time almost any quantity may be made per day (many tons—simply depending upon the apparatus), at a cost of less than 1s. per lb., and upwards, according to the quality required.

The two specimens shown illustrate the great difference between the two productions. The small piece is made from the collodion of commerce, at a cost of 130s. per lb.; and the sheet is the result of that made from the present mode of manufacture, at a cost of less than 1s. per lb. This will at once establish in your minds the practical value of the invention.

Having then satisfied himself as to the possibility of producing the material at such a price as would allow of its application in the arts, his next step was, by the combination of various substances, to counteract the inflammability of the material—to produce various colours—and to modify its hardness, toughness, and elasticity; and although from the above statements the object sought to be obtained may appear exceedingly simple, it nevertheless necessitated an enormous amount of application to arrive at the knowledge which has enabled him to produce the specimens which are now laid before you. . . .

The applications of this material to manufacturers appear almost unlimited, for it will be available for spinners' rolls and bosses, for pressing rolls in dyeing and printing works, embossing rolls, knife handles, combs, brush backs, shoe soles, floor-cloth, whips, walking sticks, umbrella and parasol handles, buttons, brooches, buckles, pierced and inlaid work, book-binding, tubes, chemical taps and pipes, photographic baths, battery cells, philosophical instruments, waterproof fabrics, sheets, and other articles for surgical purposes, and for works of art in general.

There is one application of the parkesine which, as far as experiments have gone, promises to be of great importance, viz., insulating telegraph wires. It will be at once evident, from the nature of the ingredients used, that by simple mechanical and chemical processes, perfect freedom from impurities or foreign ingredients can be attained; a most important property in a material which it is intended to employ for electrical and insulating purposes. The difficulty of producing a pure and homogeneous article, has, there is reason to believe, resulted in the total failure of some thousands of miles of submarine cables and underground wires. Parkesine is placed upon the wire by being forced through a die in successive coatings with the same facility as gutta-percha, and the author believes it to be far less liable to faults than india-rubber.

A few specimens of the application of the material for electrical and telegraphic purposes are exhibited. It is, however, deemed advisable to state, that extensive experiments have been made under the direction of Mr. Owen Rowland (Electrician to the late Joint Committee of the Board of Trade and the Atlantic Telegraph Company, appointed to inquire into the construction of submarine telegraph cables, &c.), with a view of ascertaining to the fullest extent the electrical properties and applicability of the material for the above important purposes. The results of those experiments leave no reason to doubt that, on the completion of the necessary machinery, a most excellent and efficient insulator will be produced (and indeed it has already been produced, even by imperfect and inadequate machinery), possessing all the requirements of insulation.

Short specimens of insulated wire (made by hand) in this substance for underground and aerial lines are exhibited, the latter in the form of a multiple cable, according to the valuable patent of Professor Wheatstone, containing an insulated sustaining iron wire, and 79 insulated conducting copper wires, both insulation and protecting envelope being effected by parkesine, which, possessing great strength and flexibility, and being a non-oxidising material, is extremely well adapted for the latter purpose. This cable is believed to be capable of bearing its own weight in air for a distance of upwards of one mile.

In its hard and solid form, by virtue of its high insulating and non-oxidising properties, this material is peculiarly well adapted for electrical instruments, terminal boards, testing boxes, batteries, insulators for poles and, many other philosophical purposes, and the advantages to be derived from the employment of a material which remains free from oxidation under all conditions, will be duly appreciated by electricians and experimentalists in their daily operations and investigations. Its tensile strength is considerably above that of either gutta percha, india rubber, or any other insulating material. Joints can be made with the greatest facility and perfection.

A. Parkes, 'On the Properties of Parkesine, and its Application to the Arts and Manufactures', *Journal of the Society of Arts*, 1865, **14**, 81.

13b 'Bakelite'

. . . A careful study of the condensation process of phenols and formaldehyde, made me discover that this reaction instead of occurring in two stages can be carried out in three distinct phases. This fact is much more important than it appears at first sight. Indeed it has allowed me to prepare a so-called *intermediate condensation product*, the properties of which simplify still further my methods of molding and enlarge very much the scope of useful applications of my process.

The three phases of reaction can be described as follows:

First phase. The formation of a so-called *initial condensation product* which I designate as *A*.

Second phase. The formation of a so-called *intermediate condensation product*, which I designate as *B*.

Third phase. The formation of a *final condensation product*, which I designate as *C*.

As to the properties of each of these condensation products I can define them in a few words:

A, at ordinary temperatures, may be liquid, or viscous, or pasty, or solid. Is soluble in alcohol, acetone, phenol, glycerine and similar solvents; is soluble in NaOH. *Solid A* is very brittle and melts if heated. All varieties of *A* heated long enough under suitable conditions will change first into *B* then finally into *C*.

B is solid at all temperatures. Brittle but slightly harder than solid *A* at ordinary temperatures: insoluble in all solvents but may swell in acetone, phenol or terpineol without entering into complete solution. If heated, does not melt but softens decidedly and becomes elastic and somewhat rubber-like, but on cooling becomes again hard and brittle. Further heating under suitable conditions changes it into *C*. Although *B* is infusible it can be molded under pressure in a hot mold to a homogeneous, coherent mass, and the latter can be further changed into *C* by the proper application of heat.

C is infusible, insoluble in all solvents; unattacked by acetone, indifferent to ordinary acids, or alkaline solutions; is destroyed by boiling concentrated sulphuric acid, but stands boiling with diluted sulphuric acid; does not soften to any serious extent if heated, stands temperatures of 300°C.; at much higher temperatures begins to be destroyed and chars without entering into fusion. It is a bad conductor of heat and electricity.

The preparation of these condensation products *A* and *B* and their ultimate transformation in *C* for technical purposes constitute the so-called Bakelite process. This can be described easily:

I take about equal amounts of phenol and formaldehyde and I add a small amount of an alkaline condensing agent to it. If necessary I heat. The mixture separates in two layers, a supernatant aqueous solution and a lower liquid which is the initial condensation product. I obtain thus at will, either a thin liquid called *Thin A* or a more viscous mass, *Viscous A* or a *Pasty A*, or even if the reaction be carried far enough, a *Solid A*.

Either one of these four substances are my starting materials and I will show you now how they can be used for my purposes.

If I pour some of this *A* into a receptacle and simply heat it above 100°C., without any precaution, I obtain a porous spongy mass of *C*. But bearing in mind what I said previously about dissociation, I learned to avoid this, simply by opposing an external pressure so as to counteract the tension of dissociation. With this purpose in view, I carry out my heating under suitably raised pressure, and the result is totally different.

This may be accomplished in several ways but is done ordinarily in an apparatus called a Bakelizer. Such an apparatus consists mainly of an interior chamber in which air can be pumped so as to bring its pressure to 50 or better 100 lbs. per square inch. This chamber can be heated externally or internally by means of a steam jacket or steam coils to temperatures as high as 160°C. or considerably higher, so that the heated object during the process of Bakelizing may remain steadily under suitable pressure which will avoid porosity or blistering of the mass.

For instance if I pour liquid *A* into a test tube and if I heat in a Bakelizer at say 160–180°C., the liquid will change rapidly into a solid mass of *C* that will take exactly the shape of its container; under special conditions it may affect the form of a transparent hard stick of Bakelite. It is perfectly insoluble, infusible, and unaffected by almost all chemicals, an excellent insulator for heat and electricity and has a specific gravity of about 1.25.

It is very hard, cannot be scratched with the finger nail; in this respect it is far superior to shellac and even to hard rubber. It misses one great quality of hard rubber and celluloid, it is not so elastic nor flexible. Lack of flexibility is the most serious drawback of Bakelite. As an insulator, and for any purposes where it has to resist heat, friction, dampness, steam or chemicals it is far superior to hard rubber, casein, celluloid, shellac and in fact all plastics. In price also it can splendidly compete with all these.

Instead of pouring liquid *A* into a glass tube or mold I may simply dip an object into it or coat it by means of a brush. If I take a piece of wood, and afterwards put it into a Bakelizer for an hour or so, I am able to provide it rapidly with a hard brilliant coat of Bakelite, superior to any varnish and even better than the most expensive Japanese lacquer. A piece of wood thus treated can be boiled in water for hours without impairing its gloss in the slightest way. I can dip it in alcohol or other solvents, or in chemical solutions and yet not mar the beautiful brilliant finish of its surface. But I can do better, I may prepare an *A*, much more liquid than this one, and which has great penetrating power, and I may soak cheap, porous soft wood in it, until the fibres have absorbed as much liquid as possible, then transfer the impregnated wood to the Bakelizer and let the synthesis take place in and around the fibres of the wood. The result is a very hard wood, as hard as mahogany or ebony of which the tensile- and more specially the crushing strength, has been considerably increased and which can stand dilute acids or water or steam; henceforth it is proof against dry rot. I might go further and spend a full evening on this subject alone and tell you how we are now bringing about some unexpected possibilities in the manufacture of furniture and the wood-working industry in general. But I intend to devote a special evening to this subject and show you then how with cheap soft wood we are able to accomplish results which never have been obtained even with the most expensive hard wood.

In the same way I have succeeded in impregnating cheap ordinary cardboard or pulp board and changing it into a hard resisting polished material that can be carved, turned and brought into many shapes. I might take up much more of your time by simply enumerating to you the applications of this impregnation method, with wood, paper, pulp, asbestos, and other fibrous and cellular materials; how it can be applied for fastening the bristles of shaving brushes, paint brushes, tooth brushes, how it can be used to coat metallic surfaces with a hard resisting protecting material; how it may ultimately supplant tin in canning processes; but I have no doubt that your imagination will easily supply you a list of possible technical uses even if I defer this subject for some other occasion.

As to Bakelite itself, you will readily understand that it makes a substance far superior to amber for pipe stems and similar articles. It is not so flexible as celluloid, but it is more durable, stands heat, does not smell, does not catch fire and at the same time is less expensive.

It makes excellent billiard balls of which the elasticity is very close to that of ivory, in short it can be used for similar purposes like knobs, buttons, knife handles, for which plastics are generally used. But its use for such fancy articles has not much appealed to my efforts as long as there are so many more important applications for engineering purposes.

Bakelite also acts as an excellent binder for all inert filling materials. This means, that it can be compounded with sawdust, wood pulp, asbestos, coloring materials, in fact with almost anything the use of which is warranted for special purposes. I cannot better illustrate this than by telling you that here you have before you a grindstone made of Bakelite and on the other hand a self-lubricating bearing which has been run dry for nine hours at 1800 rev. per minute without objectionable heating and without injuring the quickly revolving shaft.

If I mix Bakelite with fine sand or slate dust I can make a paste of it which can be applied like a dough to the inside of metallic pipes or containers, or pumps, and after Bakelizing, this gives an acid proof lining very useful in chemical engineering.

Valve seats, which are unaffected by steam, steam-packing that resists steam and chemicals, have been produced in a similar way.

Phonograph records have been made with it, and the fact that Bakelite is harder than rubber, shellac, or kindred substances indicates advantageous possibilities in that direction.

For the electrical industry, Bakelite has already begun to do some useful work. There too its possible applications are numerous. Armatures or fields of dynamos and motors, instead of being varnished with ordinary resinous varnishes, can simply be impregnated with *A*, then put into a Bakelizer and everything transformed into a solid infusible insulating mass; ultimately this may enable us to increase the overload in motors and dynamos by eliminating the possibility of the melting or softening of such insulating varnishes as have been used until now. But the subject of dynamos and motor construction is only at its very modest beginnings and I prefer to mention to you what has been already achieved in the line of molded insulators of which you will find here several interesting samples.

This brings me to the subject of molding Bakelite.

For all plastics like rubber, celluloid, resins, etc., the molding problem is a very important one. Several substances which otherwise might be very valuable are useless now because they cannot economically be molded. The great success of celluloid has mainly been due to the fact that it can easily be molded. Nitrated cellulose alone, is far superior in chemical qualities to celluloid, but until Hyatts' discovery, it could only be given a shape by an evaporation process and its applications were very limited. The addition of camphor and a small amount of solvent to cellulose nitrate was a master-stroke, because it allowed quick and economic molding.

In the same way white sand or silica would be an ideal substance for a good many purposes, could it be easily compressed or molded into shape and into a homogeneous mass. But it *cannot*; and therefore remains worthless. And that is the main difference between a *plastic* and a *non-plastic*. It so happens that Bakelite in *C* condition does not mold; it does not weld together under pressure even if heated; only with much effort is it possible to shape some kind of an object out of it, but someway or another the particles do not stick well together; in other terms it is not a true plastic. Therefore the molding problem has to be solved in the anterior stages of the process. We have seen how Smith, Luft, and Story tried to solve a similar problem by the admixture of solvents and subsequent evaporation, but we know now that these very solvents imply most serious drawbacks.

I have already shown you how I am able to mold and harden quickly by pouring liquid *A* into a mold and heating it in a Bakelizer. But even that method is much too slow for most purposes. Furthermore, molds cost money; any rubber or celluloid manufacturer will tell you that the item of molds represents a big portion of the cost of his plant. If an order for 10,000 pieces has to be delivered and it takes an hour for molding, it will require between three and four years to fill this order with one mold and if the mold costs $100 it will require $5000 for molds alone if the order has to be finished within 20 days. For that very reason I have devised my molding methods so as to use the molds only during the very minimum of time. I have succeeded in doing so in several ways. One of the simplest ways is the following:

As stated before, the use of bases permits me to make a variety of *A* that is solid although still fusible. The latter is as brittle as ordinary rosin and can be pulverized and mixed with suitable filling materials. A mixture of the kind is introduced in a mold and put in the hydraulic press, the mold being heated at temperatures preferably about or above 160–200°C. The *A* melts and mixes with the filler, impregnating everything; at the same time it is rapidly transformed into *B*. But I have told you that *B* does not melt, so the molded object can be expelled out of the mold after a very short time and the mold can again be refilled. All the molded articles are now in *B* condition; relatively brittle but infusible. At the end of the day's work or at any other convenient time all the molded articles are put in the Bakelizer and this of course without the use of any molds; in this way they are finally transformed in '*C*' Bakelite of maximum strength and hardness and resisting power.

The process can still be further simplified. Instead of using *A*, we can use *B* and mold it in the hot press where it welds and shapes itself. After a very short time, the *B* begins to transform into *C* and can now be expelled from the mold. If the transformation in *C* is not complete, a short after-treatment in the Bakelizer will finish everything. I have succeeded thus in reducing the molding to less than two minutes for small objects.

The valuable properties of *B* may be used in many other ways; for instance *A* may be poured into a large container and be heated slowly at 70°C. until it sets to a rubber-like mass and shows that it is transformed into *B*. This block of *B* if warm has very much the consistency of printers' roller-composition, but is brittle when cold. The warm flexible mass can now be removed from its container or, divided, cut, or sawed to any desired shape and the so-shaped articles can be simply placed in a Bakelizer; no melting nor deformation can occur, so we need no mold while maximum heat is applied to bring everything in condition *C*.

I could multiply these examples by numerous other modifications of my process but I believe that what I have said will be enough to convince you of its many uses; we are studying now applications of Bakelite in more than forty different industries on some of which I shall report on some future occasion. . . .

L. H. Baekeland, 'The Synthesis, Constitution, and Uses of Bakelite', *Journal of Industrial and Engineering Chemistry*, March 1909, 155.

13c Nitrocellulose

From my earliest experiments in nitrocellulose, incited by accidentally finding a dried bit of collodion the size and thickness of my thumb nail, and by my very earnest efforts to find a substitute for ivory billiard balls, it was apparent that a semi-liquid solution of nitrocellulose three-fourths of the bulk of which was a volatile liquid and that the final solid was less than one-fourth the mass of the original mixture, was far from being adapted to the manufacture of solid articles, and that I must initially produce a solid solution by mechanical means. The only useful solvent known to me at that time was a mixture of alcohol and sulphuric ether; with the old formula (about equal parts of ether and alcohol) I mixed in a closed mill a thick paste of soluble cotton, made by the old firm of Charles Cooper and Co. After mixing, the cover of the mill was removed and the mass evaporated down to a thick dough, which was then forced accurately around the ball (made of another substance), and allowed to dry. This resulted in a rather brittle coating, owing, as I found, to the unequal evaporation, leaving only the alcohol, a poor solvent by itself, at the final shrinkage. This was remedied by using only the least amount of alcohol (five parts ether to one alcohol) necessary as a solvent. Even this coating shrunk to less than half of its original thickness and required to be dried immersed in water under several hundred pounds' pressure per square inch to insure its solidity and freedom from bubbles. It became necessary to strain the mass by forcing it through a very fine sieve to exclude the unnitrated fibres. All these difficulties stood in the way of success, except in high-priced articles like billiard balls. Other seriously objectionable features became apparent. In order to secure strength and beauty only colouring pigments were added, and in the least quantity: consequently a lighted cigar applied would at once result in a serious flame, and occasionally the violent contact of the balls would produce a mild explosion like a percussion guncap. We had a letter from a billiard saloon proprietor in Colorado, mentioning this fact and saying he did not care so much about it, but that instantly every man in the room pulled a gun.

I next had a small beating engine made, to pulp the nitrocellulose. In one experiment I placed a small quantity of this dried pulp in a strong cylinder and forced quickly a minimum quantity of liquid solvent into it with high pressure. The result was a solid substance, as hard as sole-leather, which soon dried, leaving a fine hard disc about three-fourths of an inch in diameter and one-half an inch thick. While no explosion ever resulted from this, there was real danger to be feared, and the matter was dropped.

My brother, I. S. Hyatt, joined with me at that time, and finding it stated in some patents to which I was referred, that a little camphor added to the liquid solvent was beneficial, we conceived the idea that it might be possible to mix solvents mechanically with the pulp and colouring matter while wet, then absorbing the moisture by blotting papers under pressure, and finally submitting the mass to heat and pressure. To our surprise, we noticed a slightly solvent action of the precipitated and washed camphor upon the pulp, even before the heat and pressure, and without other solvent we succeeded in producing a transparent slab one-fourth of an inch thick, fine, and as hard as a piece of wood. My brother took some samples to the American Hard Rubber Company, with the view of interesting them. They employed the late Professor Charles A. Seeley, who had made collodion for the Government during the Civil War, to investigate the matter. He came to our place in Albany, N.Y., and we conducted the whole process for his inspection, very successfully. He remarked that he had come prepared to detect some chicanery, but could see no deception, and expressed himself as satisfied. He kindly advised us that if, accidentally or otherwise, we were to apply a little too high temperature, the quantity we were dealing with would inevitably destroy us with the building and adjacent property. While we did not accept this as true, it was disturbing. The following day between 12 and 1, when all were out, I rigged up a four-inch plank used as a vice-bench, braced it between the floor and ceiling, between the hydraulic press and the hand pump, intending it to shield me from possible harm. I then prepared the mould, heating it to about 500 degrees Fahrenheit, knowing it would certainly ignite the nitrocellulose and camphor, and would abide by the result. The gases hissed sharply out through the joints of the mould, filling the room with the pungent smoke. The mould, press, building and contents were there, including myself, very glad that I did not know as much as the Professor.

The next step was to make a small stuffing machine, consisting of a cylinder, four-inch bore, about a foot long, terminating in a tube three-fourths of an inch bore, ten inches long, immersed in an open oil-jacket, with a gas-burner and thermometer. A cap-nut with one and one-fourth inch screw forced a plunger upon the cakes of incipient celluloid, which were heated at the outlet end of the cylinder passing through the heated tube into moulds, and also through nozzles forming rods, tubes, etc. We found it advisable to lower the melting point of the camphor by adding a small quantity of alcohol or other solvent of camphor. A much larger hydraulic stuffing machine was then made. Dental plate blanks were the principal part of our early business.

My brother interested some New York capitalists—principally composed of General Marshall Lefferts, Joseph Larocque and Tracey R. Edson—and we removed to Newark, N.J., in the winter of 1872–73. We had nearly all the appliances and machinery to design and build, which progressed with some halts and mistakes for two or three years, when a fire occurred in our four-storey and basement building, utterly destroying all our stock and machinery, and pushing out the whole front of the building (which was very weak), and severely injuring several of our men, who all finally recovered, although one—Mr. Charles S. Lockwood—was in great danger for a long time. We then started our present works, acquiring knowledge and experience, my brother starting a separate factory for making brushes, mirrors and combs. Early in 1878 my brother went to Paris and started the French Celluloid Company.

Before we left Albany we began the use of tissue paper in place of carded cotton or bleached flax, which finally resulted in getting tissue paper of definite thickness and materials, and breaking it up into small separate pieces by a machine I constructed, drying these separate pieces in a loose mass and nitrating them instead of the carded cotton.

In 1876 I was allowed to employ a chemist for determining our acids and to systematize our nitration, instead of merely using hydrometers and thermometers, and Mr. Frank Vanderpoel was selected by me. So far as I know the analyses of mixed spent acids were not applied to practical work. Professor Seeley proposed that we separate the nitric and sulphuric acids by distillation, and then bring these to the proper strength before mixing; this was not adopted. Vanderpoel first removed the gelatinous flock by adding finely powdered natural barium sulphate and allowing the barium and flock to subside, which was a great success. I had previously tried filtration, which was impractical. After much labour Vanderpoel perfected a quick and accurate method of determination of the spent acids, and then added the required fresh acids. An elaborate and extensive battery of tanks was constructed, the acids being moved through copper pipes, where required, by means of air pressure. The tanks for mixing the comminuted tissue paper with the prepared acids were arranged on a turntable supporting a dozen or more tanks, each holding hundreds of pounds. A rotating stirrer was arranged to stir in the requisite amount of paper in one tank, the turntable being revolved to the next tank, and so on. At a distance embracing about half a dozen tanks from the stirrer was located a centrifugal machine, which separated out most of the acids. The nitrated paper was then thrown into a washing device, and the partly washed paper then fell into large tanks of fresh water and was thoroughly washed. A very considerable number of tons of paper per day were and are treated in this way.

The next procedure was to grind this nitrated paper in a beating engine.

The water was then mainly removed by a centrifugal machine. The mixture of pigments, camphor and nitrocellulose was then made, according to the results required, and ground together in a mill, then pressed into slabs or cakes, and still further deprived of moisture by great pressure, and finally deprived entirely of moisture by blotting papers, repeatedly changed. The cakes were then broken up and alcohol or other desired solvents added and allowed to soak in. The partly dissolved mass was next masticated in heated rolls, turned out into necessarily rough sheets, cut into strips and rolled up into cylinders, ready for the stuffing machine.

The great want was to furnish celluloid in solid perfect sheets, and this neither the rolls nor the stuffing machines could possibly do, with all our experience. I constructed a special nozzle for the stuffing machine, to form a slab about one and one-half inches thick, seven inches wide and several feet long, which was pressed between steam plates, one of them with undercut grooves, to hold the slab. A machine was made with shaving knife and reciprocating bed, and by it the slab was shaved into sheets of the desired thickness. These slabs were necessarily imperfect, and were too small. After showing this result to the other directors, I proposed building a hydraulic pressure steam-heated mould, to make a slab or block about two feet by five, and three inches thick, in which to weld and press a pile of rough sheets from the masticating rolls. To those who knew the least about celluloid my suggestion looked pretty good, but to others came a great fear of possible danger and failure to weld up such a mass. At last they gave me authority to build the apparatus, limiting me to less than half its cost if fairly estimated. With this authority, which consisted more of the privilege of trying such an apparatus than in providing for its cost, I ordered the machine, agreeing to pay the machinist all above the appropriation if it failed to work. The machine was built and most carefully tried in the presence of only the new president, who knew little about celluloid, and Mr. Stevens and Mr. Lockwood among the employees. The mould was cooled and opened, and nice sheets were at once made, considering that only scrap material was used, though less than half the block was welded. This meant re-heating, which was at once done, and the whole was known to be a success.

I dwell upon this because of its vast importance to us. It multiplied the value of celluloid many times, making a much stronger and finer material than was produced by the former means.

About this time we were precluded from bleaching the iron out of the pulp, caused by the iron-beating knives, and it was necessary to remedy the serious difficulty. The mixture of nitrated paper, camphor, and pigments was placed in a large revolving drum of copper, through the axis of which was a shaft with beating arms rotating at a very high speed, and the whole was beaten to a pulp without the discoloration of the iron and water, and without the loss in washing and bleaching the fine pulp from the beating engine.

In order clearly to differentiate my work in celluloid from that of others in the nitrocellulose industry, I will summarize as briefly as I can. First, the idea of combining with the nitrocellulose only the exact or approximate amount of solvent required for a solid solution. This required a nearly perfect mechanical mixture before very much solvent action could take place. Second, completing the solution by means of heat and pressure. Third, eliminating the practically unnitrated fibres by pulping. Fourth, employing tissue paper in place of carded fibre. Fifth, avoiding the terrible danger of drying out the moisture by exposure in drying room. Many years ago I patented the process of first pressing out all the water possible and displacing the remainder by forcing alcohol through under pressure. This, I understand, is now used by other nitrocellulose workers, but we prefer the method herein described. Sixth, the stuffing machine process. Seventh, the sheeting process, most essential.

'Presentation of the Perkin Medal to Mr. John Wesley Hyatt', *J. Soc. Chem. Ind.*, 1914, **33**, 226.

13d Polythene

The Story of Polythene. M. W. Perrin, *Wellcome Foundation Ltd, London*

AMONG the many important developments that have occurred in chemical technology during this century few are more striking than the production and use, on a large scale, of synthetic plastics. The high molecular weight polymer of ethylene, which may be regarded as the simplest member of this series of synthetic plastics, was discovered relatively late and was, in fact, produced in the course of research initiated with no such aim and carried out by those who had, at the time, no specialized knowledge of the plastics field or appreciation of the rather obvious gap existing in it.

World production of polyethylene, or polythene as it is called for convenience in the United Kingdom, has increased very rapidly in recent years, and is now approaching the 100,000 tons a year mark. While the bulk of this production is in the United States, the United Kingdom is also an important producer.

The story of its initial discovery, however, provides an unusually clear-cut instance of the unexpected results that may come from research and of the importance of the role of chance in such work. At the present time, when so much thought is being given to the proper organization of research in industry, it may be of value to record the full story of this particular discovery which, as time passes, naturally tends to become idealized and presented in a way which suggests a steady and logical growth from the start of the general research programme to the discovery, recognition and development of the particular product.

This article, therefore, relates the facts in historical sequence and, necessarily, involves an expression of personal opinion and reference to individuals by name. The facts quoted are, to the best of the author's knowledge, correct; but the opinions may not all be accepted by those whose names are quoted and it has also been impossible to mention all those who played an important part.

EARLY BACKGROUND

The Research Department of Brunner Mond and Co., which in 1926 became the Alkali Division of Imperial Chemical Industries Limited, had acquired a considerable reputation for the quality and variety of its work by the end of the 1914–18 war. A strong tradition was developed at Winnington of the value of technique in research and of the mutual benefit to be derived from research in the universities. The building up of this tradition is largely due to Dr F. A. FREETH who, in this formative period, was in charge of research at Winnington. In the early twenties there was close contact with Professor KAMMERLINGH ONNES's laboratory at Leiden and J. C. SWALLOW and R. O. GIBSON, who had both done post-graduate research there, joined the Winnington laboratory. The latter, while at Leiden, met Professor A. MICHELS of Amsterdam University at a time when he was just starting to develop his technique for the very accurate measurement of the effects of pressure on the physical properties of matter.

Based on a talk given to the Research Directors' Conference organized by the Federation of British Industries.

I.C.I. gave Michels financial support and Gibson was sent to work in his laboratory. After a visit to Amsterdam by J. C. SWALLOW, W. R. D. MANNING and E. W. COLBECK of the Winnington staff in 1928, certain specialized equipment to Michels's design was ordered for the laboratory. This apparatus was delivered at the beginning of 1931 when Gibson returned to Winnington and the programme of high pressure work there can properly be regarded as having started.

In its initial phases, however, the problems studied were more of a physical than chemical character. The results were of no particular practical value but the work did provide valuable experience in handling organic liquids under pressure and in the design of apparatus for the study of chemical reactions.

HIGH PRESSURE CHEMISTRY

At the end of 1929 the author joined I.C.I. and was sent to work with Michels at Amsterdam. By this time his high pressure school was expanding, both in numbers of research workers and in experimental facilities, and it was possible to turn to the measurement of the effect of pressure on properties other than the more simple ones such as compressibility, viscosity and electrical resistance. One of these was the specific heat at constant volume. Changes in this quantity with pressure would measure changes in internal energy in the molecules and such changes, it was recognized, might profoundly affect chemical reactions. More thought was therefore given to the possible chemical effects of high pressure.[1] It is also very relevant to the story that this line of approach laid emphasis on study of the highest attainable pressures. These would be needed if the internal energy was to be changed significantly while the more obvious effects, produced by changes in concentration, in equilibrium constant and in viscosity or degree of ionization, would play their part in the lower pressure range which was being studied in a growing number of laboratories in Europe and the United States.

As this interest in the theoretical possibilities of high pressure chemistry grew, close contact was maintained with the Winnington laboratory and many discussions took place with J. C. Swallow and other members of the Alkali Division. In 1930 and 1931 the country was facing an economic depression and, naturally, specially anxious thought was given to the kind and the amount of effort that should be devoted to research. In the summer of 1930 the Research Manager at Winnington, H. E. COCKSEDGE, and other senior members of the Alkali Division staff put forward, to the Main Board of I.C.I., considered views for a long-term research programme in relation to their own laboratory and the other research interests of I.C.I. It was evidence of a particularly courageous and imaginative outlook that, under the prevailing economic conditions, thought should be given to the selection of those long-term research projects which might only hold promise of reward at some future unspecified date.

One of the selected projects was the effects of high pressures on chemical reactions and, with J. C. Swallow, the author submitted a report to the local Board in January 1932 making a case for this work. The proposal was accepted immediately.

The following sentence from the introduction of the report may be quoted:

> In this report an attempt is made to cover the chemical possibilities up to 20,000 atmospheres, and arguments are made for working in this field rather than doing further work in the lower range already covered by so many investigators engaged in academic and industrial research. By this it is not meant that the lower range is not a fruitful one for a laboratory of I.C.I. interested in organic syntheses, but that the authors of this report are concerned only in making out a case for the higher range. . . .

The report summarized published work on the predominantly chemical effects of high pressure and set out the theoretical reasons for expecting that pressure would influence a chemical reaction. Such experimental results as had been published were examined in the light of these theories and particular emphasis was placed on the work of P. W. BRIDGMAN at Harvard. While Michels had concentrated on extreme accuracy in his work at Amsterdam, Bridgman had developed a technique for exploratory work in a much higher pressure range and J. B. CONANT, working with his apparatus, had studied some organic reactions in the liquid phase up to 20,000 atmospheres. This work showed that polymerization reactions were markedly affected, and Conant reported the formation of unknown polymers of butyraldehyde when this material was subjected to a pressure of 12,000 atmospheres for 40 hours. This result, together with Bridgman's observation that an allotropic form of phosphorus, stable when the pressure is released and which cannot be prepared by other means, is produced at a pressure of 12,000 atmospheres and a temperature of 200°C, inspired some confidence in the hope that there really were novel effects to be discovered. Extracts from the conclusions of the report can, again, usefully be quoted:

A 9 l. capacity vessel in furnace used for pilot scale experiments on polythene production (By courtesy of Imperial Chemical Industries Ltd)

> We have attempted, therefore, to answer the question whether results of chemical interest are likely to be obtained by working at pressures which, at the moment, can only be controlled on an experimental scale. The answer is that owing to the absence of experimental and theoretical data, we can only say—probably. . . . We feel, however, that although this survey does not lead to any great expectations at the moment, yet an attempt to control a variable such as pressure over the whole range as it is known today, will be worth while provided that no artificial restrictions are made on the scope of the work. We feel also that such work, carried out along a considered line and not empirically, may be of real scientific interest in providing experimental data which may control or lead to new advances in the theory of the mechanism of chemical reaction. There is no doubt that a clearer understanding of pressure effects in chemical reactions will be of value, firstly in preventing useless work being carried out and secondly in that it will inevitably couple up with other work in the Winnington Research Department.

The recommendations were that work, 'as detached as possible in its initial stages', should be carried out for a year, when the results should be reviewed. It was also recommended that 'if work of an essentially organic character be embarked upon which is considered to be of value to B.D.C. (now I.C.I. Dyestuffs Division) in any connection, then an organic chemist from that company should collaborate with those engaged on the work in the Winnington laboratories'.

This recommendation reads somewhat strangely out of character with what has been said about the main line of argument in the report. Its inclusion has, however, a rational explanation and its acceptance turned out, by chance, to have a major bearing on the discovery of polythene.

While intensive discussions had been taking place at Winnington on the choice of long-term research projects, Michels had also aroused the interest of the Dyestuffs Division of I.C.I. in the possible effects of high pressure on chemical reactions. The Division's Research Committee included several university professors among whom Sir ROBERT ROBINSON was particularly keen to try a number of organic reactions under pressure. He put forward a detailed scheme of typical reactions. These were not selected because they might be expected, on theoretical grounds, to be markedly affected by pressure but rather it was hoped that compounds, not normally reactive, might become so or that, at high pressures, reactions which normally need vigorous catalysts might occur without them.

This independent approach, based on a different philosophy, led to the decision that the chemical work at Winnington should be started on the lines of the Dyestuffs Division's scheme and a member of the research staff, E. W. Fawcett, was transferred to work with Gibson while the author continued with Michels in Amsterdam.

The chemical programme at Winnington was started in February 1932. The earlier work there and the acquisition of equipment from Holland, together with Gibson's experience in the technique, made a rapid start possible. To begin with, liquid phase reactions from the Dyestuffs Division's list were examined as this could be done more easily than experiments with gases. A range of pressure up to 3000 atmospheres was employed and the temperature of the small reaction vessels could be raised to 180°C. Some fifty reactions were tried under these conditions during 1932 and 1933 but the results were all disappointing. In some cases a positive pressure effect was discovered but this was never strikingly large or of any practical significance. Meanwhile, W. R. D. Manning, in charge of the engineering aspects of the programme, and E. W. Colbeck, the metallurgist, had developed the technique in many respects, and more elaborate equipment was made available for the study of reactions involving gases. It was obvious that, with these, much larger pressure effects due to change in concentration could be expected than with the relatively incompressible liquids. The Dyestuffs Division's list of reactions included several with ethylene and carbon monoxide, both of which would, it was hoped, react with other organic compounds to give useful products. The first gas-phase reactions studied, however, were those between carbon dioxide and phenol and its derivatives. These were unaffected by pressures of 2000 atmospheres at 150°C and the experiments were switched to a mixture of ethylene and carbon monoxide. The two were found to react and, at 1400 atmospheres, gave a yield of about 30 per cent of a solid polymer of acrolein. This experiment was followed, in March 1933, by one in which an attempt was made to react ethylene with benzaldehyde.

DISCOVERY OF POLYTHENE

Fawcett and Gibson, when this experiment was done, had the reaction vessel in an oil thermostat at 170°C, nearly the maximum available with this equipment. The ethylene pressure was raised to 1400 atmospheres. There was no indication from change in pressure, that reaction had occurred and when the vessel was dismantled the benzaldehyde was recovered unchanged. The walls of the vessel were, however, found to be coated with a thin layer of a 'white, waxy solid', to quote from Gibson's notebook record of 27 March. This material was analysed and found to contain no oxygen which confirmed the observation that the benzaldehyde, present in the vessel, had taken no part in the reaction. The solid was recognized as a hydrocarbon and, apparently, a polymer of ethylene. A similar result was obtained when ethylene alone was subjected to this pressure but the amount of product formed was always extremely small. When the pressure was increased a very different result was obtained. The ethylene decomposed with great violence to give hydrogen and methane, while the reaction vessel was left filled with amorphous carbon. This decomposition reaction, moreover, produced a very large and rapid rise in pressure and on several occasions led to bursts in the joints, tubing or gauges attached to the apparatus. Such accidents might have caused serious damage in the laboratory as it was then laid out and it is not surprising that the Research Manager at Winnington decided that no more work with ethylene should be done until more suitable equipment was ready and had been installed in a special laboratory with adequate safety arrangements. On 7 April it was reported to the Dyestuffs Group Research Committee that 'the work on the reaction between ethylene and benzaldehyde at 2000 atmospheres has been abandoned. The first experiment gave a wax-like substance, probably polymerized ethylene, but a repetition resulted in an explosion which smashed the gauges'.

A critical point in the story of polythene had been reached. The polymer had been discovered and, in a limited sense, recognized in March 1933. This discovery was a chance result of an attempt to use high pressure to bring about another reaction which, if successful, might have been of technical interest. It would be very satisfactory to have to record that this discovery was followed by an interval for the introduction of improved technical facilities in order that its full implications could be examined and that the ethylene polymerization reaction and the alternative decomposition could be studied in detail. Unfortunately this ideal sequence of events did not take place and, if there are any lessons to be learnt from the story, it is very desirable to have a full record of what did happen.

The problems associated with technique certainly were an important consideration. Work with liquid phase reactions in the range up to 3000 atmospheres did not involve any very difficult or novel problems in metallurgy or engineering design and, owing to the small compressibility of liquids, the pressure was relatively easily produced and the equipment could be used in an ordinary open laboratory. During 1932 and 1933 efforts were made to develop this technique along two lines. The first was an increase in the liquid phase pressure range to 12,000 atmospheres, and the second the design and construction of equipment to handle gases at pressures up to 3000 atmospheres and, at the same time, to increase the available temperature range by the use of electrically heated furnaces in place of oil thermostats. Important progress was made, including the introduction of a new type of joint, by Manning, which greatly simplified the design and assembly of reaction vessels and made it practicable to increase their size.

These developments had already pointed the need to work in a more specialized laboratory and plans were in hand to take over the space that had been used for early work on the hydrogenation of coal and oil which had been transferred from Winnington to Billingham. The potential dangers disclosed by the first experiments with ethylene merely added urgency to the plan and the transfer to the new laboratory was, in fact, completed by the autumn of 1933 although, of course, additional equipment was being installed in it for several years afterwards.

It was now possible to segregate individual reaction vessels in brick cubicles, while the provision of semi-automatic pressure intensifiers for both oil and gas greatly simplified the experimental procedure. Equipment was now also available in which, for example, the reactants (either gas or liquid) could be stirred under pressure, and insulated electrical leads could be taken through the wall of a pressure vessel for internal heating or temperature measurement. Various methods had also been developed for dealing with corrosive liquids or gases. Thought was also being given to the problems in the design of liquid and gas compressors for the pressure range above normal industrial practice.

REVISED HIGH PRESSURE CHEMISTRY PROGRAMME

When the author returned from Amsterdam to Winnington at the end of 1933 the high pressure chemistry programme that had been started at the beginning of 1932 seemed to be coming to a halt and, early in 1934, serious discussion took place on the advisability of closing it down. The interest of the Dyestuffs Group Research Committee in the work had fallen off. None of the reactions that had been examined seemed to be of practical interest and they were of a kind from which it was almost impossible to deduce anything about the reason for such pressure effect as had been observed. The main conclusion drawn, at the time, was that a pressure of 3000 atmospheres would accelerate the rate of those reactions, which normally took place slowly, about 5 to 10 times, and that those reactions which did not take place at atmospheric pressure (in the absence of catalysts) could not be made to do so at 3000 atmospheres. In only one of the many reactions studied was a fairly detailed attempt made to study the kinetics and the mechanism.[2]

It is significant that such further reactions as were suggested, at the end of 1933, as worth examination did not include the polymerization of ethylene or any closely related chemical compounds and were still mainly based on the general philosophy that it would be useful if they should be found to be affected.

However, during 1934, the decision was taken at Winnington to make a different kind of attack on the problem of high pressure chemistry and, with the improved facilities then available, to revert to the proposals in the 1932 report which has already been quoted. An attempt was to be made to find the real reason for any specialized pressure effects.

Fawcett was transferred to other work at about this time and with Gibson, who also left this project towards the end of the year, the author started work on a systematic study of the reaction between carbon monoxide and water to give formic acid and also on the effect of pressure, up to the extreme range of 12,000 atmospheres, on the kinetics of different types of reaction in dilute solution. This latter work, in which E. G. WILLIAMS later joined, continued through to 1937. It is only mentioned here because its results were of considerable academic interest and led to active cooperation with several of the leading schools of physical chemistry in the country at a time when they were deeply concerned with the development of modern theories of reaction mechanism.[3] This cooperation was to prove of great value when, later, an attempt had to be made to understand the complex polymerization reaction by which polythene is formed.

The work on carbon monoxide and water continued till the autumn of 1935 but, meanwhile, a start had been made with the study of another type of reaction. Fawcett, when he left the high pressure work, moved to another long-term research project in progress at Winnington, the study of molecular distillation, and became interested in its application to the 'bodying' or polymerization of linseed oil and other similar materials used in the paint industry. Pressure was known to have a marked influence on polymerization reactions and its effects on linseed oil were therefore examined. It soon looked as though results of practical value might be obtained in this field. In addition to the small laboratory-scale batch experiments, it became necessary to design and build a semi-technical scale unit to test the possibility of a continuous high pressure polymerization process. This work, in its turn, provided very valuable opportunities for testing new ideas in technique.

It so happens that, in May 1935, J. C. Swallow and the author prepared a second report for the I.C.I. (Alkali) Division Board with the title *High Pressure Research 1931 to 1935*. In the body of the report the results obtained to the end of 1933 were recorded and, among them, the early experiments with ethylene were briefly mentioned, but no special significance was attached to them.

The change in outlook, and the next important stage in the polythene story, can be dated to the end of 1935. After the reaction of carbon monoxide with water had been studied in considerable detail it seemed advisable to examine other reactions with this gas. The late J. G. PATON had recently joined the Winnington staff and he found that carbon monoxide reacted with aniline under pressure to give a high molecular weight solid compound that was, apparently, hitherto unknown. The reaction between ethylene, as opposed to carbon monoxide, and water was also considered for systematic study, together with more work on the reaction between ethylene and carbon monoxide. This, as had been noted by Fawcett and Gibson at the beginning of 1933, gave rise to a solid polymer of acrolein.

RE-DISCOVERY OF POLYTHENE

It was only at this point that the earlier experiments with ethylene were clearly recalled and it was at once obvious that further experiments should be made. By this time, however, the background was entirely different. Improvements in technique and in the lay-out of the laboratory made the work easier and also removed any danger from an explosive decomposition. But even more important was the different outlook on the problem due, largely, to the contribution made by the general knowledge that had been acquired in the interval about the nature of the pressure effect and its significance in polymerization reactions.

In December 1935, therefore, another experiment was done with ethylene alone in the reaction vessel. The pressure was applied with the semi-automatic gas intensifier and the author, personally, was not at all surprised to see the gauge pointer slowly fall back. It seemed that polymerization was taking place at a steady rate, and the pressure was raised again. This process was repeated till the charge of gas in the intensifier was exhausted; the reaction vessel was then cooled and dismantled. Again it was with no great surprise, but considerable pleasure, that the vessel was seen apparently to be filled with a white powder. The disappointment only came when it was found that the total weight of solid that had been produced was about 8 grams. Polymerization had certainly not accounted for all the pressure drop during the experiment and there must have been a leak at one of the joints in the apparatus. Once again the element of chance played an important role. It took some months of intensive work by all those in the team to elucidate the full reasons as to why, if the leak had *not* occurred, the experiment would probably have been far less spectacular than it was, and might well only have been a repetition of the earlier ones.

A mechanically driven compressor designed by Dr Michels, capable of working to pressures of 3000 atmospheres, used in the high pressure research programme (By courtesy of Imperial Chemical Industries Ltd)

As things were, however, this 're-discovery' of polythene in December 1935 gave rise to a completely different sequence of events than had the initial discovery in March 1933. The polymer was recognized as a new compound which could, perhaps, only be formed at high pressures and this alone prompted an effort to examine it in detail. Sufficient of it was available to show, in a very short time, that its properties were of considerable scientific interest. It was crystalline and, by X-ray methods, its molecular structure could be accurately elucidated. A high molecular weight polymer of ethylene would be the simplest compound of this type that could exist, consisting only of long chains of methylene ($-CH_2-$) groups. It showed the phenomenon of molecular orientation, or 'cold-drawing', that had recently been observed by W. H. CAROTHERS of du Pont's in his original work on poly-esters and nylon.

Furthermore, there was immediate recognition of the possible industrial value of the material. It could be moulded, when heated under a slight pressure, and formed into films and threads. It did not melt in boiling water. It was chemically resistant, as would be expected with a pure hydrocarbon. Also, because of its chemical composition, it would be sure to have outstandingly good properties as an electrical insulator.

Already by February 1936 the first British provisional patent specifications had been filed to cover its preparation and principal applications and by April a brand name, *Alketh* (later changed to *Alkathene*), had been registered for the material made by I.C.I. But before serious development work could be started two obstacles had to be surmounted. The conditions must be found for the polymerization reaction to take place without risk of the alternative violent decomposition of the ethylene, and equipment must be devised for the production of the polymer in reasonable quantities.

The first six months of 1936 were devoted to laboratory scale study of the reaction and this work disclosed the importance of traces of oxygen in the ethylene as a catalyst for the initiation of the polymerization chain reaction and of removal of the heat of reaction. The success of the experiment in the previous December was, in fact, due to the series of additions to the reaction vessel of fresh ethylene to replace that which had leaked out. This ethylene contained, by chance, about the right amount of oxygen to catalyse the formation of successive amounts of the polymer.

Arrangements were also made for the construction, in Holland, of a high pressure mechanically-driven gas compressor to an original design of Dr Michels which had been under discussion for some time past as a result of the work on carbon monoxide and water. The Winnington team was joined, at this stage, by E. HUNTER who had also spent some time at Amsterdam and who now became primarily responsible for the semi-technical development of polythene.

By the end of 1936 the size of the reaction vessel in use had been increased from about 80 to 750 ml and two of these were made. The importance of stirring their contents to remove the heat of the reaction was recognized. With these 750 ml vessels polythene was made in pound quantities, as compared with grams previously, and a heated valve was fitted to the bottom of one of them in order to prove the feasibility of extruding the molten polymer continuously as it was produced. Early in 1937 a still larger vessel, of 9 litres capacity, was delivered as also was the compressor from Holland.

SEMI-TECHNICAL AND PILOT PLANTS

With this additional equipment it was possible to put up a semi-technical unit in a special cubicle built on to the high pressure laboratory and this was used, between August 1937 and March 1938, to develop a practical operating process and to produce polythene on the hundredweight scale, for evaluation and development work. Even when the mechanical gas compressor and the 9 litres capacity vessel were ordered, however, in the middle of 1936, it was clear that it would be inconvenient to work on a continuous scale in the research laboratory. Plans were accordingly made for the erection of a small plant to produce ethylene from alcohol and of a building to house the high pressure equipment on the works site. Permission was obtained, in June 1937, for the execution of this programme and the work was finished by the end of March 1938 when the equipment was transferred from the laboratory and the polythene pilot plant came into existence. By the end of 1938 a ton of polythene had been made in this plant.

The subsequent history of the building of full-scale production units does not fall within the scope of this paper but the early story would not be complete without reference to the start of development work on polythene.

DEVELOPMENT

During 1936 and the beginning of 1937, as ounces and then pounds of the new material became available, visits were paid by members of the research team to all the divisions in I.C.I. who might be interested and, as far as possible, samples were supplied. The Research Department of Metropolitan-Vickers Ltd carried out a detailed examination of its electrical properties on the understanding that the results obtained would be freely available to anyone interested. In

June 1937 it was decided that the moulding uses of polythene should be developed by the Plastics Group of I.C.I., the textile uses by the Dyestuffs Group and the electrical and other unspecified uses by the Alkali Group.

There was a firm belief that the electrical, mechanical and chemical properties of polythene must lead to an important field of application for it as a dielectric in cables and efforts were made to interest individual firms in that industry as well as the Cable Makers' Association which, in November 1937, was supplied with all the information then available about this new product. Some interest was certainly shown but it was again due to chance that this field was really opened up early in 1938.

A member of the I.C.I. (Dyestuffs) staff, B. J. Habgood, who had recently had first-hand knowledge of the cable industry, then pointed out the similarity between some of the properties of polythene and those of gutta-percha. The Telegraph Construction and Maintenance Co. Ltd had many years' experience of the specialized problems in the production of submarine cables insulated with the latter material which is of natural origin and closely related to rubber. In May 1938 this firm received a sample of polythene, and with its experience and specialized equipment, was able to evaluate it rapidly. Already in September 1938 100 tons were ordered to be delivered by the middle of 1939. The outbreak of war, however, had a great effect on the subsequent history of polythene, both as regards development and production, since its application in the high-frequency equipment used in radar, for which its electrical and mechanical properties made it peculiarly suitable, provided a priority over all other uses. This, again, is another story and falls outside the scope of this paper which has only attempted to cover the birth and development of the *discovery* of polythene and does not touch the development, in the general sense, of the material itself.

CONCLUSION

In concluding this story of polythene, an attempt must be made to point out any lessons that can be learnt from it. Sometimes discoveries are made by those who deliberately set out to achieve a result that has been foreseen in general terms, and of which the potential value can be predicted. Polythene is not an example of this type, but of the more frequent chance result that may come unexpectedly from research done for other reasons. Such discoveries may result from fundamental work, designed only to increase scientific knowledge, or from an applied research programme devoted to an entirely different object. Under such conditions it would seem that a factor of outstanding significance is the *recognition* of the discovery, and this may be even more important than the discovery itself.

Since a chance discovery may, and probably will, lie outside the field of immediate interest to those who make it, rapid and effective recognition is more likely if the research man concerned has as wide a knowledge as possible of science and its application in industry. The contribution of the specialist is certainly needed and this, again, is clearly shown up in the polythene story but it also serves to emphasize the importance of collaboration in, and appreciation of, the work of others in widely different fields of academic and applied research.

If discoveries, such as that of polythene, may come from any research programme, it might be argued that there is no limit, in range or amount, to the effort that industry can usefully devote to research. This is obviously untrue and must be limited on financial grounds as well as by the availability of suitable staff. But it does lead to consideration of the kind of research programme that should be undertaken, in addition to that directly connected with the products of the company. It would seem wise, if such detached work is to be done, to base it on a broad thesis, such as was the effect of high pressure on chemical reactions, where there is reason to expect new and significant results. It cannot be claimed with certainty that the formation of polythene is entirely, or even mainly, due to the particular effects that were believed, in 1932, to be of most interest in high pressure chemistry; but the fact that they were being looked for does account for some part of the interest shown in the polymer on its re-discovery at the end of 1935.

Choice of a thesis such as this, and for these reasons, immediately raises the question of a semi-empirical approach, in which many trial experiments are made, as compared with a more systematic study designed to prove the underlying principles of the thesis. The polythene story provides an example of both methods of attack. It is, perhaps, not too serious an over-simplification to say that the first led, by chance, to the discovery in March 1933 while the second was required before its significance was properly recognized in the re-discovery in December 1935. The most difficult problem is that of blending the two methods. Either, alone, is easily subjected to criticism if interesting results do not quickly materialize or can be defended enthusiastically in the event of success and each usually appeals to a different type of research worker.

Sufficient indication has, I hope, also been given in this paper of the importance of advances in technique. Without these there would have been a much longer delay before the mechanism of the ethylene polymerization reaction could have been elucidated or, still more, a production process devised.

With regard to the initial stages of development of a discovery arising, by chance, from an unrelated research programme, the story of polythene shows the importance of the research team taking an active, and indeed leading, part in preliminary evaluation and in defining, in broad terms, the lines for the later commercial development which will be associated with the erection of production plants. Particularly when the discovery is novel, and far outside the normal interests of the research department or the firm concerned, personal enthusiasm and conviction of its importance are essential.

REFERENCES

[1] MORGAN, G. T. *Proc. roy. Soc. A* 127 (1930) 240
[2] FAWCETT, E. W. and GIBSON, R. O. *J. chem. Soc.* 53 (1934) 386
[3] GIBSON, R. O., FAWCETT, E. W. and PERRIN, M. W. *Proc. roy. Soc. A* 150 (1935) 223
WILLIAMS, E. G., PERRIN, M. W. and GIBSON, R. O. *ibid* 154 (1936) 684
PERRIN, M. W. and WILLIAMS, E. G. *ibid* 159 (1937) 162
— *Trans. Faraday Soc.* 34 (1938) 144

M. W. PERRIN, 'The Story of Polythene', *Research*, 1953, 111.

13e The Mond Nickel Process

The History of My Process of Nickel Extraction. By Ludwig Mond, F.R.S.

SINCE I promised to your chairman, shortly after my arrival, to read a paper to you before leaving this country, I have considered what subject I could bring before you that would be of sufficient interest, considering that I am unprepared to illustrate it by experiments or diagrams.

I have decided to give you an account of the history of my process of extracting nickel from its ores, as an instance of an investigation undertaken in pursuit of pure science, which has led unexpectedly, in a few years, to an important industrial application.

I have often been asked by scientific men how I came to discover nickel carbonyl. I will tell you the story.

Soon after I had satisfied myself that the ammonia-soda process was far cheaper than the Leblanc process for producing carbonate of soda pure and simple, it became evident to me that a time would come when the Leblanc process would produce carbonate of soda as a by-product of bleaching powder manufacture, which it would have to dispose of at any price it would fetch. I therefore undertook a series of experiments with the object of producing bleaching powder as a by-product of the ammonia-soda process. You all know that the usual form of this process consists in treating a solution of common salt in which caustic ammonia has been dissolved by carbonic acid, with the result that bicarbonate of soda is precipitated while a solution of ammonium chloride is formed. In the ordinary course of things this solution of chloride of ammonium is distilled with caustic lime, yielding gaseous ammonia, which returns to the process, and a solution of calcium chloride. Some investigators had endeavoured to produce chlorine from the chloride of calcium obtained by evaporating these solutions; others had proposed to substitute magnesia for the lime in distilling the solution of ammonium chloride, and to produce chlorine or hydrochloric acid from the $MgCl_2$ obtained by evaporating its solution.

I gave my attention to producing the chlorine direct from the ammonium chloride, separated from the solution, in which it is originally obtained, by refrigeration.

Ammonium chloride when vaporised is, as is well known, dissociated into ammonia and HCl; thus, if the vapour of NH_4Cl is passed over a metallic oxide this oxide is converted into a chloride, and the ammonia passes on together with the steam formed by the reaction.

I found that nearly all metallic oxides, with the exception of the alkalis and alkaline earths, and even a large number of metallic salts, were acted upon in this way by NH_4Cl vapour, and that a large number of the metallic chlorides so formed would give up their chlorine and be re-converted into oxides by submitting them to the action of air of a suitable temperature. I found that of all the substances investigated oxide of nickel yielded the best results, and that the next best were obtained by magnesia mixed with a certain amount of chloride of potassium. For obvious reasons I preferred the latter in working the process industrially, which has now been successfully at work for several years on a very considerable scale in our Winnington Alkali Works.

In developing this process I had to construct a plant for volatilising NH_4Cl, which I found an extremely difficult problem, as the vapour of this substance not only acts on oxides and salts, but also violently attacks the large majority of metals. I succeeded in lining iron vessels with glazed tiles in such a way that they would withstand this action, but I required valves for changing from the current of ammonium chloride vapour to hot air and *vice versa*, which had to be very tight to prevent a large loss of ammonia.

I found that nickel was one of the few substances suitable for the construction of these valves, and that it was not at all attacked by ammonium chloride vapour. On the laboratory scale these nickel valves worked perfectly, but when I applied them on a manufacturing scale I found them to be acted upon and very soon to become leaky. The faces became covered with a black crust, which, on examination, was found to contain carbon.

The source of this carbon seemed mysterious, the only difference between the work on the small scale and that on the large scale being that on the small scale we swept the ammonia out of the apparatus before admitting the hot air by means of pure CO_2, while on the large scale we used the gases from a lime kiln, containing a few per cent. of CO. This led us to study the action of CO on nickel.

We found this metal to have the remarkable property of splitting off carbon from CO at a moderate heat, transforming it into CO_2.

About the same time I was engaged with my collaborator, Dr. Langer, in trying to find a method for eliminating the CO from hydrogenous gases, which we wanted for use in our gas battery. We tried to avail ourselves for this purpose of the remarkable property of nickel we had just discovered, and found to our satisfaction that by passing gases containing hydrogen, CO and a certain quantity of steam over finely-divided nickel at a temperature of 400° C. we could completely convert the CO into CO_2, obtaining its equivalent of hydrogen, which was just what we wanted.

This led to a more elaborate study of the action of CO upon nickel, with a view of determining whether a definite compound of nickel and carbon was formed. We found that a small quantity of nickel decomposed a very large quantity of CO, so that we could obtain a product containing only 15 per cent. of nickel and 85 per cent. of carbon, the nickel in which was only partially soluble in acids.

In the course of these experiments finely-divided nickel, formed by reducing nickel oxide at 400° C. by hydrogen, was treated with pure CO in a glass tube, at varying temperatures, for a number of days, and was then cooled down in a current of CO before it was removed from the tube. In order to keep the poisonous CO out of the atmosphere of the laboratory, we simply lit the gas escaping from the apparatus. To our surprise we found that, while the apparatus was cooling down, the flame of the escaping gas became luminous and increased in luminosity as the temperature got below 100° C. On a cold plate of porcelain put into this luminous flame, metallic spots were deposited similar to the spots of arsenic obtained with a Marsh apparatus; and on heating the tube through which the gas was escaping we obtained a metallic mirror, while the luminosity disappeared.

At the first moment we thought that there must be an unknown element in our nickel giving rise to the production of this effect, but when we examined the mirrors we found them to consist of pure nickel. As it seemed so very improbable that so heavy a metal as nickel should form a readily-volatile compound with CO, we purified our CO as perfectly as possible, but still obtained the same results.

We now endeavoured to isolate this curious and interesting substance by preparing the nickel with great care at the lowest possible temperature, and treating this nickel with CO at about 50° C., and thus we gradually increased the amount of the volatile nickel compound in the gases passing through the apparatus. We absorbed the excess of CO by cupreous chloride solution, and thus obtained a residue of several cubic centimetres, containing the volatile nickel compound mixed with a little nitrogen. By passing this gas through a heated tube we separated the nickel, obtaining an increased volume of gas, and found in this a quantity of CO corresponding to about four equivalents for one equivalent of nickel.

By further improving our method of preparing the finely-divided nickel and by passing the resulting gas through a refrigerator, cooled by snow and salt, we at last succeeded in liquefying this compound, and were able to produce it with ease and facility in any quantity we desired.

Thanks to the kindness of Dr. Torry, of Harvard University, I am able to show you a small quantity of this nickel carbonyl, which he prepared within a few days after he saw my first publication on the subject. It is a colourless liquid boiling at 43°C. and which solidifies at −25°C., forming needle-shaped crystals. It is soluble in alcohol, petroleum, and chloroform; it is not acted upon by dilute acids or alkalies, and can be readily distilled without decomposition. But on heating the gas to 150°C. it is completely dissociated into its components, pure CO being obtained and the nickel being deposited in a dense metallic film upon the sides of the vessel in which it is heated.

For a long time, while we were engaged in investigating the physical and chemical properties of this interesting substance—which was without parallel in the history of chemistry—and while we were endeavouring to obtain other similar compounds with other metals, I had myself no suspicion that this substance, which was until then only obtainable by very careful and elaborate laboratory manipulations, should ever become available for industrial purposes. But the longer we went on preparing it for our investigations, the more easy we found it to prepare it in quantity, after we once knew exactly the best conditions for so doing. After that I came to the conclusion that it ought to be possible to make use of the ease with which nickel is converted into a volatile gas by CO, while practically all other metals, and notably cobalt (which is so difficult to separate from nickel by other methods), was not acted upon by this gas, for separating nickel from cobalt and other metals on a manufacturing scale, and for obtaining it in a very pure state.

I erected a plant on a large scale near Birmingham, and after several years of hard work, during which the apparatus has had to be several times reconstructed so as to fulfil all the conditions of this rather delicate process, we have succeeded in our object, and now have for some time produced nickel at the rate of a ton and a half per week from the Canadian nickel copper matte imported into England. This matte, which contains about 40 per cent. of nickel, and an equal quantity of copper, is carefully roasted to drive out the sulphur as far as possible, and is then subjected to the action of hydrogenous gases, either water-gas or producer-gas, rich in hydrogen, in an apparatus which is called the 'reducer,' the temperature of which is under perfect control, so that 400°C. is never exceeded. From this apparatus the substance, which is now reduced to the metallic state, is taken through air tight conveyors and elevators into another apparatus, called the 'volatiliser,' in which it is subjected, at a temperature not exceeding 80°C., to the action of CO gas.

This apparatus consists of an iron cylinder, divided into numerous compartments by shelves, and provided with a stirring device, which gradually moves the material from the top to the bottom, while the CO gas passes through in an opposite direction. The CO gas, which should be as rich as practicable, we prepare by passing pure CO_2 through incandescent coke; the pure CO_2 we make by passing the flue gas of a boiler or of a fire through a solution of carbonate of potash, and subsequently boiling the solution. The CO gas, charged with nickel carbonyl, leaving the volatiliser, is passed through a series of tubes or chambers, heated to about 180°C., in which the nickel is deposited in various forms, according to the speed of the gas current, the richness of the gas, and the existing temperature. The CO gas, thus almost completely freed from the nickel, is taken back by means of a blower into the volatiliser where it takes up a fresh quantity of nickel and is constantly used over and over, so that the quantity consumed is limited to the very small amount of unavoidable loss through leakage of the plant.

The material under treatment is repeatedly taken from the volatiliser to the reducer and *vice versa*, by means of air-tight conveyors and elevators, until the amount of nickel volatilised begins to fall off. It is then roasted again to remove the sulphur, which it still contains, and is treated by sulphuric acid to dissolve part of the copper. The residue, containing nickel, some copper, and the other impurities of the matte, is again subjected to the previously described treatment until the nickel has been extracted as far as practicable; and the ultimate residue, still containing a few per cent. of nickel, is melted up into matte again.

I have here a few samples of the various forms in which the nickel is obtained by this process. If the nickel is allowed to deposit slowly at a carefully regulated temperature, it can easily be obtained from the gas as a coherent metallic film, so that it is possible to coat any substance which can stand heating to 150°C. with a perfect covering of metallic nickel, and also to make articles of metallic nickel for direct use. I have here a knife-handle made in this way, and the finest delineaments of a mould can be perfectly reproduced in this manner.

I believe that the great facility with which hollow nickel goods can be made in this way, which at present either cannot be made at all, or only by the use of very powerful hydraulic machinery, will give a great impetus to the manufacture of nickel utensils for domestic purposes, the use of which is so very desirable from a sanitary point of view. The cost of the process, if carried out on a sufficiently large scale, is, as you will readily perceive, inconsiderable, as the consumption of material is very small, and the expenditure consists mainly in carefully superintending the various operations.

What gives me the greatest satisfaction in relation to this investigation is, that I believe I have succeeded in working out a purely chemical process for extracting nickel from its ores, which will be cheaper and simpler than any electrolytic process that can be used for the same purpose. Of late years, there has been a tendency to take chemical operations out of the hands of the chemist and to thrust them into those of the electrician, in the belief that the simplest way of obtaining a chemical change consists in the pulling apart a chemical compound by electrical energy, and subsequently putting its constituents together again in the forms that are desired. I know there are many chemical operations which will always be carried out to much greater advantage by the old chemical methods, and I have no doubt that newer methods will be found, of which nobody thinks at present, based upon purely chemical reactions, such as the process which I have brought before you to-night, which will effect the chemical changes we want to produce at a smaller expense of energy than can be done by electrolysis.

L. W. Mond, 'The History of My Process of Nickel Extraction', *J. Soc. Chem. Ind.*, 1895, **14**, 945.

13f 'Special Steels'

INTRODUCTION

The author regards it as a privilege to be invited to review in these columns the general progress of special or alloy steels during the first twenty-five years of His Majesty's reign. His pleasure in accepting this invitation is further increased by the fact that he had last year the honour of being one of the deputation from the Iron and Steel Institute to present His Majesty with the Bessemer Gold Medal, which he graciously accepted. In doing so, His Majesty was following a precedent established by Queen Victoria and continued by King Edward VII, to whom in 1906 the author himself had the honour of presenting the Medal on behalf of the Iron and Steel Institute. Thus for three generations the Royal Family has demonstrated its concern and interest for the welfare and progress of the iron and steel industry, upon which the prosperity of our country and its position in the world so largely depends.

Within the limits of space here available, it is not possible to give anything approaching a complete survey of advances in the vast field of special and alloy steels during the past quarter of a century. The story is in fact one which begins much earlier and shows during this notable period of industrial activity the vigorous expansion of the era of alloy steels, which at its commencement had become well established but was then comparatively young. In 1929 the world's output of steel, then at its maximum production, was 120 million tons, of which no less than about 6 million tons represented alloy steels.

In reviewing progress in this period of twenty-five years it is necessary, therefore, to take account both of those newer alloy steels which it has seen initiated and also of the developments which have taken place in the use of those which had already found practical application.

The author's discovery and invention of manganese steel in 1882, as expressed by so many well-known metallurgists, marked the dawn of the age of high alloy steels. The extraordinary and valuable results then obtained led him to investigate in many other directions this hitherto practically unknown field, and later he was followed by numerous other investigators. To-day manganese steel, with modern improvements in manufacture, is used throughout the world in increasing quantities and for a greater variety of purposes than ever before. In addition, there have been developed corrosion- and heat-resisting steels, high-speed tool steels, special steels for armaments and ordnance, special structural steels, high tenacity steels, and steels with remarkable magnetic properties, whether as regards low energy losses in electrical machinery and apparatus, specially high permeability at low magnetisations, or special suitability for permanent magnets, as the case may be. The arts of peace have been served in no less measure than the needs of national defence, and progress has been astonishingly rapid in all the branches of metallurgical knowledge concerned.

MANGANESE STEEL

Much progress has been made, and many difficulties overcome, in the production and applications of manganese steel since the original specimens of this material were exhibited by the author at the reading of his first paper on this subject before the Institution of Civil Engineers in 1887, that is, almost exactly forty-eight years ago. This alloy, now known throughout the world as the Hadfield manganese steel, consists of a non-magnetic alloy of iron with from 11 to 14 per cent of manganese and about 1·25 per cent carbon, and remains one of the most remarkable ferrous materials yet produced, whether from the point of view of practical application or scientific interest.

Whilst the general progress of metallurgy has brought about improvement in the actual manufacture of the material, as also its heat treatment which plays such an essential part in the full development of its qualities, this important alloy, manganese steel, remains exactly the same as when originally invented in 1882.

SILICON STEEL

The history of the development of this remarkable steel, invented by the writer in the latter portion of the nineteenth century, and how the early difficulties attending its introduction and production, including the best and most suitable proportions to be used of the element silicon, also heat treatment, were surmounted, are given very fully in his book 'Metallurgy and its Influence on Modern Progress'.

Since its first employment early in the present century, silicon steel has become of the highest importance in the electrical industry, by its use in transformers, generators, motors and many other purposes. Without the use of silicon steel the efficiency of electrical generation and transformation would be seriously lower than at present and many modern designs of electrical plant would be physically impossible owing to the increased energy losses.

The hysteresis loss of this steel is initially about two thirds that of unalloyed iron, and actually decreases during a period of years in service. Its high electrical resistance, about six times that of iron, is also of great advantage in minimising losses due to eddy currents, while its maximum magnetic permeability is about 25 per cent higher than that of iron.

CORROSION-RESISTING STEELS

The remarkable developments in corrosion-resisting steels, as also in heat-resisting steels to be referred to later, have taken place entirely within the past twenty-five years. Several varieties of specialised types have been developed to meet different requirements. There are many applications in chemical and other industries for materials combining resistance to corrosion with the mechanical properties of steel, that is, apart from the use of some of these special alloy steels for domestic and ornamental applications. No single steel yet developed meets all possible requirements in this field, but many special steels have been perfected, and from these it is generally possible to choose one which is effective against any particular one or more of the many corrosive agencies encountered in industry.

The simpler and original type of corrosion-resisting steel, usually termed 'stainless' steel, established by Brearley, containing 12–14 per cent of chromium and from about 0·05 per cent up to about 0·40 per cent of carbon, also still higher percentages such as 16–18 per cent of chromium, according to the purpose for which they are intended, have valuable mechanical properties, and are specially suitable for applications in which the corrosive action is of the ordinary kind, that is, due to air and water. For chemical and other applications where a wider range of resistance to corrosion is required, the steels most generally used are of the nickel-chromium type, low in carbon and with from about 12–20 per cent of chromium and about 7–12 per cent of nickel, and in certain cases also small additions of elements such as tungsten, molybdenum and titanium, and other elements are being investigated such as columbium. These materials are now available in castings, forgings, rolled sheets or bars, tubes and wire; also by machining, pressing, welding and other methods of assembly they can be fabricated into practically any form desired.

The development of non-rusting steel covers an exceedingly wide field and those specially interested in the history describing its origin and rise are recommended to refer to Chapter ix of the author's book 'Faraday and his Metallurgical Researches' where the later development of alloy steels is dealt with.

Steels of this type are now used widely for tanks, pans and other parts of chemical or other manufacturing equipment exposed to corrosive media too numerous to be mentioned individually, including many of the most commonly encountered acids and corrosive salts. There is also large employment for these steels for decorative purposes where high polish is required. For other applications, further and more special types of corrosion-resisting steels have been developed.

One of these, for example, is useful in the sulphate house of coal and coke by-product recovery plants, owing to its resistance to dilute sulphuric acid. Also it finds many applications in sugar refineries owing to its resistance to acid calcium phosphate and other chemicals employed; and it is specially valuable in food industries, being unaffected by the dilute solutions of phosphoric acid contained in chemical foods.

In a very different field, the use of a corrosion-resisting special steel for the reinforcements of the masonry in the preservation of St. Paul's Cathedral, London, suggests that such material will be employed to an increasing extent, ensuring the permanence of our historic buildings as necessity arises. It will doubtless be a satisfaction to all concerned with the metallurgy of iron that the special steels so employed and supplied by three Sheffield firms are associated in such an important capacity with the safety of the cathedral in which the Royal Silver Jubilee Thanksgiving Service is to be held.

In the construction of new buildings also, high tensile non-corrodible steel is being used, as in the case of certain structural members in the new library of the University of Cambridge.

HEAT-RESISTING STEELS

Almost contemporary with the development of corrosion-resisting steels has been the recognition and utilisation of the merits of heat-resisting steels. There are, in general, ferrous alloys with high nickel and chromium contents, variable in amount according to the properties required in the products. In some cases, the iron content barely exceeds 50 per cent.

The term 'heat-resisting', as applied to steels, implies resistance to oxidation and other forms of corrosion at high temperatures, while retaining a high degree of mechanical strength. Materials fulfilling these conditions find many important applications, including the valves of internal combustion engines, rotors of gas turbines, mechanical parts of furnaces of all kinds, reaction vessels for coal-hydrogenation, and many other uses requiring mechanical reliability combined with freedom from oxidation and scaling. Developments in this field have been so numerous that it is impossible here to do more than indicate their variety and importance. The following examples are, therefore, to be taken merely as representative and by no means exhaustive.

Cast iron parts formerly used in mechanical stokers in the Canadian Pacific Steamship Company's vessels of the *Beaver* class, operating with air preheated to 310°F., were found to require replacement after every second round trip. The parts exposed to the severest conditions, such as tuyères, grate plates, slicing bars and extension plates, were, therefore, made of heat-resisting steel, and the grates were then found to be in perfect condition after a year's continuous service. Similar examples are provided in many other directions and, to take one at the other end of the scale, claw bars of heat-resisting steel in a certain type of domestic hot-water boiler are still in perfect condition after two years service, where new cast iron bars were formerly required every three or four months. This is only one of many instances.

The exhaust valves of internal combustion engines are, specially where high efficiency is aimed for, exposed to most difficult conditions as regards temperature, mechanical stress and liability to scaling and erosion. Here again, special heat-resisting steels come to the rescue, and exhaust valves of such material have contributed to successive world speed records on land, water and in the air. Similar conditions apply in the case of the rotors of exhaust gas turbines, in which applications equal success has been attained.

For the vessels used in modern high-pressure high-temperature chemical processes, heat-resisting qualities must be combined with resistance to chemical action, and here again the conditions are being successfully met.

The applications of these new alloy steels are, in fact, so numerous and important that they will probably have as important an influence on the progress of engineering and the economics of industry as that exerted by the earlier alloys, manganese steel and silicon steel, in their particular spheres.

SPECIAL STEELS FOR ARMOUR AND ORDNANCE

Without ceasing for a moment to regret the waste and folly of war, it is impossible to ignore the important contributions to national defence resulting from the applications of special steels in armour, projectiles and other munitions during the War. Then, and, in fact, throughout the particular period under review, the rivalry between armour and projectile continued with steady improvement in each. However, the reference made here to this subject can only be brief, though without doubt our nation must continue to supplement its work by continued efforts if national security is to be preserved.

One of the most striking applications of alloy steel in the War, and one to which it is very appropriate to make present reference owing to its humanitarian aspect, was the use of manganese steel for the helmets of our soldiers and most of our allies. Though commonly known as 'tin hats', these helmets were of the much sterner material known universally as the Hadfield manganese steel, and there can be no question that very large numbers of our brave soldiers owed their lives to the remarkable toughness and resistance to shrapnel fire of this extraordinary material with its non-magnetic qualities most useful in many cases.

SPECIAL STEELS FOR OTHER PURPOSES

At least a passing reference must also be made to high tenacity steels, large steel forgings of extraordinary weight and size, special steels for steam turbines and steam fittings, high-speed tool steels, in which great advances have been made, special magnetic steels and alloy steels for general engineering and constructional purposes. Nor should it be forgotten that there have been important advances in the quality of carbon steels, which are strictly speaking alloy steels, for, of all the elements alloyed with iron, carbon is in many respects the most powerful and remarkable in its effects.

Alloy steels possessing high breaking strengths up to 100 tons per sq. in. and above have been used by engineers for some time past, and there has been steady progress during recent years in the improvement of the quality of such high-tenacity alloy steels, specially as regards toughness and resistance to shock and vibration.

In many of the largest steam turbines on land and sea, and even in more ordinary installations where there may seem to be no special liability to erosion and corrosion, a special alloy of iron high in its nickel and chromium contents is used for the blading, with great advantage in the maintenance of high steam economy, this alloy being highly resistant to erosion and completely non-corrodible under all the conditions experienced by steam turbine blading. For example, already turbines of no less than 7 million horse power in total capacity are now bladed in whole or in part with such special steel.

One of the most interesting developments in modern high-speed tool steels, specially in view of the fact that manganese steel was formerly considered to be practically unmachinable, is the discovery of alloy steels which will drill and cut water-toughened manganese steel, thus making possible a further extension in the many applications of this exceedingly useful material.

Great progress has been made in special steels for permanent magnets and in steels of very high permeability at low inductions; also in new non-magnetic alloys of iron. Again, there have been important developments in the use of alloy steels for general constructional purposes, for example in the hulls of large vessels where reduction of weight is an important consideration. Many other instances might be mentioned, if space permitted, of improvements in alloy steels which have greatly advanced engineering and industrial practice since the year 1910.

IMPORTANCE OF RESEARCH

If there is one outstanding lesson that has been brought out by metallurgical progress in general and made possible the rise of alloy steels in particular, it is the importance of continual research. Also this research, specially where alloy steels are concerned, must be over as wide a field as possible.

As an instance it may be mentioned that the laboratories of the author's firm, as well as those of most modern steel works, have for many years used for routine testing as well as original investigations, equipment for every kind of test—chemical, mechanical and physical—with special apparatus for carrying out tensile, bending, endurance, impact, hardness, electrical conductivity, magnetic permeability, hysteresis, measurement of temperatures from the lowest to the highest, the determination of thermal change points and other tests. In a single week as many as 12,000 pyrometric observations have been made, and equipment for the microscopic examination and photomicrography of steels at all powers up to 5,000 and even 8,000 diameters is continually in use.

In any appreciation of the great and increasing importance of metallurgical research during the past twenty-five years account must be taken of the part played by our universities in the training of metallurgical students, and in many instances conducting valuable research investigations. The special duties and responsibilities of the research laboratories maintained by firms, and working in direct touch with the conditions and requirements of commercial development, can never be fully undertaken by other institutions or organisations. At the same time, metallurgists owe a great debt of gratitude to the universities. As an example, it may be mentioned that the training and research work carried out in the laboratories of the Applied Science Department of the University of Sheffield are worthy of the highest traditions of the city, and the establishment of metallurgy as an independent faculty with, quite recently, a special founding course and degrees in foundry science is but one instance of the way in which the authorities are keeping in the forefront of metallurgical progress. The importance of research facilities is equally recognised at other universities, each in its own sphere, but in regard to ferrous metallurgy the University of Sheffield naturally occupies a special position and is second to none in the world.

Outside the many and valuable researches carried out at the various universities of Great Britain reference must certainly be made to the important work being carried on continuously by the National Physical Laboratory, originally under the able guidance of Sir Richard Glazebrook, and now that of Sir Joseph Petavel, to whom and his staff, composed of men of scientific eminence, great credit is due. This great national establishment is now supported by large annual grants, aided by Parliament through the Royal Society and the Department of Scientific and Industrial Research.

THE FUTURE

Finally, as will have been gathered in the reading of this article, the continued and increasing use of alloy steels in their many forms is most certainly assured.

The great value of such products can almost be said to be beyond computation in view of the services they render, and modern civilisation could not be sustained by any other means; they may, therefore, be justly said to be beyond any adequate expression in terms of money.

It is scarcely realised how largely further advances of present modern civilisation depend upon alloy steels Take these away and our modern civilisation could certainly not exist, at any rate, on anything like its present high plane, and we should in many respects be cast back to the comparatively unsatisfactory conditions of more than a century ago.

It is entirely the production and use of alloy steels which have rendered practicable the bringing into being of engineering and other constructions which would not otherwise have been possible, whether in their employment on land or sea or in the air.

One useful alloy after another, new or improved, takes its place in the civilisation which it advances and thereafter it is indispensable unless indeed a still better material can be discovered. Progress far beyond the bounds which could be foreseen twenty-five years ago has already been achieved, and the rate of such advance appears likely to continue undiminished. Therein lie challenge, inspiration and encouragement for those of the younger generation.

R. Hadfield, 'Special Steels', *Nature*, May 1935, p. 741.

14 COMMUNICATIONS EXPLOSION

14a Prince Albert Inspecting the Galvanic Telegraph

. . . The whole party then left the interior of the college, and passed by the eastern archway to the terrace in front of Somerset House, where a very large crowd of persons had assembled. Here, too, was stationed a band of the Guards, who received the Prince with 'God save the Queen.' His Royal Highness then inspected the working of Professor Wheatstone's electro-magnetic telegraph, placed on the terrace; the wires of communication being laid from thence across Waterloo-bridge, westward, to the summit of the lofty shot-tower, in Lambeth. His Royal Highness, having walked to the extreme end of the terrace, returned, and entered his carriage at half-past one o'clock. His departure was accompanied by the same expressions of loyalty and attachment as his arrival had been. The whole ceremony was managed in the best taste, and reflects much credit upon those by whom the arrangements were made.

PRINCE ALBERT INSPECTING THE GALVANIC TELEGRAPH.

By courtesy of the London Library

Illustrated London News, 1843, **3**, 61.

14b 'The Electric Parliament'

THE ELECTRIC PARLIAMENT

It being universally admitted that people die for want of breath, the great waste of that valuable article which occurs nightly in Parliament must be admitted to be as awful as it is useless. The philanthropist, therefore, no less than the Utilitarian, will hail any invention whereby a saving in that respect may be effected. Such an invention philosophy now presents to legislation through *Punch* in the Electric Telegraph. Let conducting wires be laid down at the nation's expense, from St. Stephen's to every county, borough, or other place which returns members to Parliament, each being connected with a galvanic arrangement similar to that in use on the Great Western Railway.

Thus the senatorial and personal seats of Honourable Members will be identified; for gentlemen will be enabled, sitting at home in their easy-chairs, to express themselves in electric short-hand. It will be only necessary to have a working committee to sit in each House, in order to take down what falls, or rather, radiates, from the several senators. At this employment noble lords and honourable gentlemen can take turns; relieving each other in gangs, which may go and leave town if they like, when off duty. The advantages of this arrangement will be numerous. Debaters, in the first instance, will be the better able to think before they deliver their sentiments; whereby they will probably say much less, in fact, and somewhat more to the purpose, than they generally do now. The instances in which they commit themselves, or expose themselves to be committed, will be fewer: and it will be at the committee's discretion—a virtue for which they will be selected,—what questions to put or remarks to make; so that of course they will abstain from communicating those which are silly or impertinent. This will be another large economy of words.

Whistlings, hootings, groans, and crowings will from their very nature be incommunicable; collisions between Brougham and Campbell impossible; time, therefore, will be saved, and respectability promoted, in the House of Commons particularly. Members are to address the House in regular rotation; whereby all disputes about precedence will be avoided. Members on their legs are at present often roughly interrupted, which will not happen with them on their wires. It is true that by this plan members at a distance will be prevented from hearing each other's arguments; but since they never attend to them as it is, this will be of no consequence. The next day's *Times* will report their speeches to the country, which is all they want. A more serious objection is, that it may be difficult to teach all of them to use the electric telegraph; and perhaps electricity must be made easy for Sibthorp. But then it is not every-body who has something worth hearing to say. On the whole, this application of the silent system to Parliamentary Debates will work admirably.

Punch, 1845, **8**.

14c 'Marvels of the Electric Telegraph'

In the United States telegraphic department I saw and heard Elisha Gray's splendidly worked-out electric telephone actually sounding four messages simultaneously on the Morse code, and clearly capable of doing yet four times as many with very moderate improvements of detail; and I saw Edison's automatic telegraph delivering 1,015 words in 57 seconds; this done by the long-neglected electro-chemical method of Bain, long ago condemned in England to the helot work of recording from a relay, and then turned adrift as needlessly delicate for that. In the Canadian department I heard 'To be or not to be, . . . there's the rub,' through an electric telegraph wire; but, scorning monosyllables, the electric articulation rose to higher flights, and gave me passages taken at random from the New York newspapers:—'S.S. Cox has arrived' (I failed to make out the S.S. Cox); 'The City of New York,' 'Senator Morton,' 'The Senate has resolved to print a thousand extra copies:' 'The Americans in London have resolved to celebrate the coming 4th of July.' All this my own ears heard, spoken to me with unmistakable distinctness by the thin circular disc armature of just such another little electro-magnet as this which I hold in my hand. The words were shouted with a clear and loud voice by my colleague-judge, Prof. Watson, at the far end of the telegraph wire, holding his mouth close to a stretched membrane, such as you see before you here, carrying a little piece of soft iron, which was thus made to perform in the neighbourhood of an electro-magnet in circuit with the line motions proportional to the sonorific motions of the air. This, the greatest by far of all the marvels of the electric telegraph, is due to a young countryman of our own, Mr. Graham Bell, of Edinburgh and Montreal, and Boston, now becoming a naturalised citizen of the United States. Who can but admire the hardihood of invention which devised such very slight means to realise the mathematical conception that, if electricity is to convey all the delicacies of quality which distinguish articulate speech, the strength of its current must vary continuously and as nearly as may be in simple proportion to the velocity of a particle of air engaged in constituting the sound?

'Marvels of the Electric Telegraph', *Nature*, 1876, 14 September, p. 427.

14d 'Researches in Electric Telephony'

Mr. President and Gentlemen of the Society of Telegraph Engineers. It is to-night my pleasure, as well as duty, to give you some account of the telephonic researches in which I have been so long engaged. Many years ago my attention was directed to the mechanism of speech by my father, Alexander Melville Bell, of Edinburgh, who has made a life-long study of the subject. Many of those present may recollect the invention by my father of a means of representing, in a wonderfully accurate manner, the positions of the vocal organs in forming sounds. Together we carried on quite a number of experiments, seeking to discover the correct mechanism of English and foreign elements of speech, and I remember especially an investigation in which we were engaged concerning the musical relations of vowel sounds. When vowel sounds are whispered, each vowel seems to possess a particular pitch of its own, and by whispering certain vowels in succession a musical scale can be distinctly perceived. Our aim was to determine the natural pitch of each vowel; but unexpected difficulties made their appearance, for many of the vowels seemed to possess a double pitch—one due, probably, to the resonance of the air in the mouth, and the other to the resonance of the air contained in the cavity behind the tongue, comprehending the pharynx and larynx.

I hit upon an expedient for determining the pitch which at that time I thought to be original with myself. It consisted in vibrating a tuning-fork in front of the mouth while the positions of the vocal organs for the various vowel sounds were silently taken. It was found that each vowel position caused the reinforcement of some particular fork or forks.

I wrote an account of these researches to Mr. Alex. J. Ellis, of London, whom I have very great pleasure in seeing here to-night. In reply he informed me that the experiments related had already been performed by Helmholtz, and in a much more perfect manner than I had done. Indeed, he said that Helmholtz had not only analysed the vowel sounds into their constituent musical elements, but had actually performed the synthesis of them.

He had succeeded in producing, artificially, certain of the vowel sounds by causing tuning-forks of different pitch to vibrate simultaneously by means of an electric current. Mr. Ellis was kind enough to grant me an interview for the purpose of explaining the apparatus employed by Helmholtz in producing these extraordinary effects, and I spent the greater part of a delightful day with him in investigating the subject. At that time, however, I was too slightly acquainted with the laws of electricity fully to understand the explanations given; but the interview had the effect of arousing my interest in the subjects of sound and electricity, and I did not rest until I had obtained possession of a copy of Helmholtz' great work,* and had attempted, in a crude and imperfect manner it is true, to reproduce his results. While reflecting upon the possibilities of the production of sound by electrical means, it struck me that the principle of vibrating a tuning-fork by the intermittent attraction of an electro-magnet might be applied to the electrical production of music.

I imagined to myself a series of tuning-forks of different pitches, arranged to vibrate automatically in the manner shown by Helmholtz, each fork interrupting at every vibration a voltaic current; and the thought occurred, 'Why should not the depression of a key like that of a piano direct the interrupted current from any one of these forks, through a telegraph wire, to a series of electro-magnets operating the strings of a piano or other musical instrument, in which case a person might play the tuning-fork piano in one place and the music be audible from the electromagnetic piano in a distant city?'

The more I reflected upon this arrangement the more feasible did it seem to me; indeed, I saw no reason why the depression of a number of keys at the tuning-fork end of the circuit should not be followed by the audible production of a full chord from the piano in the distant city, each tuning-fork affecting at the receiving end that string of the piano with which it was in unison. At this time the interest which I felt in electricity led me to study the various systems of telegraphy in use in this country and in America. I was much struck with the simplicity of the Morse alphabet, and with the fact that it could be read by sound. Instead of having the dots and dashes recorded upon paper, the operators were in the habit of observing the duration of the click of the instruments, and in this way were enabled to distinguish by ear the various signals.

It struck me that in a similar manner the duration of a musical note might be made to represent the dot or dash of the telegraph code, so that a person might operate one of the keys of the tuning-fork piano referred to above, and the duration of the sound proceeding from the corresponding string of the distant piano be observed by an operator stationed there. It seemed to me that in this way a number of distinct telegraph messages might be sent simultaneously from the tuning-fork piano to the other end of the circuit, by operators each manipulating a different key of the instrument. These messages would be read by operators stationed at the distant piano, each receiving operator listening for signals of a certain definite pitch, and ignoring all others. In this way could be accomplished the simultaneous transmission of a number of telegraphic messages along a single wire, the number being limited only by the delicacy of the listener's ear. The idea of increasing the carrying power of a telegraph wire in this way took complete possession of my mind, and it was this practical end that I had in view when I commenced my researches in Electric Telephony. . . .

. . . I have before alluded to the invention by my father of a system of physiological symbols for representing the action of the vocal organs, and I had been invited

* Helmholtz. *Die Lehre von dem Tonempfindungen.* (English Translation by Alexander J. Ellis, *Theory of Tone.*)

by the Boston Board of Education to conduct a series of experiments with the system in the Boston school for the deaf and dumb. It is well known that deaf mutes are dumb merely because they are deaf, and that there is no defect in their vocal organs to incapacitate them from utterance. Hence it was thought that my father's system of pictorial symbols, popularly known as visible speech, might prove a means whereby we could teach the deaf and dumb to use their vocal organs and to speak. The great success of these experiments urged upon me the advisability of devising methods of exhibiting the vibrations of sound optically, for use in teaching the deaf and dumb. For some time I carried on experiments with the manometric capsule of Koenig, and with the phonautograph of Léon Scott. The scientific apparatus in the Institute of Technology in Boston was freely placed at my disposal for these experiments, and it happened that at that time a student of the Institute of Technology, Mr. Maurey, had invented an improvement upon the phonautograph. He had succeeded in vibrating by the voice a stylus of wood about a foot in length which was attached to the membrane of the phonautograph, and in this way he had been enabled to obtain enlarged tracings upon a plane surface of smoked glass. With this apparatus I succeeded in producing very beautiful tracings of the vibrations of the air for vowel sounds. Some of these tracings are shown in fig. 20. I was much

By courtesy of the Royal Institution

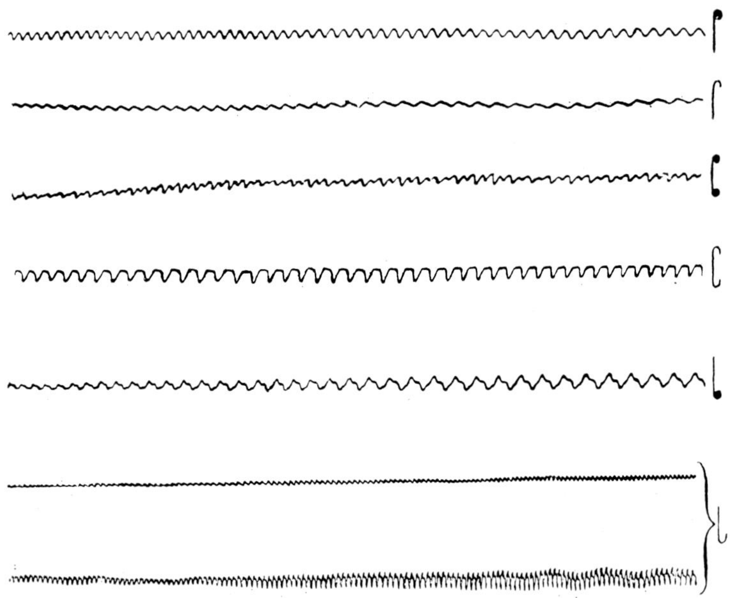

Fig. 20.

struck with this improved form of apparatus, and it occurred to me that there was a remarkable likeness between the manner in which this piece of wood was vibrated by the membrane of the phonautograph and the manner in which the *ossiculæ* of the human ear were moved by the tympanic membrane. I determined therefore to construct a phonautograph modelled still more closely upon the mechanism of the human ear, and for this purpose I sought the assistance of a distinguished aurist in Boston, Dr. Clarence J. Blake. He suggested the use of the human ear itself as a phonautograph, instead of making an artificial imitation of it. The idea was novel and struck me accordingly, and I requested my friend to prepare a specimen for me, which he did. The apparatus, as finally constructed, is shown in fig. 21. The *stapes* was removed and a stylus of hay about an inch in length was attached to the end of the incus. Upon moistening the membrana-tympani and the ossiculæ with a mixture of glycerine and water, the necessary mobility of the parts was obtained; and upon singing into the

By courtesy of the Royal Institution

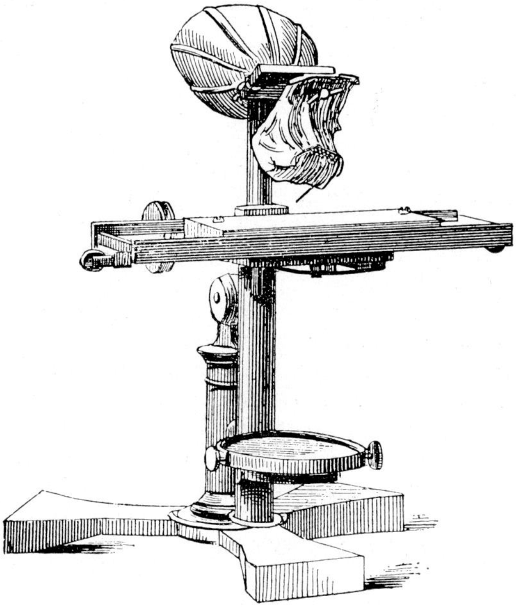

Fig. 21.

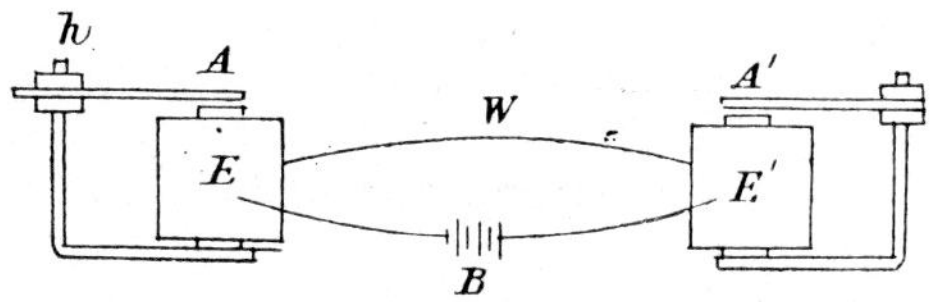

Fig. 22.

external artificial ear the stylus of hay was thrown into vibration, and tracings were obtained upon a plane surface of smoked glass passed rapidly underneath. Whilst engaged in these experiments I was struck with the remarkable disproportion in weight between the membrane and the bones that were vibrated by it. It occurred to me that if a membrane as thin as tissue paper could control the vibration of bones that were, compared to it, of immense size and weight, why should not a larger and thicker membrane be able to vibrate a piece of iron in front of an electro-magnet, in which case the complication of steel rods shown in my first form of telephone . . . could be done away with, and a simple piece of iron attached to a membrane be placed at either end of the telegraphic circuit.

Fig. 22 shows the form of apparatus that I was then employing for producing undulatory currents of electricity for the purposes of multiple telegraphy. A steel reed A was clamped firmly by one extremity to the uncovered leg *h* of an electro-magnet E, and the free end of the reed projected above the covered leg. When the reed A was vibrated in any mechanical way, the battery current was thrown into waves, and electrical undulations traversed the circuit B E W E′, throwing into vibration the corresponding reed A′ at the other end of the circuit. I immediately proceeded to put my new idea to the test of practical experiment, and for this purpose I attached the reed A (fig. 23) loosely by one extremity to the uncovered pole *h* of the magnet, and fastened the other extremity to the centre of a stretched membrane of goldbeaters' skin *n*. I presumed that upon speaking in the neighbourhood of the membrane *n* it would be thrown into vibration and cause the steel reed A to move in a similar manner, occasioning undulations in the electrical current that would correspond to the changes in the density of the air during the production of the sound; and I further thought that the change of the intensity of the current at the receiving end would cause the magnet there to attract the reed A′ in such a manner that it should copy the motion of the reed A, in which case its movements would occasion a sound from the membrane n′ similar in *timbre* to that which had occasioned the original vibration.

By courtesy of the Royal Institution

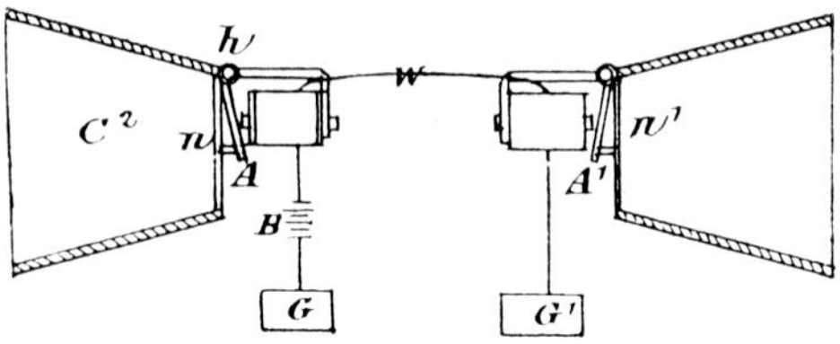

Fig. 23.

The results, however, were unsatisfactory and discouraging. My friend Mr. Thomas A. Watson, who assisted me in this first experiment, declared that he heard a faint sound proceed from the telephone at his end of the circuit, but I was unable to verify his assertion. After many experiments attended by the same only partially-successful results, I determined to reduce the size and weight of the spring as much as possible. For this purpose I glued a piece of clock spring, about the size and shape of my thumbnail, firmly to the centre of the diaphragm, and had a similar instrument at the other end (fig. 24); we were

By courtesy of the Royal Institution

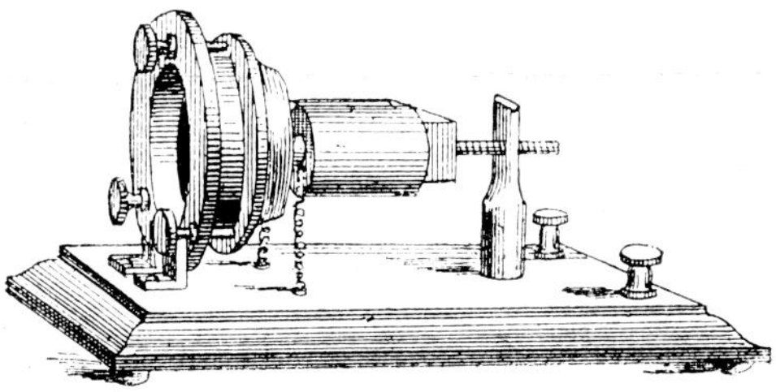

Fig. 24.

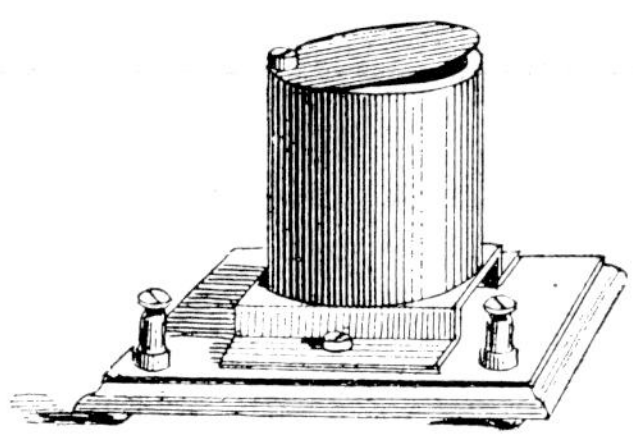

Fig. 25.

then enabled to obtain distinctly audible effects. I remember an experiment made with this telephone, which at the time gave me great satisfaction and delight. One of the telephones was placed in my lecture-room in the Boston University, and the other in the basement of the adjoining building. One of my students repaired to the distant telephone to observe the effects of articulate speech, while I uttered the sentence, 'Do you understand what I say?' into the telephone placed in the lecture-hall. To my delight an answer was returned through the instrument itself, articulate sounds proceeded from the steel spring attached to the membrane, and I heard the sentence, 'Yes, I understand you perfectly.' It is a mistake, however, to suppose that the articulation was by any means perfect, and expectancy no doubt had a great deal to do with my recognition of the sentence; still, the articulation was there, and I recognised the fact that the indistinctness was entirely due to the imperfection of the instrument. I will not trouble you by detailing the various stages through which the apparatus passed, but shall merely say that after a time I produced the form of instrument shown in fig. 25, which served very well as a receiving telephone. In this condition my invention was exhibited at the Centennial Exhibition in Philadelphia. The telephone shown in fig. 24 was used as a transmitting instrument, and that in fig. 25 as a receiver, so that vocal communication was only established in one direction.

Another form of transmitting telephone exhibited in Philadelphia intended for use with the receiving telephone (fig. 25) is represented by fig. 26.

A platinum wire attached to a stretched membrane completed a voltaic circuit by dipping into water. Upon speaking to the membrane, articulate

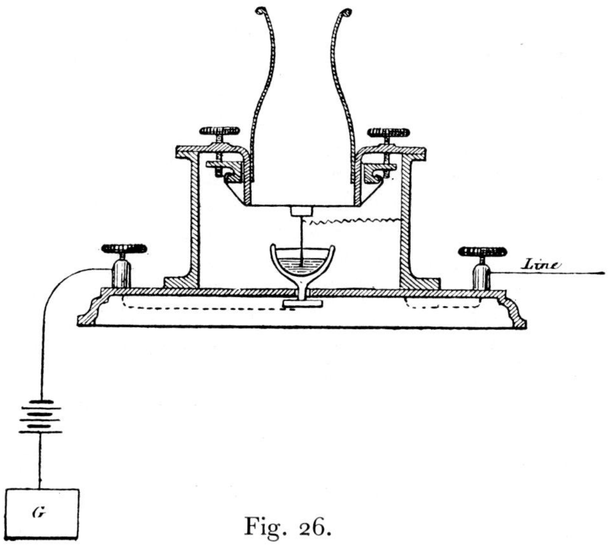

By courtesy of the Royal Institution

Fig. 26.

sounds proceeded from the telephone in the distant room. The sounds produced by the telephone became louder when dilute sulphuric acid, or a saturated solution of salt, was substituted for the water. Audible effects were also produced by the vibration of plumbago in mercury, in a solution of bichromate of potash, in salt and water, in dilute sulphuric acid, and in pure water.

The articulation produced from the instrument shown in fig. 25 was remarkably distinct, but its great defect consisted in the fact that it could not be used as a transmitting instrument, and thus two telephones were required at each station, one for transmitting and one for receiving spoken messages.

It was determined to vary the construction of the telephone shown in fig. 24, and I sought by changing the size and tension of the membrane, the diameter and thickness of the steel spring, the size and power of the magnet, and the coils of insulated wire around their poles, to discover empirically the exact effect of each element of the combination, and thus to deduce a more perfect form of apparatus. It was found that a marked increase in the loudness of the sounds resulted from shortening the length of the coils of wire, and by enlarging the iron diaphragm which was glued to the membrane. In the latter case, also, the distinctness of the articulation was improved. Finally, the membrane of gold-beaters' skin was discarded entirely, and a simple iron plate was used instead, and at once intelligible articulation was obtained. The new form of instrument is that shown in fig. 27, and, as had been long anticipated, it was proved that the only use of the battery was to magnetize the iron core of the magnet, for the effects were equally audible when the battery was omitted and a rod of magnetized steel substituted for the iron core of the magnet.

It was my original intention . . . and it was always claimed by me, that the final form of telephone would be operated by permanent magnets in place of

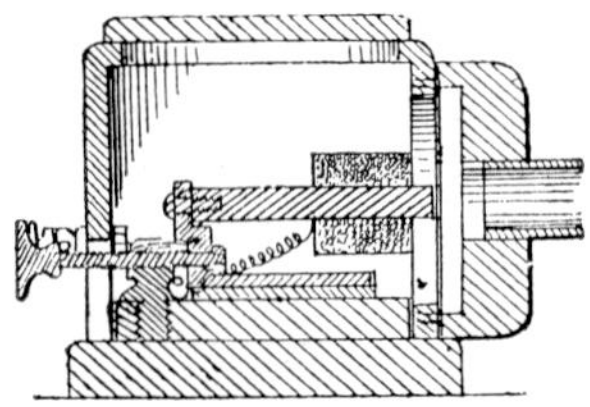

Fig. 27. *By courtesy of the Royal Institution*

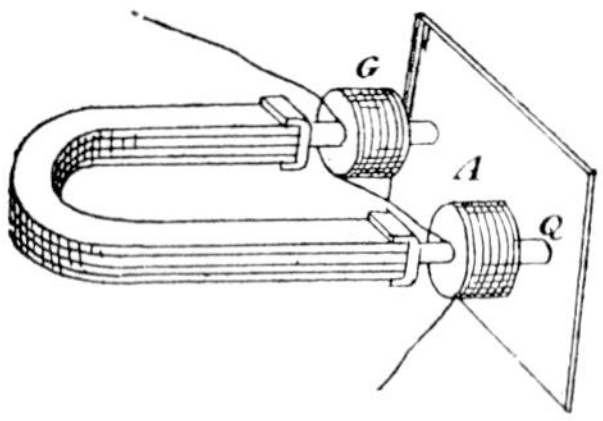

Fig. 28. *By courtesy of the Royal Institution*

batteries, and numerous experiments had been carried on by Mr. Watson and myself privately for the purpose of producing this effect.

At the time the instruments were first exhibited in public the results obtained with permanent magnets were not nearly so striking as when a voltaic battery was employed, wherefore we thought it best to exhibit only the latter form of instrument.

The interest excited by the first published accounts of the operation of the telephone led many persons to investigate the subject, and I doubt not that numbers of experimenters have independently discovered that permanent magnets might be employed instead of voltaic batteries. Indeed one gentleman, Professor Dolbear, of Tufts College, not only claims to have discovered the magneto-electric telephone, but I understand charges me with having obtained the idea from him through the medium of a mutual friend.

A still more powerful form of apparatus was constructed by using a powerful compound horse-shoe magnet in place of the straight-rod which had been previously used (see fig. 28). Indeed the sounds produced by means of this instrument were of sufficient loudness to be faintly audible to a large audience, and in this condition the instrument was exhibited in the Essex Institute, in Salem, Massachusetts, on the 12th Feb. 1877, on which occasion a short speech shouted into a similar telephone in Boston, sixteen miles away, was heard by the audience in Salem. The tones of the speaker's voice were distinctly audible to an audience of 600 people, but the articulation was only distinct at a distance of about 6 feet. On the same occasion, also, a report of the lecture was transmitted by word of mouth from Salem to Boston, and published in the papers the next morning.

From the form of telephone shown in fig. 27 to the present form of the instrument (fig. 29) is but a step. It is in fact the arrangement of fig. 27 in a portable

By courtesy of the Royal Institution

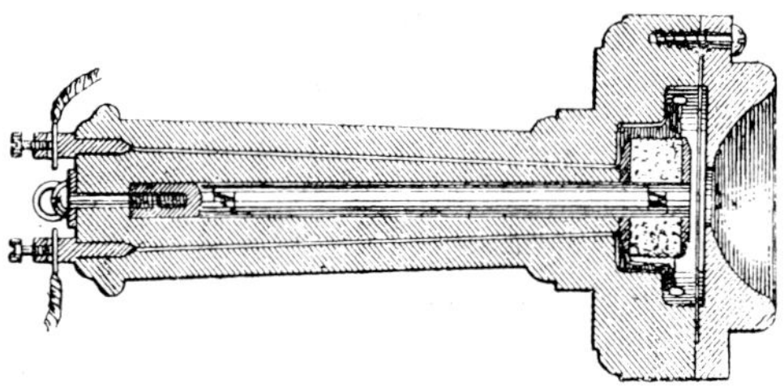

Fig. 29.

By courtesy of the Royal Institution

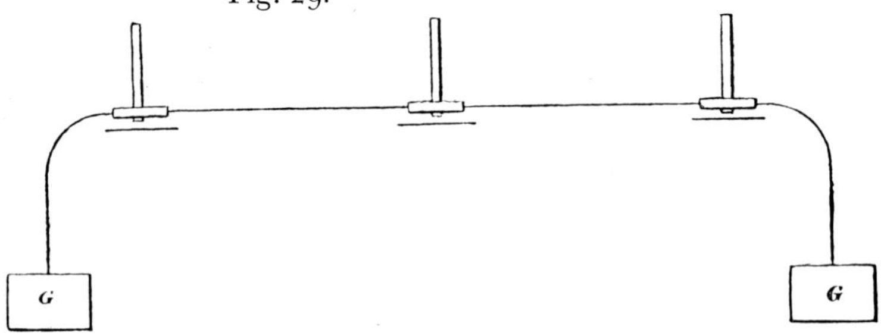

Fig. 30.

form, the magnet F H being placed inside the handle and a more convenient form of mouthpiece provided. The arrangement of these instruments upon a telegraphic circuit is shown in fig. 30.

The question will naturally arise, Through what length of wire can the telephone be used? In reply to this I may say that the maximum amount of resistance through which the undulatory current will pass, and yet retain sufficient force to produce an audible sound at the distant end, has yet to be determined; no difficulty has, however, been experienced in laboratory experiments in conversing through a resistance of 60,000 ohms, which has been the maximum at my disposal. On one occasion, not having a rheostat at hand, I may mention having passed the current through the bodies of sixteen persons, who stood hand in hand. The longest length of real telegraph line through which I have attempted to converse has been about 250 miles. On this occasion no difficulty was experienced so long as parallel lines were not in operation. Sunday was chosen as the day on which it was probable other circuits would be at rest. Conversation was carried on between myself, in New York, and Mr. Thomas A. Watson, in Boston, until the opening of business upon the other wires. When this happened the vocal sounds were very much diminished, but still audible. It seemed, indeed, like talking through a storm. Conversation though possible could be carried on with difficulty, owing to the distracting nature of the interfering currents.

I am informed by my friend Mr. Preece that conversation has been successfully carried on through a submarine cable, sixty miles in length, extending from Dartmouth to the Island of Guernsey, by means of hand telephones similar to that shown in fig. 30.

A. G. BELL, 'Researches in Electric Telephony', *Journal of the Society of Telegraph Engineers*, 1877, **6**, 385.

14e Troubles with the Telephone, 1898

Select Committee on Telephones. 16 *June* 1898. Mr Benn

CHAIRMAN

2422. What other statistics have you to show that the service is inadequate in London?—I do not propose to submit any other figures than those on inadequacy.

2423. Now as to inefficiency?—Inefficiency, we took some trouble in 1895 to send circular letters to the subscribers to the National Telephone Company: we received 2,536 replies; 1,305 of those expressed dissatisfaction, various grades and descriptions.

2424. In what does the inefficiency consist?—We carefully inquired into the nature of the complaints, and it would appear that the most usual cause of complaint was failure to get connection on the ground of 'engaged,' many subscribers stating that they had reasons for saying that in many cases this excuse for not connecting them was absolutely incorrect. Delay in getting connection is another cause of complaint, and also cutting off before conversation is finished. Indistinctness caused by cross talk is also much complained of. The arrangements relating to the provision of keys for public call offices are also much complained of. It is suggested that there is a breach of faith on the part of the company with regard to the provision of those keys.

B.P.P., 1898 (12).

15 SCIENCE AND ENGINEERING

15a 'Science in Engineering'

LECTURE V. On the Necessity of Incorporating with the Practice of the Mechanical and Industrial Arts a Knowledge of Practical Science*

If I possessed a command of language capable of rendering what I have to communicate tolerably attractive, I should probably have less difficulty in making the subject, on which I venture to address you, clearly and explicitly understood. I appear before you at the request of the Directors of this Institution, and I hope, notwithstanding the defects under which I labour, to be able to engage your attention on a subject of some importance, inasmuch as its true appreciation may lead to results in every sense calculated to improve our practice and enlarge our conceptions in the pursuit of our respective callings.

In attempting to investigate the present state of our knowledge of the constructive arts, I have been induced to believe, that, although we may be considered to rank amongst the first in the art of mechanical design and construction, we are, nevertheless, not the first in some other departments of the useful arts. To make our position more clearly known, I shall endeavour to point out wherein our superiority consists, and wherein I consider we are behind the inventive talent and practice of other countries.

There cannot exist a doubt that we are far from perfect and have much to learn in this respect. The Great Exhibition of last year directed public attention to our position as an industrial people, and one of the beneficial results of that immense and varied collection of constructive art is observable in the notice which the subject is now beginning to receive in quarters best calculated to direct the public taste, and encourage a better and purer style of construction. It is a singular but an important fact, in countries where the industrial arts are cultivated with the greatest success, that the principles on which they are founded should be so imperfectly understood. How very few of our best practitioners in Architecture, Civil and Mechanical Engineering, are acquainted with the rudiments, or even with the simplest theoretical rules of their professions; and how often have they to depend upon chance, instead of sound elementary knowledge, for the various constructions on which they elaborate defective, if not abortive results! I have myself laboured, and still labour, under these disadvantages, and it is from a consciousness of this deficiency that I now address you. I do so under the impression that our successors may attain greater distinction, and greater certainty, in the strength, beauty and proportion of their constructions, than we have hitherto been able to accomplish. At the present moment, and for many years past, we have suffered severe inflictions on our national pride, in being called upon to witness failures and abortions in the art of construction, which a cultivated taste, superior skill and extended knowledge

* Delivered, at the request of the Directors, to the Members of the Manchester Mechanics' Institution, March 1852.

By courtesy of the British Museum

USEFUL INFORMATION

FOR

ENGINEERS;

BEING

A SERIES OF LECTURES

DELIVERED TO

THE WORKING ENGINEERS OF YORKSHIRE AND LANCASHIRE;

TOGETHER WITH

A SERIES OF APPENDICES,

CONTAINING

THE RESULTS OF EXPERIMENTAL INQUIRIES INTO THE STRENGTH OF MATERIALS, THE CAUSES OF BOILER EXPLOSIONS, ETC.

BY

WILLIAM FAIRBAIRN, F.R.S., F.G.S.,

PRESIDENT OF THE LITERARY AND PHILOSOPHICAL SOCIETY OF MANCHESTER,
CORRESPONDING MEMBER OF THE NATIONAL INSTITUTE OF FRANCE; CHEVALIER OF THE LEGION OF HONOUR, ETC.

LONDON:
LONGMAN, BROWN, GREEN AND LONGMANS.
1856.

would have prevented. This is a national reproach, and the example of the past will continue to be the rule of the future, so long as we have to grope our way in the dark, under the guidance of prejudice and ignorance, instead of being governed by sound principles and correctly determined laws calculated to produce very different and much happier results.

To render the subject on which I purpose to address you as intelligible and as explicit as possible, I shall adopt the useful practice of dividing it into heads, as follows:—

I. On the necessities which exist for a knowledge of science in union with practice.

II. On the importance of national schools and institutes for imparting scientific knowledge to foremen and managers of works.

III. Educational institutions established on the continent of Europe and America.

IV. Differences of scholastic institutions in this as compared with those of other countries.

V. On the comparative state of our manufacturing industry as applied to cotton, during the last and the present centuries.

VI. The increase of other branches of industry in a similar ratio to that of cotton.

Lastly. Self-acting machines, and their application to constructive science.

I. *On the necessities which exist for a knowledge of science in union with practice.*

It is extraordinary, that in this country—which, above all others, is famed for the extent of its manufactures, mechanical skill, and extensive practice in the useful arts—there should be no institute, nor any establishment whatever to teach and instruct the rising generation in the elementary rules of their respective professions: these, of all others, are the most important to the community, and the best calculated to enhance the value, and extend the influence of our industrial resources. In my opinion, every one should be taught the rudiments and higher branches of their professions upon the same principles as barristers or physicians are taught, and, with this difference only, that no man should be ineligible to practise his profession, whenever he finds himself competent to the discharge of its duties; any enactment to the contrary would militate against the freedom of labour and the employment of capital, and if carried out would paralyse the best interests of a manufacturing and commercial community. Every person intended for professional pursuits in connexion with the art of construction, should have a theoretical as well as a practical education. They should be taught the fundamental rules connected with their respective professions, or at least so much of theory as would enable them to enter upon the practice with some degree of certainty, and that more especially in the practical development of those principles on which the safety of the public and the success of their professional career depend.

It is absurd to talk against theory, as if a knowledge of the exact sciences was a dangerous and a useless attainment; nothing can be more erroneous than this impression, as on close inspection there is no practice without theory, any more than there is no effect without a cause. In the useful arts, theory can only be considered dangerous when it is not reducible to practice, and the real meaning of the term *theory*,—which creates so much alarm in the minds of practical men,—is neither more nor less than a series of definite rules by which practice is governed, and through which we derive, from fixed and definite laws, those sound and definite results, which of all others it is the primary object of practice to accomplish. In the Mechanical Arts how difficult, precarious and unsatisfactory are the thoughts of men unacquainted with first principles, and how very often does that deficiency lead them into mal-construction and those errors which a knowledge of science would teach them to avoid! It is true, that some of our first engineers and some of our most ingenious mechanicians have been men of limited education—men of humble origin, but how much more perfect would have been their labours had the emanations of their minds and their subsequent constructions been based upon the unerring laws of natural science!

A knowledge of the exact sciences must be valuable under every circumstance of life, and this knowledge when united to sound judgement is irrevocably the forerunner of a sound and perfect construction. I could multiply examples where ignorance—as a pretender to knowledge—has been productive of the most untoward results, not only in abortive attempts at construction, but in those on which the lives and property of individuals depend. It is not an uncommon occurrence to witness in works of this kind the most glaring imperfections, a waste of material, and a total want of proportion arising from the absence of this knowledge; and in order to lessen the number of those discrepancies, our practical men should be educated, and that education should be accompanied with the conviction that sound practice can never be attained without some definite rule for its guidance. Fully impressed with these views, and the advantages to be derived from theory in the exercise of a well-founded practice, I shall endeavour to prove from evidence which I possess, that theory and practice are the twin sisters of science, and cannot be separated without endangering the connexion, or destroying the beauty, harmony and solidity of construction.

II. *On the importance of national schools and institutes for imparting scientific knowledge to foremen and managers of works.*

Before I direct your attention to the facts which I propose to lay before you, I would first advert to the present inefficient state of our educational resources, the imperfect means we have at our disposal, and the advantages we might derive from well-conducted and well-organized institutions, such as are now established and hold out encouragement to aspirants for scientific distinction in other countries. It is essential to the prosperity of every country,—and to none more so than our own,—that the industrial arts should be carefully nursed and cultivated with energy in every department, and by the introduction of increased intelligence, every endeavour should be made to perpetuate to succeeding generations an improved and healthy progression.

To effect these desirable objects, we must rear for future service a more intelligent and better educated class of foremen, managers and workmen. We must enlarge their sphere of action, and prepare them by suitable instruction for their respective callings. We must offer them, at a cheap rate, such rudimentary and theoretical knowledge as will qualify them for the due and faithful performance of their duties, and thus raise them to a higher standard of character, both as individuals, or members of the community. To elevate the mind and improve the intellectual character of the rising generation must be a work of time, but the necessities of the age demand increased intelligence in our workshops, and in every other department of our national industry; and any attempt to improve the condition and enlarge the mental capacities of that important class, must eventually lead to the best results.

It is evident to every reflecting mind that society never stands still; there is a restless activity abroad; and if ever we become stationary, it is easy to foresee that others will reap the harvest of our past labours, and be first in the race of national supremacy.

It is true we have our Mechanics' Institutes, Athenæums, Lyceums, and many similar institutions, which do honour to the country; but they are so constituted, or perhaps the persons for whom they were instituted are so constituted, as to justify the observation that they have failed to serve the particular classes for whom they were originally intended, and they have probably been still less efficient as regards the advancement of the industrial arts. We are still in want of institutions and museums for scientific and industrial tuition, and until such institutions are established in the metropolis and in some of the large towns, and in active co-operation with the middle and working classes themselves, we may look in vain for that combination of talent, in theory and practice, the want of which must sooner or later seriously affect the destinies of our manufacturing and mechanical industry.

William Fairbairn, *Useful Information for Engineers*, London, 1856, p. 93.

15b 'Mechanical and Scientific Knowledge'

The extension of experimental methods of investigation, has caused even manual skill in practical mechanics, when scientifically exercised, to be duly honoured, and not (as in ancient times) to be regarded as beneath the dignity of science.

As a systematically avowed doctrine, there can be no doubt that the fallacy of a discrepancy between rational and practical mechanics came long ago to an end; and that every well-informed and sane man, expressing a deliberate opinion upon the mutual relations of those two branches of science, would at once admit that they agree in their principles, and assist each other's progress, and that such distinction as exists between them arises from the difference of the *purposes* to which the same body of principles is applied.

If this doctrine had as strong an influence over the actions of men as it now has over their reasonings, it would have been unnecessary for me to describe, so fully as I have done, the great scientific fallacy of the ancients. I might, in fact, have passed it over in silence, as dead and forgotten; but, unfortunately, that discrepancy between theory and practice, which in sound physical and mechanical science is a delusion, has a real existence in the minds of men; and that fallacy, though rejected by their judgments, continues to exert an influence over their acts. Therefore it is that I have endeavoured to trace the prejudice as to the discrepancy of theory and practice, especially in Mechanics, to its origin; and to show that it is the ghost of a defunct fallacy of the ancient Greeks and of the mediæval schoolmen.

This prejudice, as I have stated, is not to be found, at the present day, in the form of a definite and avowed principle: it is to be traced only in its pernicious effects on the progress both of speculative science and of practice, and sometimes in a sort of tacit influence which it exerts on the forms of expression of writers, who have assuredly no intention of perpetuating a delusion. To exemplify the kind of influence last referred to, I shall cite a passage from the same historical work which I recently quoted for a different purpose. Lord Macaulay, in treating of the Act of Toleration of William III., compares, metaphorically, the science of politics to that of mechanics, and then proceeds as follows:—

> The mathematician can easily demonstrate that a certain power, applied by means of a certain lever, or of a certain system of pulleys, will suffice to raise a certain weight. But his demonstration proceeds on the supposition that the machinery is such as no load will bend or break. If the engineer who has to lift a great mass of real granite by the instrumentality

of real timber and real hemp, should absolutely rely on the propositions which he finds in treatises on Dynamics, and should make no allowance for the imperfection of his materials, his whole apparatus of beams, wheels, and ropes, would soon come down in ruin, and with all his geometrical skill, he would be found a far inferior builder to those painted barbarians who, though they never heard of the parallelogram of forces, managed to pile up Stonehenge.*

It is impossible to read this passage without feeling admiration for the force and clearness (and I may add, for the brilliancy and wit) of the language in which it is expressed; and those very qualities of force and clearness, as well as the author's eminence, render it one of the best examples that can be found to illustrate the lurking influence of the fallacy of a double set of mechanical laws, rational and practical.

In fact, the mathematical theory of a machine,—that is, the body of principles which enables the engineer to compute the arrangement and dimensions of the parts of a machine intended to perform given operations,—is divided by mathematicians, for the sake of convenience of investigation, into two parts. The part first treated of, as being the more simple, relates to the motions and mutual actions of the solid pieces of a machine, and the forces exerted by and upon them, each continuous solid piece being treated as a whole, and of sensibly invariable figure. The second and more intricate part relates to the actions of the forces tending to break or to alter the figure of each such solid piece, and the dimensions and form to be given to it in order to enable it to resist those forces: this part of the theory depends, as much as the first part, on the general laws of mechanics; and it is, as truly as the first part, a subject for the reasonings of the mathematician, and equally requisite for the completeness of the mathematical treatise which the engineer is supposed to consult. It is true, that should the engineer implicitly trust to a pretended mathematician, or an incomplete treatise, his apparatus would come down in ruin, as the historian has stated: it is true also that the same result would follow, if the engineer was one who had not qualified himself, by experience and observation, to distinguish between good and bad materials and workmanship; but the passage I have quoted conveys an idea different from these; for it proceeds on the erroneous supposition, that the first part of the theory of a machine is the whole theory, and is at variance with something else which is independent of mathematics, and which constitutes, or is the foundation of, practical mechanics.

The evil influence of the supposed inconsistency of theory and practice upon speculative science, although much less conspicuous than it was in the ancient and middle ages, is still occasionally to be traced. This it is which opposes the mutual communication of ideas between men of science and men of practice, and which leads scientific men sometimes to employ, on problems that can only be regarded as ingenious mathematical exercises, much time and mental exertion that would be better bestowed on questions having some connection with the arts, and sometimes to state the results of really important investigations on practical subjects in a form too abstruse for ordinary use; so that the benefit which might be derived from their application is for years lost to the public; and valuable practical principles, which might have been anticipated by reasoning, are left to be discovered by slow and costly experience.

But it is on the practice of mechanics and engineering that the influence of the great fallacy is most conspicuous and most fatal. There is assuredly, in Britain, no deficiency of men distinguished by skill in judging of the quality of materials and work, and in directing the operations of workmen,—by that sort

* Vol. iii, p. 84.

of skill, in fact, which is purely practical, and acquired by observation and experience in business. But of that scientifically practical skill which produces the greatest effect with the least possible expenditure of material and work, the instances are comparatively rare. In too many cases we see the strength and the stability, which ought to be given by the skilful arrangement of the parts of a structure, supplied by means of clumsy massiveness, and of lavish expenditure of material, labour, and money; and the evil is increased by a perversion of the public taste, which causes works to be admired, not in proportion to their fitness for their purposes, or to the skill evinced in attaining that fitness, but in proportion to their size and cost.

With respect to those works which, from unscientific design, give way during or immediately after their erection, I shall say little; for, with all their evils, they add to our experimental knowledge, and convey a lesson, though a costly one. But a class of structures fraught with much greater evils exists in great abundance throughout the country:—namely, those in which the faults of an unscientific design have been so far counteracted by massive strength, good materials, and careful workmanship, that a temporary stability has been produced, but which contain within themselves sources of weakness, obvious to a scientific examination only, that must inevitably cause their destruction within a limited number of years.

Another evil, and one of the worst which arises from the separation of theoretical and practical knowledge, is the fact that a large number of persons, possessed of an inventive turn of mind and of considerable skill in the manual operations of practical mechanics, are destitute of that knowledge of scientific principles which is requisite to prevent their being misled by their own ingenuity. Such men too often spend their money, waste their lives, and it may be lose their reason, in the vain pursuit of visionary inventions, of which a moderate amount of theoretical knowledge would be sufficient to demonstrate the fallacy; and for want of such knowledge, many a man who might have been a useful and happy member of society, becomes a being than whom it would be hard to find anything more miserable.

The number of those unhappy persons—to judge from the patent-lists, and from some of the mechanical journals—must be much greater than is generally believed. The most absurd of all their delusions,—that commonly called the perpetual motion, or to speak more accurately, the inexhaustible source of power,—is, in various forms, the subject of several patents in each year.

The ill success of the projects of misdirected ingenuity has very naturally the effect of driving those men of practical skill who, though without scientific knowledge, possess prudence and common sense, to the opposite extreme of caution, and of inducing them to avoid all experiments, and to confine themselves to the careful copying of successful existing structures and machines: a course which, although it avoids risk, would, if generally followed, stop the progress of all improvement. A similar course has sometimes, indeed, been adopted by men possessed of scientific as well as practical skill: such men having, in certain cases, from deference to popular prejudice, or from a dread of being reputed as theorists, considered it advisable to adopt the worse and customary design for a work in preference to a better but unusual design.

Some of the evils which are caused by the fallacy of an incompatibility between theory and practice having been described, it must now be admitted, that at the present time those evils show a decided tendency to decline. The extent of intercourse, and of mutual assistance, between men of science and men of practice, the practical knowledge of scientific men, and the scientific knowledge of practical men, have been for some time steadily increasing; and that combination and harmony of theoretical and practical knowledge—that skill in the application of scientific principles to practical purposes, which in former times was confined to a few remarkable individuals, now tends to become more generally diffused. With a view to promote the diffusion of that kind of skill, Chairs were instituted at periods of from fifteen to ten years ago, in the two Colleges of the University of London, in the University of Dublin, in the three Queen's Colleges of Belfast, Cork, and Galway, and in this University of Glasgow.

For the sake of a parallel, it may be here worth while to refer to another branch of practical science—that of Medicine. From the time of the first establishment of Medical Schools in Universities, there have existed, not only Chairs for the teaching of the purely scientific departments of Medical Science, such as Anatomy and Physiology, but also Chairs for instruction in the art of applying scientific principles to practice, such as those of Surgery, the Practice of Physics, and others. The institution of a Chair of Mechanics and Engineering in a University where there have long existed Chairs of Mathematics and Natural Philosophy, is an endeavour to place Mechanical Science on the same footing with that of Medicine.

Another parallel may be found in an Institution, which, though not a University, and though established as much for the advancement as for the diffusion of knowledge, has had a most beneficial effect in promoting the appreciation of science by the public,—I mean the British Association. When that body was first instituted, both the theoretical advancement and the practical application of Mechanics, and the several branches of Physics, were allotted to a single section, called Section A. The business before that Section soon became so excessive in amount, and so multifarious in its character, that it was found necessary to institute Section G, for the purpose of considering the practical application of those branches of science to whose theoretical advancement Section A was now devoted; and notwithstanding this separation, those two Sections work harmoniously together for the promotion of kindred objects; and the same men are, in many instances, leading members of both. What Section G is to Section A in the British Association, this class of Engineering and Mechanics is to those of Physics and Mathematics in the University.

It being admitted, that Theoretical and Practical Mechanics are in harmony with each other, and depend on the same first principles, and that they differ only in the purposes to which those principles are applied, it now remains to be considered, in what manner that difference affects the mode of instruction to be followed in communicating those branches of science.

Mechanical knowledge may obviously be distinguished into three kinds: purely scientific knowledge,—purely practical knowledge—and that intermediate knowledge which relates to the application of scientific principles to practical purposes, and which arises from understanding the harmony of theory and practice.

The objects of instruction in purely scientific mechanics and physics are, first to produce in the student that improvement of the understanding which results from the cultivation of natural knowledge, and that elevation of mind which flows from the contemplation of the order of the universe; and secondly, if possible, to qualify him to become a scientific discoverer. In this branch of study exactness is an essential feature; and mathematical difficulties must not be shrunk from when the nature of the subject leads to them. The ascertainment and illustration of truth are the objects; and structures and machines are looked upon merely as natural bodies are:—namely, as furnishing experimental data for the ascertaining of principles, and examples for their illustration.

Instruction in purely practical knowledge is that which the student acquires by his own experience and observation of the transaction of business. It enables him to judge of the quality of materials and workmanship, and of questions of convenience and commercial profit, to direct the operations of workmen, to imitate existing structures and machines, to follow established practical rules, and to transact the commercial business which is connected with mechanical pursuits.

The third and intermediate kind of instruction, which connects the first two, and for the promotion of which this Chair was established, relates to the application of scientific principles to practical purposes. It qualifies the student to plan a structure or a machine for a given purpose, without the necessity of copying some existing example, and to adapt his designs to situations to which no existing example affords a parallel. It enables him to compute the theoretical limit of the strength or stability of a structure, or the efficiency of a machine of a particular kind,—to ascertain how far an actual structure or machine fails to attain that limit,—to discover the causes of such shortcomings,—and to devise improvements for obviating such causes; and it enables him to judge how far an established practical rule is founded on reason, how far on mere custom, and how far on error.

There are certain characteristics in the mode of treating the subjects, by which this practical-scientific instruction ought to be distinguished from instruction for purely scientific purposes.

In the first place it will be universally admitted, that as far as is possible, mathematical intricacy ought to be avoided.

In the original discovery of a proposition of practical utility, by deduction from general principles and from experimental data, a complex algebraical investigation is often not merely useful, but indispensable; but in expounding such a proposition as a part of practical science, and applying it to practical purposes, simplicity is of the first importance:—and, in fact, the more thoroughly a scientific man has studied the higher mathematics, the more fully does he become aware of this truth,—and, I may add, the better qualified does he become to free the exposition and application of scientific principles from mathematical intricacy. I cannot better support this view than by referring to Sir John Herschel's *Outlines of Astronomy*—a work in which one of the most profound mathematicians in the world has succeeded admirably in divesting of all mathematical intricacy the explanation of the principles of that natural science which employs the higher mathematics most.

In fact, the symbols of algebra, when employed in abstruse and complex theoretical investigations, constitute a sort of thought-saving machine, by whose aid a person skilled in its use can solve problems respecting quantities, and dispense with the mental labour of thinking of the quantities denoted by the symbols, except at the beginning and end of the operation. In treating of the practical application of scientific principles, an algebraical formula should only be employed when its shortness and simplicity are such as to render it a clearer expression of a proposition or rule than common language would be, and when there is no difficulty in keeping the thing represented by each symbol constantly before the mind.

Another characteristic by which instruction in practical science should be distinguished from purely scientific instruction, is one which appears to me to possess the advantage of calling into operation a mental faculty distinct from those which are exercised by theoretical science. It is of the following kind:—

In the theoretical science, the question is—*What are we to think?* and when a doubtful point arises, for the solution of which either experimental data are wanting, or mathematical methods are not sufficiently advanced, it is the duty of philosophic minds not to dispute about the probability of conflicting suppositions, but to labour for the advancement of experimental inquiry and of mathematics, and await patiently the time when these shall be adequate to solve the question.

But in practical science the question is—*What are we to do?*—a question which involves the necessity for the immediate adoption of some rule of working. In doubtful cases, we cannot allow our machines and our works of improvement to wait for the advancement of science; and if existing data are insufficient to give an exact solution of the question, that approximate solution must be acted upon which the best data attainable show to be the most probable. A prompt and sound judgment in cases of this kind is one of the characteristics of a PRACTICAL MAN, in the right sense of that term.

In conclusion, I will now observe, that the cultivation of the Harmony between Theory and Practice in Mechanics—of the application of Science to the Mechanical Arts—besides all the benefits which it confers on us, by promoting the comfort and prosperity of individuals, and augmenting the wealth and power of the nation—confers on us also the more important benefit of raising the character of the mechanical arts, and of those who practise them. A great mechanical philosopher, the late Dr. Robison of Edinburgh, after stating that the principles of Carpentry depend on two branches of the science of Statics, remarks—'It is this which makes Carpentry a liberal art.'

So also is Masonry a liberal art,—so is the art of working in Iron, so is every art, when guided by scientific principles. Every structure or machine, whose design evinces the guidance of science, is to be regarded not merely as an instrument for promoting convenience and profit, but as a monument and testimony that those who planned and made it had studied the laws of nature; and this renders it an object of interest and value, how small soever its bulk, how common soever its material.

For a century there has stood, in a room in this College, a small, rude, and plain model, of appearance so uncouth, that when an artist lately introduced its likeness into a historical painting, those who saw the likeness, and knew nothing of the original, wondered what the artist meant by painting an object so unattractive.

But the artist was right; for ninety-one years ago a man took that model, applied it to his knowledge of natural laws, and made it into the first of those steam engines that now cover the land and the sea; and ever since, in Reason's eye, that small and uncouth mass of wood and metal shines with imperishable beauty, as the earliest embodiment of the genius of James Watt.

Thus it is that the commonest objects are by science rendered precious; and in like manner the engineer or the mechanic, who plans and works with understanding of the natural laws that regulate the results of his operations, rises to the dignity of a Sage.

W. J. M. RANKINE, *A Manual of Applied Mechanics*, London, 1858, p. 4.

15c Bridge Construction

By courtesy of the British Museum

GENERAL THEORY

OF

BRIDGE CONSTRUCTION:

CONTAINING

DEMONSTRATIONS OF THE PRINCIPLES OF THE ART
AND THEIR APPLICATION TO PRACTICE;

FURNISHING THE MEANS OF

CALCULATING THE STRAINS UPON THE CHORDS, TIES, BRACES,
COUNTER-BRACES, AND OTHER PARTS OF A BRIDGE
OR FRAME OF ANY DESCRIPTION.

WITH PRACTICAL ILLUSTRATIONS.

BY

HERMAN HAUPT, A.M.,

CIVIL ENGINEER.

NEW YORK:
D. APPLETON AND COMPANY,
346 & 348 BROADWAY.
M.DCCC.LVI.

PREFACE

From the great importance of the art of bridge construction, it might be supposed that its principles would be familiarly understood by all whose occupations require an acquaintance with it. If, however, any work exists, containing an exposition of a theory sufficient to account generally for the various phenomena observed in the mutual action of the parts of trussed combinations of wood or metals, the author has neither seen or heard of it. When his attention was first directed to the subject of properly proportioning the parts of bridges, by being called upon in the discharge of professional duties to superintend their construction, he found it impossible to procure satisfactory information, either from engineers and builders, or from books. In fact, if he may be permitted to judge from the contradictory opinions that are even yet entertained, and from the errors frequently committed by practical men in the construction of bridges, he would be inclined to infer, that of few arts of equal practical importance, are the principles so little understood.

The best works on the subject of construction, that have fallen into the hands of the writer, contain but little that will furnish the means of calculating the strains upon the timbers of a bridge truss, or of determining their relative sizes; and they do not furnish information as to what constitute the elements of a framed truss, or the most advantageous disposition of these elements to attain the maximum strength and stiffness with a given quantity of material.

In the following pages, the object has been, not so much to detail particular plans, as to establish general principles.

An attempt has been made to explain the mode of action of the parts of structures, and their mutual influence when combined; to point out the ways by which the strains can be estimated, and the relative sizes of the timbers accurately determined; new combinations of the elements in the construction of bridge trusses have been suggested, the defects of many plans in general use pointed out, and several simple means proposed for remedying these defects and adding to the strength of structures.

Unable to procure satisfactory information in any other way, the writer, in the spring of 1840, commenced a course of experiments on models, and examined many existing structures, for which his occupation at that time as a civil engineer in the service of the State of Pennsylvania, afforded great facilities.

This investigation was commenced without the most remote design of ever giving the result publicity, but solely for the purpose of enabling him to proceed intelligently in the discharge of professional duties. The facts elicited, led him to conclude that many important structures, and some of recent erection, exhibited defects which, although serious, had escaped the observation of merely practical builders; these will be pointed out in their proper places, and the writer will be amply rewarded for the labor of preparing this little treatise, if it shall prove the means of adding in the smallest degree to the stock of information already possessed on the important subject of bridge construction.

The treatise on bridge construction has been prefaced by a few pages on the resistance of solids, because a knowledge of this subject lies at the foundation of the art of construction. The mode of investigation is peculiar, and is believed to be more simple than those usually employed.

This part of the subject will not be interesting to those who are unacquainted with the application of mathematics to mechanics, but those who are, will perhaps be pleased with the simplicity of the solutions, and the novelty of representing strains by geometrical solids, deflections by parabolic areas, and the variable pressures at different points of beams by the corresponding ordinates of plane curves. The portion which refers to the principles of construction, and their application to practice, will be readily understood by the general reader. Demonstrations of propositions required in the solution of problems, and not found in any work accessible to the author, have been given in notes.

HERMAN HAUPT, *General Theory of Bridge Construction*, New York, 1856, p. 5.

15d Stress Analysis

Diagrams of the Applied Forces and of the Forces of Reaction of the Parts subjected to Stress in Framed Structures

USES OF THE METHOD

DIAGRAMS of forces must render great assistance to any one occupied in designing roofs, bridges, or other framed structures. They furnish indeed a royal road to certain results, but this power of affording help to the helpless must not be regarded as their legitimate office. By means of a diagram of forces for any arrangement of framework provided for him, a person, quite incapable otherwise of taking out the stresses in a moderately complicated structure, could get at these for any particular modification of the arrangement with some assurance of accuracy; but such a one could never properly be entrusted with the duty of choosing the design, of making the detail drawings, or of superintending the erection of the work. In the operation of drawing details judiciously, there is a constant necessity for a reasoning appreciation of the various stresses and strains* that may act in the parts, as well as a practical acquaintance with the nature of materials, their strengths, and their imperfections—acquirements that should place 'Detailing' second only among professional duties to that of determining the most economical or best design to meet the occasion.

The uses of diagrams of forces are—not to render it unnecessary that the engineer or architect should familiarize himself with the mechanical interaction of the parts of framed and other structures—but to facilitate the work of taking out the stresses, to suggest rapidly improvements in the design, and to reduce the labour of comparing the economic merits of different designs.

EARLIER EXAMPLES OF DIAGRAMS OF FORCES

The representation of interacting forces by a diagram has been long practised in its simpler forms to determine the magnitudes of the stresses at particular points or parts of framed structures. Fig. 341 i. (Plate I.) shows an application of the method that must be familiar to any one accustomed to such work; in it the line B C is drawn vertically, and so that its length will represent on a chosen scale the magnitude of the vertical reaction of the support A′A (any weight A D coming directly upon the support being neglected); then the stresses in the extreme parts of the rafter and tie-bar are at once ascertainable by measuring on the same scale the respective lengths of A B and A C, cut off by B C. This gives us one of the simplest examples of a diagram of forces. It is reproduced in

* The term *stress* expresses the condition of a part of the structure to the extremities of which are applied compressing or extending forces; the amount of the stress is measured by the magnitude of the force acting on either extremity; the *strain* is the change of length from elasticity which the part undergoes when subjected to the stress.

PLATE I

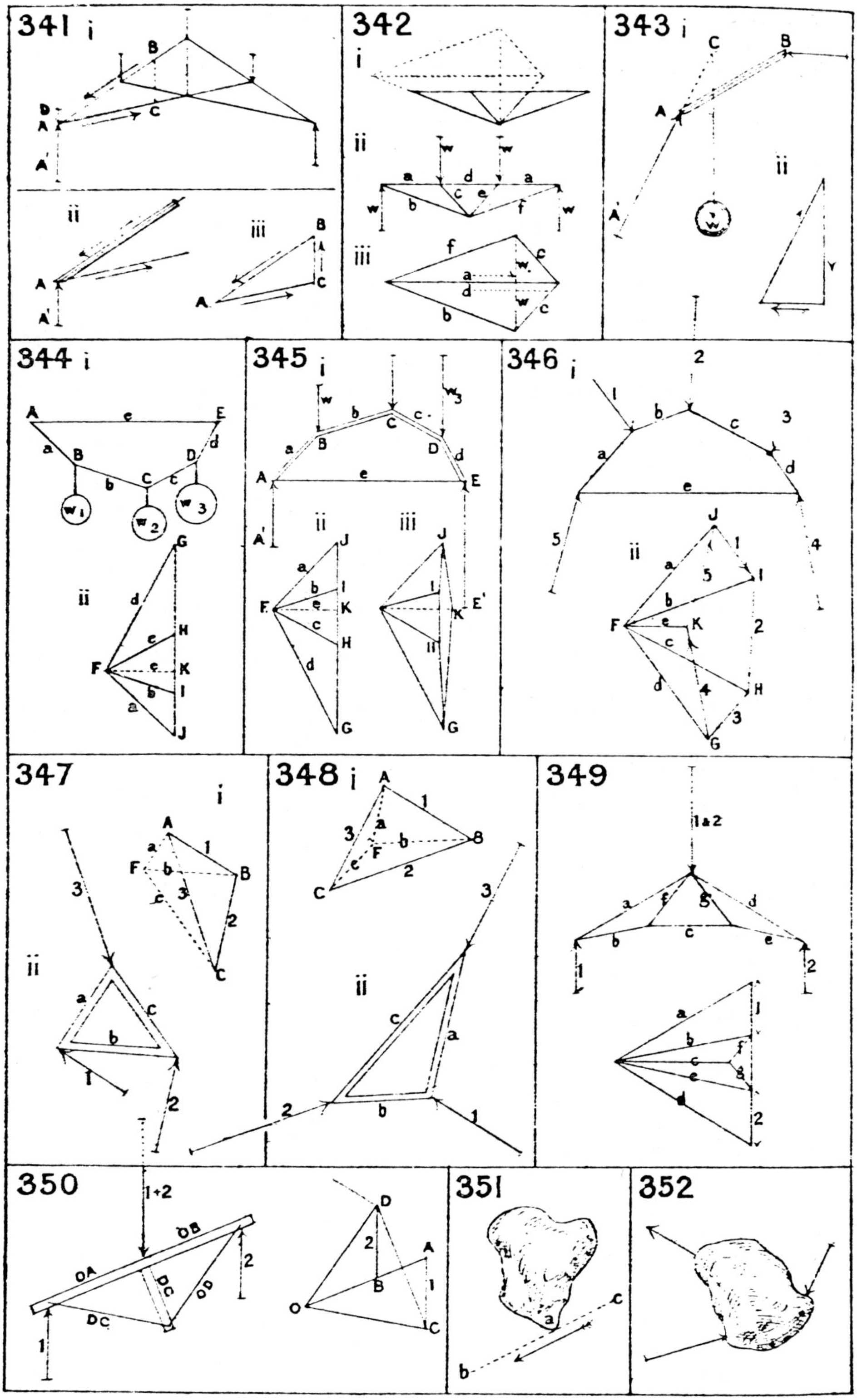

By courtesy of the British Museum

the form of two separate drawings in Figs. 341 ii. and iii.; ii. representing the end of the truss with the supporting force A′A acting upon it, and iii. the diagram of the forces. In ii. we have the three forces acting upon a *point*, while in iii. the lines parallel to and representing in magnitude these forces become the sides of a *triangle*. If A′A be oblique instead of vertical, then B C in the diagram must of course be drawn parallel to it.

Fig. 342 i. is an example taken from a paper by the late Mr. C. H. Wild, C.E., which was shown to me in the year 1854, but was of much earlier date. With the frame and diagram given separately, this is shown by Figs. 342 ii. and iii.; the stress in any part in ii. is measured by the length of the corresponding line in iii.

The next example of a 'diagram of forces,' shown by Fig. 343, is taken from Professor Rankine's *Applied Mechanics*, 1858, section 142. In this the three forces act upon a solid or frame A B, and in order that equilibrium may be produced, it is necessary that the lines of action of the forces shall intersect in some one point C; the triangle, Fig. 343 ii., formed by drawing lines parallel to the forces constitutes the diagram of forces—that is, the lengths of the lines in it give the proportionate magnitudes of the forces to which they are respectively drawn parallel.

The well-known diagram of forces for the funicular polygon is shown by Fig. 344 ii. Let A B C D E in Fig. 344 i. represent a string the weight of which may be neglected; let it be suspended from the points A and E at the same level, and let any number of weights, w_1, w_2, w_3, be hung from it as in the figure. Then the relative magnitudes of the weights which cause the string to assume the particular form shown, and of the resulting stresses in the various parts of the string, can be readily ascertained by drawing the diagram of forces Fig. 344 ii.; to construct this, we must draw from the same point F lines parallel to the different parts of the string, and intersect these by a vertical line G J. Now if the length of F J be taken to represent the stress in the part A B to which it is parallel, J K will measure its vertical component (F K being drawn horizontally); similarly the length of F I will give the stress in the part B C, and I K its vertical component, and the difference between these vertical components = I J will represent the weight w_1; in like manner H I = w_2 and G H = w_3; and the stress in C D is measured by F H, and that in D E by F G. It is obvious that the horizontal component of the stress at any part of the string must be the same, and it is measured by F K.

R. H. Bow, *Economics of Construction in relation to Frames Structures*, London, 1873, p. 45.

15e Strength of Pillars

By courtesy of Bristol Central Library

E. Hodgkinson, 'Experimental Researches on the Strength of Pillars, Cast Iron and other Materials', *Phil. Trans.*, 1840, **130**, Fig. 1.

15f Strength of Malleable Iron Tubes

Experimental Inquiry into the Strength of Malleable Iron Tubes for supporting the Chester and Holyhead Railway across the Estuary of the Conway and the Menai Straits

. . . TAKING the whole series of experiments, it was considered necessary to inquire into the powers of resistance offered,—first, by cylindrical tubes to a transverse strain; secondly, by those of the elliptical or egg-shaped form; and lastly, by those of a rectangular section, including such other modifications as the experiments might indicate during the progressive stages of the investigation.

Each experiment is recorded, with some slight alteration, in the order in which it was made; but to give an accurate conception of the results, it will be necessary not only to describe the apparatus used in the course of the experiments, but also to show in what manner and under what circumstances those results were obtained.

The tube to be experimented upon was placed upon two solid blocks, A A (fig. 59), at a sufficient height to admit weights to be freely suspended from the

By courtesy of the British Museum

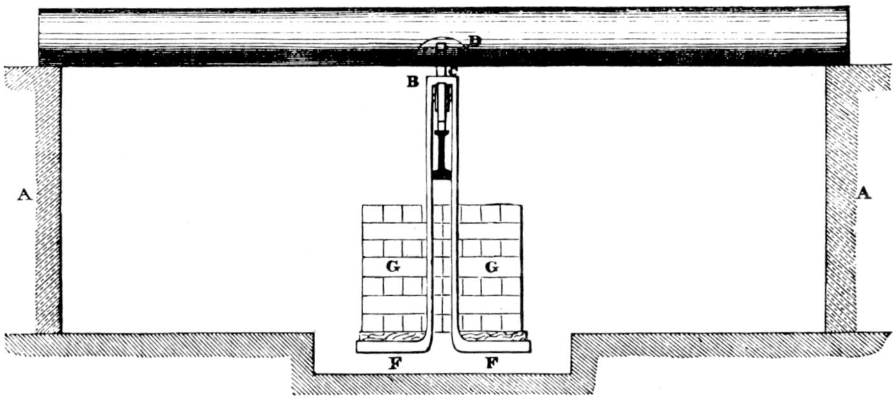

Fig. 59.

tube. In the cylindrical and elliptical tubes, the weights were suspended from a hole cut through the middle of the lower sides. The pin or shackle B was supported upon a cotter and plate resting upon a cushion of hard wood, 8 inches square, fitted into the interior on the lower side of the tube. That part of the pipe, which was cut out to receive the iron pin, was strengthened by riveting a strong plate round the outside, as shown by the dotted lines at D. After adjusting the shackle, the cross plate E (fig. 60) was attached to the iron pin by a bolt, which fitted loosely into the forked ends of the pin B. To the cast-iron plate thus attached were suspended the large cross links F, and upon the forks of these links were laid the weights, on each side, as exhibited at G.

It will be observed, that by this arrangement the tubes were subjected to the test of a *dead* weight, and the complexity and uncertainty of a system of levers were thus avoided. A lever was however found necessary, for the purpose of raising the weights off the tube, in order to ascertain the defects of elasticity, or the amount of permanent deflection which had taken place when the load was removed. These powers of restoration as well as deflection, were carefully measured and recorded every time the weights were laid on, and also when they were subsequently removed. The large lever H, with the 'screw-jack' J at the end, were used for raising and lowering the load, and for ascertaining the changes, which were going forward under the influence of the successive increase of the loading and unloading of the weights. The wooden wedges *b*, *b*, (fig. 60) were used, for the purpose of retaining the load steady during the time of raising the lever with the weights, in order to ascertain the permanent set. When the load was suspended, they were then removed.

By courtesy of the British Museum

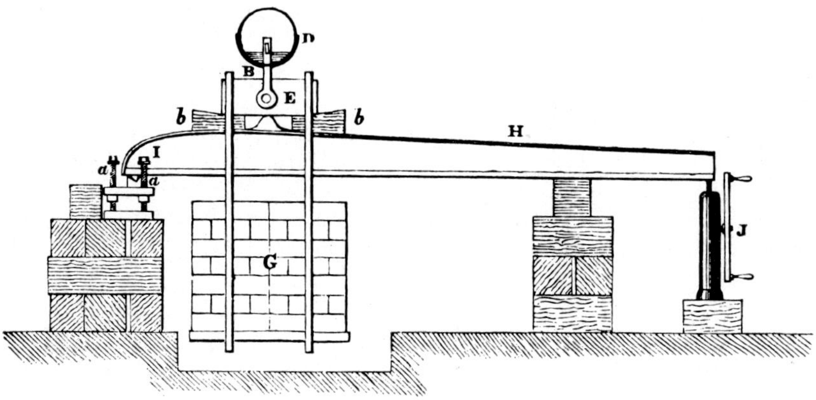

Fig. 60.

It will be noticed, that in every experiment the deflections were taken after the weights were laid on; and as some of the tubes were deflected to a considerable extent, the fulcrum I, on which the large lever rested, was supported on an iron table, which had the power of being raised or lowered by the screws *a*, *a*, *a*, to suit the deflection of the tubes.

From this description it will be seen that every facility was afforded for marking the changes which took place during the progress of the experiments; and from the accuracy with which they were conducted, the following tables will be found to contain a faithful record of the derangement of the tubes, from the time of laying on the weights till their final destruction.

SUMMARY OF RESULTS OBTAINED FROM THE EXPERIMENTS ON CYLINDRICAL TUBES

Distance between supports	Diameter in inches	Thickness of plate in inches	Ultimate deflection in inches	Breaking weight in tons	Remarks
ft. in.					
17 0	12·18	·0408	·39	3,040	Crushed top
17 0	12·00	·0370	·65	2,704	Crushed top
15 7½	12·40	·1310	1·29	11,440	Torn asunder at bottom
23 5	18·26	·0582	·56	6,400	Torn asunder at bottom
23 5	17·68	·0631	·74	6,400	Torn asunder at bottom
23 5	18·18	·1190	1·19	14,240	Torn asunder at bottom
31 3¼	24·00	·0954	·63	9,760	Torn asunder at bottom
31 3¼	24·30	·13501	·95	14,240	Torn asunder at bottom
31 3¼	24·20	·0954	·74	10,880	Torn asunder at bottom

SUMMARY OF RESULTS FROM THE EXPERIMENTS ON ELLIPTICAL TUBES

No. of experiments	Distance between the supports	Diameter congregate and transverse	Thickness of plates	Ultimate deflection in inches	Breaking weight in lbs	Remarks
	ft. in.					
19	17 0	{14·62, 9·25}	·0416	·62	2,100	Crushed on top
20	24 0	{21·66, 13·50}	1·1320	1·36	17,076	Broke by extension
21	24 0	{21·25, 14·12}	·0688	·45	7,270	By compression
22	18 6	{12·00, 7·50}	·0775	·95	6,867	By compression; this tube had a fin on the top side
24	17 6	{15·00, 9·75}	·1430	1·39	15,000	Both sides were ruptured

W. Fairbairn, *An Account of the Construction of the Britannia and Conway Tubular Bridges with a Complete History of their Progress*, London, 1849, p. 211.

15g Resistance of Ships

Observations and Suggestions on the Subject of Determining by Experiment the Resistance of Ships

[The following is a memorandum prepared by William Froude and sent to Mr. E. J. Reed, Chief Constructor of the Navy—dated December 1868]

My communication addressed to the Chief Constructor of the Navy dated 24th of April 1868 contained a lengthened report of a series of experiments made by me on the resistance at various velocities of certain models.

The results of those experiments, independently of any inference that might be founded on them favourable to the adoption of any particular forms in Naval Architecture may I think be regarded as having established the following propositions.

(i) Experiments duly conducted on a small scale will give results truly indicative of the performance of full size ships.

(ii) There are very important defects and inaccuracies in the received views concerning the resistances of different forms.

The second of these propositions shows the existence of a state of things the ill effects of which considering the magnitude of the interests involved, it is not easy to overrate; while the first indicates the applicability of what appears to me to be the only remedy, namely a system of small scale experiments; though such experiments have hitherto been considered barren and misleading.

Object of this communication

It is my purpose in this communication to illustrate these considerations more fully, as well as to carry them to a practical result.

In pursuance of this object.

Outline of Argument

(I.) Firstly, I shall place the above mentioned propositions and the grounds on which they rest in a more precise and apposite form.

(II.) Secondly, I shall point out the defects of full size trials as a means of investigation, contrasted with the effectiveness for that purpose of small scale experiments;

(III.) Thirdly, I shall describe that method of making such experiments which my experience recommends as the best;

(IV.) and Fourthly, I shall suggest a practical scheme by which such a system may be put into effect.

I. (Propositions established by former experiments)

In accordance with the above design, I begin by drawing out afresh the two propositions above mentioned.

(i) Experiments duly conducted on a small scale will give results truly indicative of the performance of full size ships.

In the former communication this proposition was shewn to be probably correct; but it may be well to recur to it here with especial reference to two objections which are commonly made to it.

The large waves seen to be made by models commonly urged as an objection to the validity of small scale experiments

(*a*) It is observed that models, when towed through the water make proportionately much larger waves than full size ships are ever seen to do, and under the impressions derived from this circumstance it is asserted that any deductions based on such experiments are misleading since it is certain that wave-genesis must form an important element of the whole resistance.

In answer to this objection I venture to assert that there is prima facie reason to suppose that full size ships would create waves of quite as formidable appearance if driven at corresponding velocities.

What is meant by 'corresponding velocities' may be better understood by referring to my former communication, where it was argued from the principles of wave motion that similar models of whatever dimensions, or what is more pertinent here, a model and a ship similar to it, would make similarly dimensioned waves when their relative velocities were as the square roots of their respective dimensions. And it was further suggested that under this condition the proportion between the resistance due to wave-genesis, and that due to other sources would be the same for both ship and model.

Principle of comparison of the performances of a model and a similar ship enunciated

To express the conclusion from this in a simple form, the diagram which exhibits to scale the resistance of a model at various successive velocities, will express equally the resistance of a ship similar to it, but of (n) times the dimension, at various successive velocities, if in applying the diagram to the case of the ship we interpret all the velocities as ($\sqrt{n}$) times, and the corresponding resistances as (n^3) times as great as on the diagram.

This principle is borne out by my experiments

The results obtained in my experiments, with the variously dimensioned, similar models when severally reduced to diagrams on this hypothesis exhibit such precise resemblance as practically to justify its adoption.

The hypothesis admits, however of definite proof, and this shall be given, but as the proof is somewhat lengthy and may be regarded as superfluous it is placed in an Appendix.

The 'rigidity' of water commonly urged as an objection to the validity of small scale experiments

(*b*) The second objection is founded upon what Professor Rankine has termed the 'rigidity' of water.

Observing that this property in any substance (such as tar or semi-fluid earth) which possesses it largely exhibits more highly marked effects when the substance is dealt with in small volumes than when in large, it is inferred that in such degree as it may exist in water, it will form an element in the resistance of a model which has no commensurable counterpart in that of a full size ship.

The degree of weight to be attached to this objection depends on whether the dimension of the model dealt with exceeds the limits of dimension within which the operation of the property, as it exists in water is practically felt.

This may be pretty fairly estimated by observing within what limits of dimension the behaviour of waves is affected by it.

In fact it is noticeable that while capillary waves are obliterated almost immediately that the originating impulse ceases to act, exhibiting in its greatest intensity the action of the property we are considering, yet waves of a few inches in length though they die out in a short distance, travel so far as to indicate that the relative intensity of its action is materially less, and waves of two or three feet in length run to such considerable distances without any apparent diminution as to indicate that they are practically exempt from it. It will be fair then to conclude that models of, say, 6 feet long and upwards will be to all intents and purposes unimpeded by such 'rigidity' as exists in water.

Models 6 feet long large enough not to be within range of the operation of the property of 'rigidity' of water

Moreover while it was found in my experiments that on comparing the performances of a 3 foot and 6 foot similar model, the former exhibited a slight excess of proportionate resistance possibly attributable to the above cause; between the 6 foot and 12 foot models no such difference was observable.

Lastly this is worth notice. Calculations based on a large number of indicator diagrams taken from the largest well-proportioned vessels of the wave-line type, at such speeds as not to create notable waves, have satisfied me that we may regard from 23 lb. to 24 lb. per square foot of midship section as the resistance of such vessels at 10 knots, this figure, reduced by the rule of the squares of the velocities, which my experiments shew to be fairly correct below such speeds as create notable waves, would give ·74 lb. to ·78 lb. per foot as the resistance due to 1·8 knots. Now, at that speed, the 6 foot wave-line model makes, according to my experiments a resistance of just ·75 lb. per square foot. Whereas if a 6 foot model were considerably affected by the action referred to, its resistance should be markedly greater.

It appears then that so far from the foregoing objections having any real foundation, we are able, in fact, to frame from the results of small scale experiments by a precise and simple method an accurate prediction of the performance of full size ships.

(ii) The existing state of knowledge as to the merits of particular forms in ships is very unsatisfactory.

Erroneous opinion concerning the resistance of forms of the 'wave-line' type

(*a*) It is generally held that Vessels of what is called the 'wave-line' type offer the least resistance, and there is nothing either in the reasonings or the statements of its advocates to limit its superiority to the condition of comparatively moderate velocity.

Now, if it is allowed as I have contended that the performance of such models as were used in my experiments is precisely similar to that of similar ships, it is clear from those experiments, that at high speeds the waveline form is markedly inferior to a form totally dissimilar to it in appearance and utterly opposed to it in principle.

Misconception of the modus operandi of Surface friction

(*b*) An opinion is beginning to gain ground and with good reason, that surface friction forms a large share of the resistance of a ship, yet concerning the nature of this force much misconception plainly exists; for by the best Authorities who have treated of the subject it has been held as the basis of calculation that a long plane surface of say one foot in width moving endways through the water encounters the same resistance of friction for each square foot throughout its length. Now it is quite clear that the first square foot in its length in experiencing frictional resistance from the particles passing it, and thus impressing force on them in return must communicate to them some velocity in the direction of its motion. The particles in contact with the second square foot having thus already received forward velocity must exert less frictional resistance upon it, unless the degree of that frictional resistance be independent of the degree of velocity with which the particles traverse the surface; a supposition manifestly erroneous, and quite as subversive of received opinions as the alternative conclusions, I advance that surface friction is not uniform for every part of the surface.

Defective state of knowledge concerning due ratio of length to breadth and to depth

(*c*) In the existing state of knowledge there is no guide for determining the influence upon a ship's resistance for the ratio of her length to her breadth and to her depth; a question of the most vital importance in the construction of ships of war, where by increasing the length though we no doubt increase the carrying power, we increase also in a greater degree the weightiest and costliest parts of the structure, adding at the same time to the difficulty of handling the ship; whereas by increasing the breadth we can add almost as largely as we please to the carrying power with an increase only to the less weighty and less costly parts of the structure, leaving the handiness unimpaired.

Defective state of knowledge concerning limiting speeds for different forms

(*d*) Though there are current very precise rules by which it is said we can determine the limit of speed beyond which a vessel of given length cannot economically be driven, these rules have been fixed by an (as it appears to me) arbitrary and not strictly relevant application of the theory of wave-motion. Moreover, the rule implies that the limit will be at the same speed for all ships of the same length, but this according to my experiments is obviously not the case, for instance in the experiments on the two 6 foot models* the same in length, breadth and displacement, one has the limit at a speed fully 8 per cent. higher than the other, making at that speed 20 per cent. the less resistance of the two though at the lower speeds it had made the greater resistance.

I have here enumerated certain deficiencies and inaccuracies in the received views on the subject. These alone are manifestly of very high importance, moreover we may reasonably expect that an extension of the inquiry will discover many other deficiencies and errors equally serious.

II.
(Full size trials compared with small scale experiments)

It needs no argument to show that correct and complete knowledge on these points is of even national importance.

Full size trials imperfect

Nevertheless, though with this view careful trials of ships have for some years been vigorously conducted yet our knowledge appears to be still in the unsatisfactory state I have been describing. Nor can we wonder at this, if we consider the defects inherent in any system of full size trials regarded as a mode of experimental inquiry.

Full size trials limited in number

In the first place the information gained is sparse, because to construct mere experimental forms of the size of large ships (ship sized models in fact) would be so costly that we are restricted to experimenting on those ships which from time to time are built for actual service, the costliness, too, of each individual experiment hinders us from so varying the trials as to make the most of even this scanty material. At best so disjointed a series of results as can be thus obtained furnishes a most imperfect basis on which to construct any kind of law.

and limited in range

It is limited in range, because the risk of signal failure incurred in deviating widely from the beaten track is too serious to be hazarded in the design of a ship intended for actual service.

also indecisive

It is moreover indecisive, because the information furnished by each individual experiment includes in one result with the resistance of the ship, the efficiency of the engines, and the efficiency of the mode of propulsion, and its

* form: comm: diagram C.

analysis is complicated by the uncertainties due to the state of the weather, the trim and other indeterminate conditions incidental to such trials.

and slow

Lastly, it is tardy, because not only is considerable time required for the preparation and completion of any one experiment on so large a scale, but before it is possible to take a single step in advance along any new line of thought which may have been suggested by the experience already acquired, it is necessary to wait during the time occupied in the design and construction of a new ship.

It has already been shewn that the results of small scale experiments are capable of a direct and accurate application to the case of full size ships and I will now shew that a system of such experiments is a mode of inquiry at once exhaustive, rapid and economical thus furnishing a marked contrast to what is in fact the only alternative, a system of full size trials.

Results of small scale experiments complete

Not only does the facility of manipulation inherent in mere smallness of dimension allow of the experiments being many, but the models being cheap of construction and specially adapted to the purpose may be multiplied and varied to an almost indefinite extent, so as to direct the inquiry to any particular question that may be opened up, and to complete all the links in each chain of results from which any individual law we are in search of is to be deduced.

Results of small scale experiments definite and precise

Again we are not perplexed by a multitude of collateral functions such as in full size trials as already has been observed must intrude themselves into the result; which result in small scale trials on the contrary is an exact and unmixed record of the resistance experienced by the form, disembarrassed from the questions of efficiency of Engine power or of propulsive agency*. The resistance moreover may be in itself freed from the adventitious influences of wind and rough waters etc., as the whole experiment may be carried on under cover.

Process of investigation by small scale experiments rapid

The experiments being under cover may be carried on in all weathers, the advantages of the rapidity thus insured to this system are too obvious to need pointing out.

III. (Description of best method of making small scale experiments)

Having thus demonstrated on general grounds the superiority of small scale experiments I will next explain what sort of arrangement my experience leads me to recommend as the best.

The water space must be devoted to the special purpose

The water space in which the experiments are to be tried should be devoted to the special purpose, so that it may be secured free from currents, as well as from the hindrances which may sometimes occur from any other uses to which the water might be put and that it may be maintained at a uniform level, and free from scum, weed etc. Further it is highly important that the water space should be housed in that the experiments may not be stopped by the weather.

Necessary dimensions of waterway

The waterway must be deep enough and broad enough not to affect the resistance of the model passing along it by constriction of the fluid and must be long enough to give the model length of run sufficient for it to attain gradually the highest speed required leaving still length enough for the experiment at a steady speed, and though I have above shewn good grounds for concluding that a model 6 feet in length (it being about half a square foot in section) is sufficiently large, yet it would be well to have arrangements capable of trying models of

* Not, of course, that these other questions are unimportant but neither can they be investigated to good purpose unless in their turn disembarrassed from the question of the resistance of the Vessel.

twice that dimension. For this, a minimum radius of 10 feet in section of waterway and a length of 250 feet* would I think be quite sufficient.

Drawing Office and Workshop should be attached

A Drawing Office and a Workshop suitable to the model making should be attached.

As to the mode of constructing the models which, if a sufficient number are made to try all the necessary variations, will form a considerable item in the expenditure it is clear that if we can by any means rapidly alter the shape of any model it is not necessary that a model of each form should survive, as long as there is a record of its shape upon paper. For purposes of alteration then, the Material for some exterior thickness should be such as can be cleanly and rapidly cut, and should be fusible for the purpose of 'melting on' additions. Stearine is perhaps the best material as it possesses sufficient firmness and is at the same time capable of being rapidly and accurately cut or pared. There must of course be an internal shell of wood which will carry the fittings for ballast, towing and handling, round this the stearine will be roughly 'melted on' in a clay or other mould to a sufficient thickness to allow for subsequent shaping.

Material best for models

Method of shaping models

The models may be cut to their primary shape, and afterwards altered to others by a revolving cutter mounted on a traversing frame (something similar to Taylor and Jordan's wood carving Machinery) save that here the tracer which regulates the path of the cutter instead of following the surface of a solid pattern will follow on a drawing the successive water-lines, the cutter being set successively at the levels corresponding to the water-lines. The intermediate excrescences left by the cutter will afterwards be pared off by hand.

Nature of experimental apparatus

Next as to the experimental apparatus itself by which we are to ascertain the relations between velocity and resistance in any given model.

Common method

The prima facie simplest way of doing this is that which has been generally adopted, namely that of applying a definite suspended weight acting through a delicate tow line as the tractive force, which force is thus in each experiment regarded as the measure of the resistance, and noting the time taken by the model to traverse certain distances.

Objections to common method

Now at the outset it is essential for the sake of avoiding 'personal error' that the results should be self-recorded, but under the above mentioned arrangement it will be found that owing to various difficulties such as the elasticity of the tow line, the dangers of causing adventitious friction etc., it is not easy to satisfy this requirement.

A still graver difficulty is that of properly guiding the model, for no model can be trusted to follow exactly the direction of the tow line, indeed some forms, in which the tendency to sheer would be least suspected, I have found to possess it to an astonishing degree. Now any guiding arrangement, such as a stretched wire, introduces a considerable yet irregular and unascertainable element of frictional resistance thus misapplying an unknown amount of the tractive force which cannot therefore be regarded as a correct measure of the resistance of the water to the model.

Improvement that I suggest on common method

If however, we substitute for the guiding wire a light railway with a truck travelling on it we may attach the model to the truck through the medium of a dynamometer, affixing the tow line to the truck instead of to the model, and the dynamometer while it guides the model inexorably will indicate purely and

* Should this seem an unnecessary distance, it must be remembered that though such a model might be started to its full speed in say 10 feet, yet the fluid motions proper to the speed take some time to become established in their normal shape, and the nature of these motions has an important effect on the resistance. For this reason it is desirable to start the model gradually.

solely the resistance experienced by the model in traversing the water. The rotation of the truck wheels will afford an exact measure of the distance as it is travelled by the model; and a 'time apparatus' such as I used in my former experiments, being added, we have all the necessary elements, which when duly combined in one apparatus will record by lines traced on a travelling sheet of paper the velocity and resistance of the model at each point of its course.

Since, under this arrangement, we have an exact record of the resistance, we may do away with the definite weight and delicate tow line; substituting for it an inexorable mechanical force, regulated by a chronometric governor which will ensure a uniform speed.

No doubt all this entails an apparatus of somewhat costly construction, but after giving the subject much trial and consideration I am confirmed in the opinion that it is the only sure way of obtaining a suitably accurate result.

What experiments should be tried and in what order

As to the experiments to be tried, it will as a preliminary be important to make a set of experiments on surface friction, chiefly in order to ascertain to what nicety it will be necessary in the construction of the models to bring them to one standard quality of surface, and these same experiments will furnish data from which the correct laws of surface friction may be determined.

Nextly before commending any methodical investigation of leading principles it will probably be well to make some experiments with existing forms adopted in the Navy, to determine whether any excessive resistance is experienced by any peculiar forms, such for instance as the double stern of some of the twin screw ships, etc., etc., and whether great length possesses that merit which has been attributed to it; inquiries of immediate practical value.

Proceeding to the methodical investigation it will perhaps be best to begin with successive variations in the forms of entrance and run, carried through every reasonable variety of form and proportion; then to examine the question of the best position for the greatest transverse section; and then to examine the effect of varying the proportion of length to breadth and to depth. It is however obvious that all these questions react on one another and that the course of investigation must consequently be in a great measure left to be directed according to the discoveries made as the inquiry proceeds.

IV.
(Scheme for carrying out experiments)

Assuming a prima facie case to have been made out in favour of establishing such appliances and apparatus and instituting such a course of experiments as I have recommended, it will become a question *where* and by *whom* the inquiry is to be conducted.

I propose to undertake them myself

And I believe it is no undue self assertion which leads me to suggest that on the score alike of efficiency and economy the work should be entrusted to myself at my residence here.

I will at the outset state that I am prepared gratuitously to devote my services to the investigation if my proposals are approved of, and if I am provided with the means of meeting the expenditure necessary for its pursuit.

With respect to the essential requirement of fitness for the work, though there are probably many men of science whose qualifications would be found in various respects superior to my own, yet, unless their services could be obtained gratuitously to place them on a footing or professional remuneration on a scale proportioned to their qualification would of itself introduce a very serious *additional* element into the cost of the inquiry.

But also it would be false delicacy on my part were I to have any scruple in asserting my entire and I may add my special competence; and I will explain more fully the grounds of the assertion.

It is a simple fact, not only that the subject is one to which I have devoted great study for very many years, but that I have devoted to it several years of almost exclusive attention, coupled with extensive experiments, carried on at a considerable expense to myself.

Facilities arising from having made previous similar experiments

I have thus acquired a large stock of apposite knowledge and matured habits of experimental inquiry. Indeed I may fairly express a doubt whether any other person has the advantage of me in this respect, or has produced as instructive a series of experimental results as that which I have already presented to the Chief Constructor.

I am besides, what is termed a good mechanic and I have a good workshop in my own house; I am now providing steam power which will be of material service in the operations I suggest.

Much of the apparatus which I have used in previous experiments is capable of being efficiently and easily adapted to the arrangement which I recommend and which I have already described, and I could manufacture at home the greater part of the additional machinery which will be required at a considerably less cost than if it were supplied in any other manner.

It will be extremely gratifying to me if the proposal I have made should be approved of but in offering my services thus gratuitously I may fairly make the following stipulations.

That I must conduct the experiments at my own residence

(i) If I carry out the experiments the establishment should be close to my own residence. This is necessary, not merely because in that way alone could I arrange to give my constant attention to the work, but also because nowhere else could I approach it with the same mechanical advantages.

(ii) As I have already pointed out there will at the outset probably be several important experiments to be tried with the various forms of ships at present in use in the Navy and in course of construction, and what is of more importance, with forms of vessels which are being designed or proposed for construction and indeed any series of experiments which may be in progress, can at any time without detriment thereto be suspended in order to try any special form of vessel, the properties of which it may be desired to determine. I should however wish to be allowed myself to have the responsible direction of the course of experiments forming the 'methodical investigation' since I could not pursue it effectively along a track divergent from my own natural course of thought.

Responsible direction to be in my own hands

(iii) During the construction of the necessary covered waterway and its accessories, which should be completed by contract under my direction, I should be provided with funds to make the requisite payments to the Contractor as well as for the tools and appliances to be used in model making and for the adaptation or completion of the experimental apparatus, and during the experiments, I should be repaid all incidental expenses including those required for the employment of an adequate staff. All such expenditure being of course properly accounted for.

Funds for original outlay and for working expenses to be provided

To place this proposal, as far as possible in a practical and definite shape, I have carefully endeavoured to ascertain the outlay which it involves.

Rental of proposed site

In the first place, as the land attached to my residence does not afford adequate level space for the formation of the waterway, I have provisionally obtained the consent of the owner of the surrounding land for the use of a sufficient portion of contiguous field on the condition that when the inquiry is concluded, I fill in the excavation and make good the land to its former surface, paying meanwhile in consideration of its use and its inevitable deterioration the sum of £12.10.0 per annum. The materials of the roof when removed will more than pay for making good the land.

Provisional tender for execution of work

In the next place I have prepared a complete set of drawings and a specification of the work to be executed comprising the waterway 250 feet long (having a depth of 10 feet, a width of 6 feet at the bottom, and of 26 feet at the top) roofed throughout, and having a drawing office and workshop attached and I have obtained from a firm on whom I can rely and for whom I would be responsible a tender for the execution of the work amounting to £584.

To fill and to keep filled, so large a water space, should the supply from my own well prove inadequate, it will be requisite to lay temporarily a three inch iron pipe for about ¾ of a mile in connection with the Torquay water service, and the cost of this will be £105.

Probable total cost

To add the necessary machinery for the model making and the completion of the experimental apparatus would probably bring the cost of the Establishment up to nearly £1,000. I will however engage if so required that that sum shall not be exceeded.

Probable working expenses

With regard to working expenses, I should require to employ probably in permanence, two competent workmen, one of them a skilled mechanic. I have two such men now in my employment who have long been engaged for me in similar work, and on whom I can rely, their united wages are 51/– per week, or if fully employed £133 per annum.

I shall also require the services of a competent draughtsman and assistant, for which purpose I should propose to employ my son, who is commencing his profession as a civil engineer and who has already served an apprenticeship in this description of work, having prepared the drawings for the models and the apparatus used in my former experiments, throughout which he was the principal operator in the experimental manipulations, and who, I may add, more thoroughly understands the work than any one else whose services I could similarly secure.

I should propose that he should be paid at a rate of say £3.3.0 per week for his time, which would probably be fully employed in the work, but which as well as that of the workmen would of course not be charged except when so employed.

These expenses in salary and wages alone would therefore amount to £289 per annum, but as other incidental expenses would inevitably be incurred, and indeed as with a view to more rapid progress it might be requisite to employ additional hands, it might be prudent to regard the working expenses as limited to, say £500 per annum, and indeed here, as in reference to the first cost of the establishment I would guarantee that the sum named should not be exceeded.

Probable duration of experiments

Though the experiments which I have contemplated would probably yield some useful results at a very early period it might occupy two years to complete the series exhaustively, so that the whole cost involved in them would be about £2,000.

Should this sum appear large certain considerations may at once be suggested to shew that the expenditure can hardly be regarded as unremunerative.

Assuming the enquiry to result in even nothing better than the saving of a permanent expenditure of Ten Horse Power in all, in the employed engine power of Her Majesty's Navy, the annual reduction in the national coal bill even thus would pay the interest of the outlay by which the saving had been effected, were it to render unnecessary half a dozen measured mile trials in the course of each year the same result in fuel saved would follow. Were it to result in beneficially taking 10 feet off the length of a single ironclad the whole outlay would probably be recouped at once.

But I think it is scarcely necessary to have recourse to such modest hypotheses for the justification of an outlay which when duly considered, promises to contribute materially to the speed and the general economy of the Navy.

(Signed) W. FROUDE

December, 1868
Chelston Cross,
Near Torquay

WM. FROUDE, *The Papers of Wm. Froude, 1810–1879*, The Institution of Naval Architects, London, 1955, p. 120.

15h Theory of Ships' Motion

Recommendation of Experiments on Rolling

THE mathematical theory of rolling is very far from easy, and leads to equations of which there is no known solution. The time of a common pendulum, for instance, depends upon an elliptic integral, and, beyond the degree of complexity involved in such a function, mathematics are in the condition of uncleared ground. Accordingly, while it is possible to give a rational account of the immediate gross results of a compound oscillation, these results cannot be expressed or measured with the requisite combination of generality and accuracy. In order to treat them, we are obliged to introduce simplifying suppositions, which do not necessarily belong to our problem—as, for instance, isochronism, or the neglect of certain elements of resistance, or the grouping of others.

Now, when this occurs with any branch of practical knowledge, the proper mode of applying mathematical investigation is to start, not from the known principles of general mechanics, but from an advanced base of observations peculiar to the science itself. In hydrodynamics, between minuteness and number, the ultimate molecular unit escapes our notice, and we are only able to observe effects in the gross; being thereby driven to a certain want of detail, both of observation and of reasoning, which allows us to trust our conclusions only when they have been made to rest on a broad experimental foundation. Whether we regard the theory of the propulsion of ships, or that of their rolling, our analysis has assuredly been pushed quite to the extreme verge to which general reasoning can be trusted; and a largely increased extent of exact observation ought to precede further attempts at inductive reasoning on these subjects. We have many exact experiments on propulsion, although, from the complicated character of the phenomena involved, it is difficult to separate the issues; and this will probably not be set right without further special investigation. With regard to rolling, however, we have much vague observation, and but little exact knowledge derived from experiment.

We are not aware of any one published experiment on the rolling of ships in waves in which the details necessary to make any mathematical use of the results are supplied. The data required are, as a minimum for each case,—

1. A draught of the ship, and her calculated elements.
2. The position of her centre of gravity.

3. Her periodic time in still water.
4. The condition of her wet surface.
5. The extent and period of her roll.
6. Was the rolling simple, or mixed with pitching?
7. The height, length, and period of the waves in which she was rolling.
8. Were these waves simple?
9. What alterations have been made in her displacement, her trim, and the position of her weights, as regards both centre of gravity and moment of inertia, previously to the trial?
10. Force and direction of wind, and condition of ship as regards resistance to it.
11. Full details as to manner in which, and the instruments or calculations by which, these data have been ascertained.

There is no doubt that for a comprehensive view of the subject, it would be necessary that these things should be ascertained with care for a large number of ships, of various classes, and under very varied conditions. But this is too much to expect to get done, although we think it would be a good thing for the Government, and other large shipowners, to keep in view as an ultimate object. Meanwhile we think it would be a very great experimental aid to science if these things could be accurately settled for even two or three ships, under different circumstances of weather and different arrangements of weight, both in amount and distribution.

Similar experiments should also be made with reference to pitching.

The trials should be made with sails furled, and as little disturbance from headway as possible. We have every wish to have parallel experiments tried under any possible conditions of sail and propulsion, and, if it may be done, on the same ships, consecutively with the simpler experiments; but it will be seen that the data are already sufficiently complex at the best, and that they must be used clear of headway and leeway before they can be discussed with reference to these.

No experiments are of use for the purpose of inductive reasoning in which any one of the data mentioned above are wanting.

We think that the Government might fairly be asked to institute such a set of calculations and experiments. We cannot find that the exact information which we have suggested is in existence anywhere. We are certain that it has not been published in any available form; and we have reason to believe that the knowledge is quite as much needed and desired by the gentlemen responsible for the construction of the navy as by merchant builders or by students of theory.

We therefore recommend that the deputation previously mentioned with reference to the experiments on resistance be also instructed to urge upon the Admiralty the importance, both practical and theoretical, of instituting such a set of experiments, of providing suitable instruments for recording exact observations, and of publishing the results. We also recommend the appointment by the Council of the Association of a committee of *three* members to confer with the officers of the Admiralty as to the drawing up of detailed instructions for conducting these experiments; and that the Lords of the Admiralty, in the event of their assenting to the proposals, be requested to nominate a committee to confer with the committee named by the Association.

In conclusion, we beg leave to recommend that this Report be officially communicated to the Councils of the Institution of Naval Architects, the Institution of Civil Engineers, and the Institution of Engineers in Scotland, and the cooperation of those bodies sought, both in applying to the Government and in making known among shipbuilders, and other persons connected with Naval Architecture, as well what is the state of our existing knowledge as what are the immediate desiderata for its extension.

CHARLES W. MERRIFIELD W. J. MACQUORN RANKINE
GEORGE P. BIDDER W. FROUDE (*subject to the following explanations*)
DOUGLAS GALTON

Mr. Froude's Explanations

The subject of a ship's resistance is one which I have for many years been independently investigating, both theoretically and experimentally; and I have been thus led to conclusions which are in very material respects at variance with those which Mr. Merrifield has placed on record for the Committee as representing the existing state of knowledge respecting it, and specially at variance with the consequent recommendations which he has drawn up, as indicating the experiments for the performance of which the assistance of Her Majesty's Government is to be sought: I thus find myself somewhat abruptly placed in a position in which I must ask permission to present, as part of our proceedings, a supplementary report explaining the reasons which oblige me to dissent from the recommendations to which I refer. Until the Draft Report was in my hands, I was unaware that 'Resistance' was regarded as included in the list of subjects submitted to the Committee; for I understood the terms 'Stability' and 'Seagoing qualities' as having reference to the theory of 'Rolling motion,' and 'Propulsion' to the theory of the Action of Propellers. The subject of Resistance appeared to belong already to the 'Steamship Performance Committee.'

Let me say at the outset that Mr. Merrifield's very full discussion of this subject appears to me to set forth most lucidly what must be called 'the existing state of knowledge' respecting it; it has evidently involved much laborious research and deep consideration.

And, on the other hand, the results at which I have arrived are in many respects so far from complete that I have hitherto hesitated to bring them before the public. But I believe I have so fully established those conclusions, to which I shall now refer, and the difference in the line of action to which they point is so serious that, under the present circumstances, I feel bound to press them on the notice of the Committee.

The Report specially recommends, as the experiment which it is important to try, the dynamometric determination of the scale of resistances for a full-sized ship.

Now, without impugning, or rather, while fully asserting that any scale of resistance, in terms of velocity, accurately determined for a full-sized ship, would be of real value and of great interest, I shall nevertheless contend (1) that experiments on the resistances of models of rational size, when rationally dealt with, by no means deserve the mistrust with which they are usually regarded, but, on the contrary, can be relied on as truly representing the resistances of the ships of which they are the models; and (2) that in order properly to open up the question, so great a variety of forms ought to be tried that it would be impossible, alike on the score of time and expenditure, to perform the experiments with full-sized ships. Both these propositions require to be drawn out at some length. The kindred proposition, that as accurate results can be obtained far more easily and rapidly in experimenting with a model than with a ship, though of great importance, is so obviously true as to require no elucidation.

The natural expectation that the ascertained resistance of a model will furnish a measure of the resistance of a ship similar to the model, depends on the *primâ facie* probability that the resistance for a given body will vary as the square of its velocity; and that in comparing similar bodies of different dimension at a given velocity, the resistance will be as the square of the dimension, since that function expresses alike the proportion of the respective midship sections and of the respective friction-bearing surfaces. Were these propositions true, the ascertained resistance of a model, at given velocity, would supply a complete scale of resistance for all velocities, both for the model and for any ship similar to the model.

Since, however, the resistance of a model or ship deviates from the law of the square of the velocity, as under certain circumstances it is known to do, in a manner dependent on its actual dimensions, it is obvious that the simple scale of comparison, which seemed *primâ facie* probable, can be no longer accepted, and it has hence been hastily concluded that no assignable scale of comparison can be found instead.

Now it appears to me to be pretty well established, and it is scarcely questioned, that, for deeply submerged bodies of tolerable size and fair shape, the resistance does follow the law of the squares with a high degree of approximation. Such deviations from this law as appear in Beaufoy's experiments are, I think, explicable by the angularity of the shapes tried and by the mode of trying the experiments, under which the considerable distance between the bodies tried and the conducting float by which they were carried involved some deviation of the body from the true axial motion, when the velocity and the consequent resistance became considerable.

That surface-friction, in particular, follows the law of the squares of the velocity very closely, is well established by the experience of the flow of water through pipes, in reference to which, I may observe, I have myself experimentally verified on a five-mile length of 9-inch pipe, the law that the delivery is almost exactly as the square root of the steepness of the hydraulic gradient. The experiments were tried with very great variations in the steepness*. Now Professor Rankine's admirable stream-line investigations have definitely established the conclusion that for symmetrically shaped bodies of 'fair' lines, not excluding by that description certain very blunt-ended ovals, when wholly submerged, the entire resistance depends on the conditions of imperfect fluidity, of which surface-friction is the only one so considerable that we need take account of if we deal with bodies of rational dimensions; and this, as I have pointed out, does follow the law of the squares. I set aside the condition of 'viscosity'; for though this defect, even as it exists in water, is certainly sufficient to affect differently the resistances of bodies of different dimensions, this is not sensibly the case unless the bodies are very minute; and having regard to the great vitality of such small surface-waves as (say) one foot in length, and to the fact that discharge of water through pipes and orifices exhibits no results indicative of this special action, unless the diameters are very small indeed, it seems extremely improbable that the resistances of bodies five or six feet in length will be affected by it. If, therefore, we were dealing with submerged bodies, we should have no reason to mistrust the *primâ facie* deductions founded on experiments with models.

When, however, we deal with a body moving at the surface, we at once meet with a *vera causa*, which alters those simple relations that exist between the resistances of differently dimensioned submerged bodies. This *vera causa* is the generation of surface-waves, which accompanies the transit of the body along the

* Though I regard these experiments as sufficiently conclusive in reference to the point to which they were directed, I am inclined to think that the theory of surface-friction in its application to a ship's resistance requires considerable revision.

surface; and it is, I believe, not merely the only known cause, but a sufficient one.

What absolute conformation and magnitude of waves a given vessel moving with a given velocity may create, and what excess of resistance may thus be developed in any individual instance, it is not necessary for the present purpose to determine; for it will appear that, whatever the excess may be, however abnormally, in virtue of it, the law of resistance for any given ship may vary in terms of her velocity, a very simple scale of comparison will express the relation between the excess as developed in a model and as developed in a ship similar to the model when moving at a corresponding velocity. I shall show, in fact, that if the velocities of the ship and the model are as the square roots, these excesses of resistance thus arising will be as the cubes of their respective dimensions, a law which, as is easily seen, expresses also the relation founded on those elements of resistance which vary as the square of the velocity and as the squares of the respective dimensions.

The principles on which Professor Rankine's stream-line investigations are founded establish generally, in relation to all wholly submerged symmetrical bodies moving in a fluid infinitely extended on all sides, that the stream-line displacements which the motion of the body imposes on the surrounding volumes of fluid are, for a given body, identical in configuration for all velocities (an identity which assigns to them always a velocity proportional to that of the body itself), and that the configuration is similar for all similar bodies.

If we now suppose that the body is moving along the surface of the fluid, and if we imagine the surface to be not under the influence of gravity or any such force, it is obvious that here also the configurations of the stream-line displacements will be identical at all velocities for the same body, and will be similar for similar bodies, including those displacements which consist of upward disturbances of the surface.

When we impose the further condition appropriate to an existing water-surface, that the replacements of the surface, when disturbed, are governed jointly by gravity and by the volumes and velocities of the original impulses of disturbance, it follows that those impulses of disturbance, being similar for all similar bodies at all velocities, will retain their similarity whenever and in the manner which the operation of gravity permits: and this will be when the similar bodies are moved with velocities proportioned to the square roots of their respective dimensions; in these a similar wave-configuration will, in each case, similarly dispose of the originally similar volume of displacement, since similar waves have their velocities so related.

These waves (as is explained by Professor Rankine), when the velocity proper to their length along the line of motion is exceeded by that of the ship, so that they cannot squarely travel with her, satisfy the conditions of their motion by travelling obliquely, and diverging into the surrounding fluid, the angle of divergence and their size forming a measure of the work constantly running away from the ship, and consequently of the resistance caused by their generation.

Now the similarity of the configuration which has been asserted involves the condition that when the velocities of similar ships are as the square roots of their respective dimensions, the angles of divergence will be equal, and therefore equal lengths of similar wave-crest will be 'run off' for equal distances travelled by the respective ships; and hence the energy abstracted in each case by these equal lengths of similar wave-crest is clearly as the cube of the dimension (since the mass elevated is as the sectional area, and the elevation is simply as the dimension); and since the forces which supply proportionate amounts of energy

By courtesy of the Royal Institution

Plate I

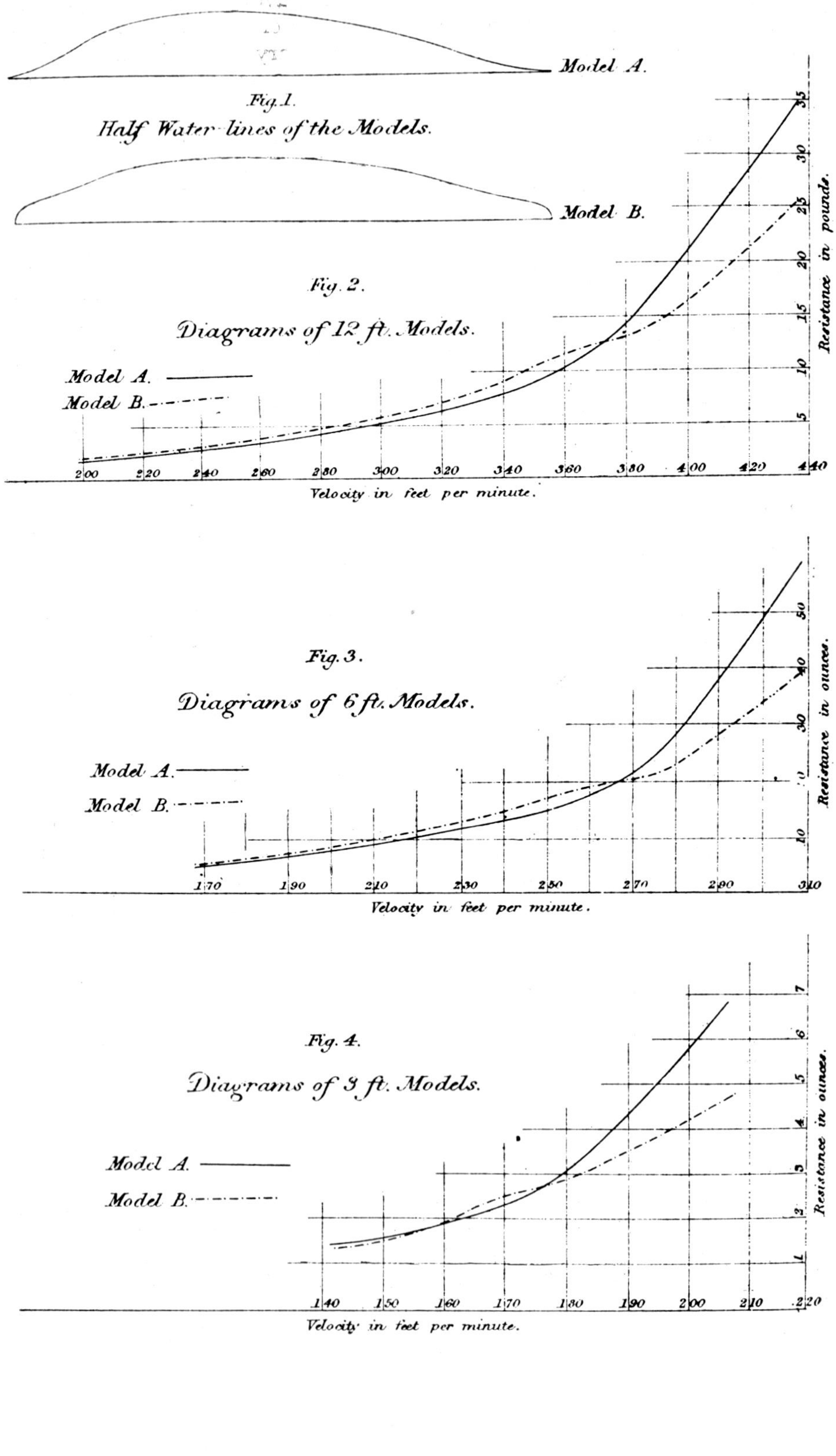

while travelling a given distance must be as the energy, it follows that the excesses of resistance thus called into existence are also as the cube of the dimension, agreeing in this respect, as has already been pointed out, with the resistances derived from the surface-friction. In fact, we are thus brought to the scale of comparison which was just now enunciated, that the entire resistances of a ship, and similar model, are as the cubes of their respective dimensions, if their velocities are as the square roots of their dimensions.

In verification and illustration of the foregoing views, I tried, in the autumn of 1867, a large number of resistance-experiments with a pair of models of contrasted forms, six feet long, by towing them simultaneously from the ends of a pair of ten-foot scale-beams connected with self-recording dynamometric apparatus, and mounted on booms projecting sideways from the nose of a steam-launch, lent me for the purpose by Mr. Bidder. The water-lines of the models are shown in Plate I. fig. 1. One was of the wave-line type, the other, having the same length, form of midship-section, and displacement, had large rounded ends. I also tried similar experiments with a pair of very nearly similar models of twice the dimensions and eight times the displacement. I had already obtained a series of experimental results of the same kind, but with less successful apparatus, from a similar pair of models, three feet long. These data enabled me to compile for each model a diagram of resistance in terms of velocity.

The three pairs of such diagrams, proper to the three pairs of models, were laid down to scales corresponding to the dimensions of the models, according to the system of comparison I have enunciated; thus the velocity-scale for the six-foot models is $\sqrt{2}$ times, and that for the three-foot models twice as open as that for the twelve-foot models; and the resistance-scales for the six-foot and three-foot are respectively 8 and 64 times as open as that for the twelve-foot. According to my proposition, were the three sets of models exactly similar the three sets of diagrams should be identical.

Reduced copies of these diagrams are shown in Plate I. figs. 2, 3, 4. Their general agreement, especially as to the position occupied in the velocity-scale by the several salient features of the curves and as to the relative resistances of the contrasted forms, is very striking. It is true that on comparing the absolute resistances, the correspondence is not so close as it at first sight appears. Thus the three-foot models exhibit throughout a decided excess of resistance as compared with the six-foot; but I think this is probably attributable to their being small enough to be within the range of viscosity. On comparing the diagrams of the twelve-foot and six-foot models, however, we find that it is the larger model that has an excess of resistance. This excess, which is slight, may be partly due to certain minor differences of form which had been introduced in the larger models. It may also have partly arisen from the fact that the twelve-foot models, owing to their greater dimensions, swam relatively nearer to the towing-boat, a circumstance which may naturally have tended to enhance their resistances.

On the whole, I think that series of diagrams supplies a very fair verification of the alleged scale of comparison.

Besides thus throwing light on the question of comparison of the performance of similar vessels of different dimensions, these experiments show very clearly that strange forms may possess merits that are entirely unknown and unexpected before experiment is made upon them; for here we find that an abnormal form (suggested simply by the appearance of water-birds when swimming), if moving with a high though not excessive velocity, experiences considerably less resistance than the wave-line form, the accredited representative of the form of least resistance, particularly at high speeds. This proves that we can have no ground for certainty that we have found even an approximation to the best form, unless we have gone experimentally over almost the whole ground and tested a very wide variety of shape. But, independently of this aspect of the question, it is, I think, certain that on very many important questions, such as, for instance, the proper ratio of length to breadth, there is no really established principle of judgment on which reliance can be placed. Yet most weighty considerations affecting economy and efficiency are involved in the settlement of even that single question. But unless we build mere experimental ship-sized models, there seems no possibility of determining the question by full-scale experiments.

It is true that the circumstances under which my experiments were tried did not admit of such exactness as to render them absolutely conclusive as the sole basis of the theory of comparative resistance in terms of dimension. Nor do I by any means pretend to be certain that there are no elements of resistance other than I have taken account of in my theoretical justification of it; but if any such do exist, they can be detected, and the laws of their operation discovered with far greater facility and completeness by small-scale than by full-scale experiments. And I contend that unless the reliability of small-scale experiments is emphatically disproved, it is useless to spend vast sums of money upon full-size trials, which, after all, may be misdirected, unless the ground is thoroughly cleared beforehand by an exhaustive investigation on small scale.

WM. FROUDE *et al.*, 'On the Stability, Propulsion, and Sea-going Qualities of Ships', *Rep. Brit. Assoc. Adv. Sci.*, 1869, p. 41.

15i Theory of Rivers

On Certain Laws relating to the Régime of Rivers and Estuaries, and on the Possibility of Experiments on a small scale. By Professor Osborne Reynolds, *F.R.S.*

[A communication ordered by the General Committee to be printed *in extenso* among the Reports]

1. THE object of this communication is to bring before Section G certain results and conclusions with respect to the action of water to arrange loose granular material over which it may be flowing. These results and conclusions were in the first instance arrived at during a long-continued investigation, undertaken with a view to bring the general theory of hydrodynamics into accord with experience, rather than with any special reference to the subject in hand, but have since been to some extent made the subject of special investigation.

2. A systematic study of the *régime* of rivers naturally divides itself under three heads, which may be stated as follows:—

(1) The more general facts observed as regards the regimen of the beds.

(2) The movements of sand consistent with these observed facts.

(3) The necessary actions of the water to produce these movements in the material of the beds.

Observed facts.—Amongst the most general facts to be observed as to the arrangement of the material forming the beds of estuaries are—

(1) The general stability or steadiness of these beds, so far as is shown by their outline or figure, while, at the same time, as is shown by the obliteration

of all footprints and markings casually placed upon them, also by the ripple mark, the material at the surface of these beds is being continually shifted.

(2) The almost absolute steadiness in figure of some of these beds.

(3) The gradual changes in the position and form of others—the growth or accumulation of sand-banks in some places, and the wasting of banks or removal of sand in others.

Movement of sand.—As regards the movement of sand consistent with these changes, in the first place the movement, whatever it may be, is one of the surface, and not one in bulk; and in the next place such movement of the surface must be continually going on, whether it produces any change in the figure of the banks or not. The invariable obliteration of footprints and marks which may have been left on the sand at low water, as well as the ripple marks, are absolute evidence of a general disturbance of the surface, and it requires but little observation to show that this disturbance is of the character of a drift of sand, in whatever direction the water may be moving.

Uniform drift.—Where the outline of the banks is not altered, this drift or motion of the sand must be uniform, as much sand being deposited at each point as is removed from that point. Although there may be a general flow of the sand in some direction, if the drift is uniform this movement will not alter the figure of the bed, which, like the balance in another kind of bank, does not depend on the rate of deposit and withdrawal, but on the excess of one of these over the other. The gradual accumulation or diminution of sand at any point is clearly not due to a simple action of deposit or removal, as they are always attended with the same evidence of the drifting of the surface, and are clearly the result of a difference in the quantities of sand deposited or removed by the drift.

Movement of water.—The manner in which a current of water acts on the granular material forming the bed of the current has been the subject of an investigation by various experimenters. It has been found that the primary action is not so much to drag the grains along the bottom, but to pick them up, hold them in a kind of eddying suspension, at a greater or less height above the bed, for a certain distance and then drop them, so that when the water is drifting the sand there is a layer of water adjacent to the bottom of a greater or less thickness charged to a greater or less extent with sand. The faster the current and the finer the sand the greater will be the thickness of the charged layer, as well as the denser is the charge in the layer.

A certain definite velocity, according to the size and weight of the grains, is required before the water will raise the grains from the bottom, and for all velocities above the minimum necessary to raise the sand the suspended charge increases with the velocity, and the rate of drift or the quantity of sand which passes a particular section increases much faster than the velocity. Attempts have been made with greater or less success to determine exact laws connecting the minimum velocities at which the sand begins to drift with the weight of the grains and other circumstances; also to determine the exact law of the rate of increase of the drift with the velocity.

For my present purpose, however, it is not necessary to enter upon such considerations.

From the facts already mentioned, it will appear that the effect of a uniform current of water over a uniform bed of sand will not be to raise or lower the bed; for, as the charge of sand in the water remains uniform, it must drop as many particles as it raises everywhere on the bed. This is the action of the water in causing a uniform drift.

It is also evident that, if the charge in the water as it comes to any particular place is less than the full charge due to its velocity, it will pick up from that place more sand than it drops, and so increase its charge at the expense of the bed, which will there be scoured or lowered. And conversely, if the water as it arrives at any place is overcharged, it will relieve itself by depositing more than it picks up, and so raise or silt up the bed.

As regards the circumstances which can cause the water to be charged to a greater or less extent than that which it would just maintain with such velocity as it has, the most important are—

(1) An increasing or diminishing velocity. When the water is moving in a stream from a point where the velocity is less to one where it is greater, the velocity of the actual water as it moves along is increasing, as will also be its normal charge of sand; hence it must be continually picking up more than it deposits. And conversely, when moving from a point of greater velocity to one of less, its normal charge will be continually diminishing through deposits on the bed.

(2) Another circumstance which affects the charge of sand with which the water may arrive at a particular point is a variation in the character of the bed. If, for instance, water flows from a rocky bed on to sand, it may arrive on the sand without charge, and immediately charges itself at the expense of the bed. Or again, where water flows from a sandy bottom on to a clean or grassy rocky bottom, it gradually loses its charge silting up the bottom.

The direction in which the sand is moved by the water is sensibly in the direction in which the water which holds the charge is moving. But, as was first pointed out by Dr. James Thomson as affording an explanation of the generally observed fact that the beds of rivers are scoured on their convex sides and silted on their concave, the layers of water adjacent to the bed do not always move in the general direction of the stream. There are often steady cross currents at the bottom, as in the case mentioned, though such cross currents do not exist except under circumstances which may be readily distinguished. The most important of these is that pointed out by Dr. Thomson—curvature in the general direction of the stream, in which case the centrifugal force of the more rapidly moving water above overbalances that of the water retarded by the bottom, and forces the latter back towards the centre of the curve.

This action is universal, where even the lateral boundaries are such as to require water to move in curved streams; the drift at the bottom does not follow the general direction of the stream, but sets towards the centre of the curve.

The result of the foregoing consideration is to lead to the conclusion that the *régime* of each part of the bed as to maintenance in steady condition, lowering or raising it any time, depends solely on the character of the motion of the water, which if straight and uniform, neither acquiring nor losing velocity, causes a uniform drift in the direction of the stream, which maintains the condition steady. If losing velocity the motion causes a depositing drift and raises the bed; if gaining velocity it causes a scouring drift and lowers the bed; while if curved, the direction of the drift is diverted towards the centre of the curve with its attendant effect to lower the convex side and raise the concave side of the bed. This conclusion seems to be of the utmost importance in dealing with this subject. For if it is correct, not only can the character of the action going on at the bed be inferred from the observed motion of the water, and *vice versâ*, but since, according to this conclusion, the character of the action is independent of the magnitude or velocity of the stream, the results will be the same on a small scale as on a large one, provided only that the character of the motion of the water is the same at all points. In this latter respect this conclusion affords an explanation of a fact that cannot fail to have struck everyone who has observed the sand-beds of the streams running over sands which have been left by the tide, viz., what an almost exact resemblance they bear to each other, whether having the size of a moderate river or of the smallest rivulet.

On the large scale of actual estuaries we can only test the conclusion by actual observation, but on a small scale we can experimentalise in whatever condition of motion we want to test, and readily observe the effects produced; a possibility of which great use has been made in this investigation, and which will be again referred to.

As applied to a *non-tidal* river, in which the direction of the motion is always the same, the foregoing conclusion would lead us to expect that the *régime* would be steady except at the bends, the sources, and the mouth, which is exactly what is observed, so that the conclusion so far agrees with experience. The most striking feature about rivers is the way they wriggle about in the alluvial valleys; a phenomenon pointed to by Lyell as one of those causes still in progress which had produced the present conditions of the valleys, and which, as already stated, was explained by Dr. Thomson. From the source of the river as the rain-water acquires the velocity, it charges itself with deposit, which charge it maintains with continual taxes and drawbacks until it reaches the ocean or lake, when its water in again losing its velocity deposits its charge, continually carrying forward the bar and extending its delta.

In non-tidal rivers, whether large or small, fast or slow, the characters of these actions are invariable, however much they may differ in intensity. The case of tidal estuaries is, however, by no means so simple. Here we have not, as in a river, a continuous progression of the same character of action at the same point. On the contrary, at every point the action is changed twice a day. For the change in the tidal current does not merely change or reverse the direction of the sand-drift at each part of the bed, but it changes and often reverses the character of this drift, changing what has been a scourging drift during the ebb-tide into a depositing drift during the flood; so that the question as to whether the *régime* is stable, depositing, or scouring is not simply a question as to whether the current at this point is uniform, accelerated, or retarded, but whether the action of the ebb to cause, say, scour is equal to, less than, or greater than the action of the flood to cause deposit.

As there is no likelihood that the resultant effect as regards the general *régime* of two opposing influences will resemble what would have been the simple effect of either of the influences acting alone, this dual control affords abundant reason why the configuration of the beds of these tidal estuaries should differ in character from the configuration of the sand-beds of continuous streams.

There is, however, another and an equally important difference between the general motion of the water in rivers and tidal estuaries.

The function of the estuary is by no means that of a simple channel to conduct the tidal water up and down. It equally discharges the function of a reservoir or basin, to be filled and emptied by each tide.

In consequence of this action as a reservoir the directions of the motions of the water during flood and ebb, and particularly towards the top of the flood and commencement of the ebb, are generally very different from what they would be were the estuary acting the simple part of a channel conducting the water from somewhere to somewhere.

When a vessel is filled by a stream entering on one side the forward motion of the water is stopped before reaching the opposite side. But if, as is always the case, the motion which the water has on entering is more than sufficient to carry it as far as is necessary, the remaining momentum is spent in setting up eddies, or a general circulation in the water, so that when the vessel is full the water within it is not by any means at rest, but may be circulating round or have any other motion. If, then, the water is allowed to flow out the initial motion will not be a steady movement towards the outlet from all parts of the vessel, but those portions of the water which are moving towards the outlet will have their motion accelerated, while those which are moving in the opposite direction will have first to be stopped before they begin to approach the outlet. And thus the ebb will begin earlier at some points in the vessel than at others.

It was the observation of such an effect as this in one of our largest estuaries that first directed my attention to the subject of this paper.

Having investigated this point sufficiently for my own satisfaction nothing further was done until 1885, when my attention was directed to the inner estuary of the Mersey.

This estuary may be described as a crescent-shaped shallow pan, eleven miles long by three broad, lying north-west and south-east, having its upper horn pointing east and its lower horn north; the northern horn, being prolonged for five miles into a narrow deep channel, runs north to the outer estuary or sandy bay of the sea. One of the most marked features presented by the configuration of the bed of this inner estuary is the invariable preference of the low-tide channels for the concave or Lancashire side; whereas, were the estuary acting merely the part of a river, whether during flood or ebb, it would be expected to follow the usual law, and have the deepest water on the convex or Cheshire side.

That this prevalence of the deepest water on the concave side must be the result of the momentum left in the water by the flood at once seemed to me probable; for if the bottom were level or deepest on the Lancashire side the effect of the curved shape would be to cause the flood entering at the northern horn to follow the south-eastern or Cheshire shore, and the momentum of this water would tend to carry it round the head of the estuary and back along the Lancashire side; would, in fact, tend to set up a circulation before the top of the flood was reached; so that on the Lancashire side the water would be moving down the estuary before the ebb commenced; whence, considering that the flood tends to raise the bottom and the ebb to lower it (for the reasons already pointed out), it seems that the stronger flood on the Cheshire side would raise this side, while the stronger ebb on the Lancashire side would lower this. This is supposing the bottom to be level.

In order to verify these conclusions a vessel was constructed having a flat bottom and a vertical boundary of the same shape as the high-tide line of the inner estuary from the rock to the same distance above Runcorn. The horizontal scale was 2″ to a mile, and the vertical scale 1 inch to 80 feet, $\frac{1}{31800}$.

A shallow tin pan was hinged on to the otherwise open channel at the rock, by raising and lowering which, when full of water, the motion of the tide could be produced throughout the model through the narrows; the true form of the bed of the channel was given to the model by means of paraffin. And in order to obtain approximately the proportional depth in the inner estuary, sand was placed level on the bottom so that the high-tide depth was reduced to the equivalent of about twenty feet. The idea in making this model was not so much to obtain a shifting of the sand as to show the circulation of the water as resulting from the flood tide with a level bottom. In the first instance the tide pan was raised and lowered by hand, but as at the first trial it became evident that the model was not only going to show the expected circulation, but was also capable of showing, by the change in the position of the sand, the effect of this circulation on the configuration of the estuary and other important effects, it was arranged that the model should be worked from a continuously running shaft. The working of the model by hand at once showed that there was only one period of working at which the motion of the water in the model would imitate the motions of the actual tide in the Mersey, which period was found to be about forty seconds; a result that might have been foreseen from the theory of wave motions, since the scale of velocities varies as the square roots of the scales of wave heights, so that the velocities in the model which would correspond to the velocities in the channel would be as the square roots of the vertical scales —about $\frac{1}{33}$—and the ratios of the periods would be the ratio of horizontal scales divided by this ratio of velocities, or

$$\frac{33}{31800} = \frac{1}{950}.$$

Hence, taking 11·25 hours 40,700 seconds as the tidal period, the period of the model

$$= \frac{40700}{950} = 42 \text{ seconds (about).}$$

This period was adopted for working the model from the shaft.

It was then found that the circulation at the top of the flood, which was very evident while the bottom was flat, caused a general rise of the sand on the Cheshire side and lowering on the Lancashire, which went on for about 2,000 tides. That during this time, owing to the increase of flood up the Lancashire side and the diminution of that on the Cheshire side which followed from the deepening of the one and the shoaling of the other, the circulation steadily diminished until its character was so changed that it could no longer be called a general circulation, and that after this, although there were further changes in detail going on in the estuary, the two sides maintained a steady condition as regards depth for low tides.

During this time banks were formed and low-tide channels, which resembled in all the principal features those actually in the Mersey; the eastern bank, with the deep sloynes on the Cheshire side, the Devil's Bank and the Garston Channel, the Ellesmere Channel and the deep water in Dungeon Bay and at Dingle Point—all these were very marked in character and closely approximate in scale.

And, what is as important, the causes of these as well as all minor features could be distinctly seen in the model.

The eastern and Devil's Bank are seen during the process of their formation to be simply an internal bar formed by carrying the sand brought down by the ebb out of the narrows and sloyne, until debouching into the broad estuary; its velocity is so far diminished that it can no longer carry its charge, just as happens at the mouth of every river. The peculiar configuration of these banks is explained by the existence of two lines of eddies from about half-tide to the top of the flood: the first of these is caused by the sharp corner at Dingle, and lies between Dingle and Garston, the eddies having their centres over the Devil's Bank; and the second, caused by the divergence of the Cheshire Bank towards Eastham, having the lines of centres over the Eastham Bank. These eddies, which during the most rapid part of the flood only affect a diminution of the velocity of the flood, cause, as the velocity slackens toward the top of the flood, back water to set in along both shores, while back waters, starting the ebb, cause this to be strongest over the Garston and Eastham Channels, which are thus kept open.

The lateral configuration of the shores at Dungeon Bay and at Ellesmere is seen to cause back waters to exist in these bays during the whole of the flood in the latter, and from one to two hours before the top of the flood in the former, which fully accounts for the deep water at these points. The existence of these back waters in the actual channel has been verified. There are many other circumstances brought to light by this model, which it is impossible for me here to notice without unduly extending the length of this paper, if, indeed, I have not already done so. I will therefore only remark that a second start was made with the sand flat in this second model, and that the result obtained was the same as regards the general features of the estuary. So interesting were these results that it was decided to try a larger scale. A model, having a horizontal scale of 6 inches to a mile, and a vertical scale of 33 feet to an inch, was therefore made, and the tide produced as before. The calculated period of this model is 80 seconds, and experiment bears this out, any variation leading to some tidal phenomena, such as bonos or standing waves, which are not observed in the estuary.

The disadvantage of the larger model is the time occupied—a little more than a minute a tide—which means about 300 tides a day, or 2,000 tides a week. On one occasion the model was kept going for 6,000 tides, and a survey was then made of the state of the sand. And this will be seen to present a remarkable resemblance in the general features to the charts of the Mersey, of which three—1861, 1871, 1881—are shown; in fact the survey from the model presents as great a resemblance to any one of these as they do to each other.

It is impossible for me to enter upon all the points of agreement. Taking into account that in both the estuary and the model there are always changes going on within certain limits, and these changes do affect the currents to a certain extent, it is not to be supposed that there will be exact agreement between the currents at all points and at all states of the tides on the model and estuary. Still there is a general agreement, and in the few verifications I have made I have found that the current found in the model at a particular point and state of tide is also to be found in the estuary.

In one respect the great difference between the model and the estuary calls for remark: this is the much greater depth of the model as compared with its length and breadth. The vertical scale being 33 feet to an inch, and the horizontal scale 880 feet to an inch, so that the vertical heights are nearly twenty-seven times greater than the horizontal distances, such a difference is necessary to get any results at all with such small scale models; and it is only natural to suppose that it would materially affect the action. As a matter of fact, however, it does not seem to do so. And, further, it would seem that, notwithstanding the general resemblance on the *régime* of the beds of large and small streams running over sand, there is in these a similar difference in vertical scale, the smaller streams not only having a greater slope, but also having greater depth as compared with their breadth and steeper banks. So far as the theory of hydrodynamics will apply, it seems that in the model the effects of the momentum of the water would be greater as compared with the bottom resistance than in the estuary, and I think that they are. But the effects of momentum in the estuary greatly preponderate on the resistances, as shown by the fact that the tide at the top of the flood rises some 2 to 3 feet higher at high spring tides than it does at the rock; nor does it do much more than this in the model. In the model it certainly seems that the general *régime* is determined by the momentum effects, and from the almost exact resemblance which this *régime* bears to that of the estuary, it would seem that, although the momentum effects may be diminished by the greater resistance on the bottom, they are still the prevailing influence in determining the configuration of the banks. Further investigation will doubtless explain this, and also determine the best proportional depths. From my present experience, in constructing another model, I should adopt a somewhat greater exaggeration of the vertical scale. In the meantime I have called attention to these results, because this method of experimenting seems to afford a ready means of investigating and determining beforehand the effects of any proposed estuary or harbour works; a means which, after what I have seen, I should feel it madness to neglect before entering upon any costly undertaking.

I have only to say that, as it was not practicable to exhibit the model to the Section, I have had it working in the new engineering laboratory through the college. Unfortunately it could not be started before Monday, and it will not yet have run more than 1,000 tides, since the sand was put in flat, so that it is not probable that the *régime* is yet quite stable; still the principal features have come out.

OSBORNE REYNOLDS, 'On Certain Laws relating to the Régime of Rivers and Estuaries', *Rep. Brit. Assoc. Adv. Sci.*, 1888, p. 555.

15j Swan's Electric Lamp

By courtesy of the Patent Office

A.D. 1880, 2nd JANUARY. N° 18.

Electric Lamps.

LETTERS PATENT to Joseph Wilson Swan, of Newcastle on Tyne, in the County of Newcastle, Chemist, for an Invention of "IMPROVEMENTS IN ELECTRIC LAMPS."

PROVISIONAL SPECIFICATION left by the said Joseph Wilson Swan at the Office of the Commissioners of Patents on the 2nd January 1880.

JOSEPH WILSON SWAN, of Newcastle-on-Tyne, in the County of Newcastle, Chemist. "IMPROVEMENTS IN ELECTRIC LAMPS."

It is well known that in that class of electric lamps in which light is produced by the incandesence of a carbon conductor contained in a closely sealed globe of glass more or less exhausted of air two difficulties have hitherto impaired their utility.

1st. The carbon conductor in process of use gradually wasted and ultimately broke.

2nd. The interior surface of the globe containing the incandescent carbon became coated with a carbonaceous deposit, which obscured the light.

I have discovered the means of avoiding both these faults (which are dependent the one upon the other) as follows, that is to say:—

I have found that if the globe or vessel of glass containing the carbon to be rendered incandescent be exhausted of air to a high degree of exhaustion by means of a "Sprengel pump," whilst the carbon contained within it is raised (by the passage of a current of electricity through it) to the same or a higher degree of incandescence as that ultimately required to be produced in using the lamp, and if the lamp be hermetically sealed whilst exhausted of air under the conditions specified the lamp so prepared is durable, both the faults above mentioned being eliminated, the incandescent carbon enduring without waste, and the interior surface of the glass globe keeping free from carbonaceous deposit.

SPECIFICATION in pursuance of the conditions of the Letters Patent filed by the said Joseph Wilson Swan in the Great Seal Patent Office on the 2nd July 1880.

JOSEPH WILSON SWAN, of Newcastle on Tyne, in the County of Newcastle, Chemist. "IMPROVEMENTS IN ELECTRIC LAMPS."

It is well known that in that class of electric lamp in which light is produced by the incandescence of a carbon conductor contained in a closely sealed globe of glass more or less exhausted of air two difficulties have hitherto impaired their utility.

1st. The carbon conductor in process of use gradually wasted and ultimately broke.

2nd. The interior surface of the globe containing the incandescent carbon became wasted with a carbonaceous deposit and obscured the light.

I have discovered the means of avoiding both these faults (which are dependent the one upon the other) as follows, that is to say:—

I have found that if the globe or vessel of glass containing the carbon to be rendered incandescent be exhausted of air to a high degree of exhaustion by means of a "Sprengel pump," whilst the carbon contained within it is raised (by the passage of a current of electricity through it) to the same or a higher degree of incandescence as that ultimately required to be produced in using the lamp, and if the lamp be hermetically sealed whilst exhausted of air under the conditions specified the lamp so prepared is durable, both the faults above mentioned being eliminated, the incandescent carbon enduring without waste, and the interior surface of the glass globe keeping free from carbonaceous deposit.

In carrying out my said Invention I take a rod, filament or lamina, of carbon of a length and thickness proportioned to the current to be employed in heating it; for example, it may be two inches in length and one fiftieth to one hundredth of an inch in sectional area. This rod, filament or lamina, I clip at its extremities by means of a pair of platina clips, to which are attached platina wires, which serve for conducting the current through the lamp. These conducting wires pass through the glass which encloses the clips and the carbon, and are sealed into it by fusion of the glass around them. I attach the globe or bulb thus arranged to the exhaust tube of a Sprengel air pump. This I do in the manner well understood and practised in the procuring of high vacua, the attachment being made by means of a small branch tube fixed on to some part of the globe or bulb by the blow pipe. This attachment being made I work the pump until not more than the one thousandth part of an atmosphere remains in the globe or bulb. I then pass a current of electricity through the carbon, so as to render it incandescent, and during the incandescence of the carbon I continue the exhaustion further until the residue of air is so small as practically not to affect the durability of the carbon when kept in a state of incandescence. By exhausting the air under these conditions a much nearer approach to an absolute vacuum is attained than otherwise would be possible.

Having thus stated the nature of my said Invention, and described the manner in which it may be carried into effect, I claim, the production of an incandescent carbon electric lamp from which the air has been exhausted while the carbon is in a state of incandescence, and which is hermetically sealed during exhaustion.

In witness whereof, I, the said Joseph Wilson Swan, have hereunto set my hand and seal, the First day of July, in the year of our Lord One thousand eight hundred and eighty.

JOSEPH WILSON SWAN. (L.S.)

J. W. SWAN, British Patent, 2 January, 1880.

15k The Diesel Engine

The Diesel Oil-engine, and its Industrial Importance, particularly for Great Britain. By Dr. Rudolph Diesel, *of Munich*
[*Translated from the German*]

THERE have been so many publications recently, and especially during the past year, in technical periodicals of all languages, on the construction of the Diesel engine and its various types, that it is hardly possible to give any fresh information on the subject. Moreover, an excellent Paper was read only last summer at the Zürich Meeting of the Institution by Mr. J. F. Schubeler. The author proposes, therefore, to discuss only questions of general importance concerning the Diesel engine, especially those questions which are brought into prominence in the title of the Paper, and to consider them only as the starting point of a thorough and stimulating discussion. He desires, further, to be exonerated if statements are made in the Paper on points which are common knowledge.

Since its first appearance about fourteen years ago the Diesel engine has been built by the thousand in the best factories of all industrial countries, and has been set up in the most remote corners of the world. It has been proved to be a most reliable engine when properly built, the working of which is quite as safe as that of any other system of prime mover; and in general it is even more simple, since it does not require any auxiliary apparatus, and since the fuel in its natural and original form, without having previously undergone any transforming process, is directly converted into work in the cylinder of the engine. As early as 1897, when, after four years of difficult experiments, the author had put the first engine into working order in the factory of the Augsburg Works, numerous engineering representatives and experts who came from various countries to examine this engine expressed the opinion that it gave better heat utilization than any known kind of heat engine. From experience gained subsequently by working many engines, by gradual improvements in the construction and manufacture, and by increasing the size, the results have been still further improved, and to-day the thermal or indicated efficiency reaches 48 per cent.* in this engine, and the effective or brake efficiency reaches, in some cases, 35 per cent. of the heat value of the fuel.

Science and technical knowledge are making continuous progress, and the time will come when even these figures will be exceeded; but, with our present scientific knowledge, any considerably higher efficiency in the process of transforming heat into mechanical work is not obtainable. Further progress seems only possible with some other method of transforming heat into work; this would mean an entirely new principle, which, however, one cannot speculate upon in the present state of science.

The Diesel engine is therefore the engine which converts the heat of the natural fuel into work in the cylinder itself, without any previous transforming process, and which utilizes it as far as the present standard of science permits; it is therefore the simplest and, at the same time, the most economical prime mover.

These two facts explain its success; it lies in the new principle of the internal working process, and not in constructional improvements or alterations of older types of engines. There is no doubt that the careful working out of all the constructional details plays a great part in the practical success of the Diesel engine. . . .

* *See* Mr. Dugald Clerk's remarks (page 236).

A further reason for this importance is that the Diesel engine has broken the monopoly of coal, and has solved the problem of using liquid fuel for power production in its simplest and most general form. It has become for all liquid fuels what the steam-engine and gas-engine are for coal, but in a much simpler and more economical way. The truth of this statement was strikingly proved at the Turin Exhibition of last year. At this Exhibition, in the large Machinery Hall, a steam-turbine and a large Diesel engine, both made by Franco Tosi, of Milan, and set up on the same stand, were worked together with the same liquid fuel. The boilers belonging to the plant were fitted with Körting nozzles for burning crude oil. The difference between the two plants was therefore this: for the working of the steam-engine the whole boiler plant with its chimney, fuel supply apparatus, purification plant for feed-water with feed-pumps, extensive steam-pipes, condensation plant with water-pumps and an enormous water consumption, had to be provided, with the final result of consuming $2\frac{1}{2}$ or more times the fuel per horse-power required by the Diesel engine standing beside it. The latter, being an entirely independent engine without any auxiliary plant, took up its crude fuel automatically and consumed it direct in its cylinders without any residue or smoke. A better proof can hardly be imagined, even for the non-technical man, that except in special cases the steam-engine cannot compete economically with the oil-engine, and from this point of view the power-plant in the Machinery Hall at Turin must be looked upon as marking a historical event. It is hardly possible for a country which produces no coal, like Italy, to develop a great industry based on the steam-engine, and this is one reason for the exhibition at Turin of about thirty Diesel engines of various types and sizes, and made in different countries.

By courtesy of the Royal Institution

FIG. 2.—*Map showing Petroleum Fields of the World in the Year* 1908.

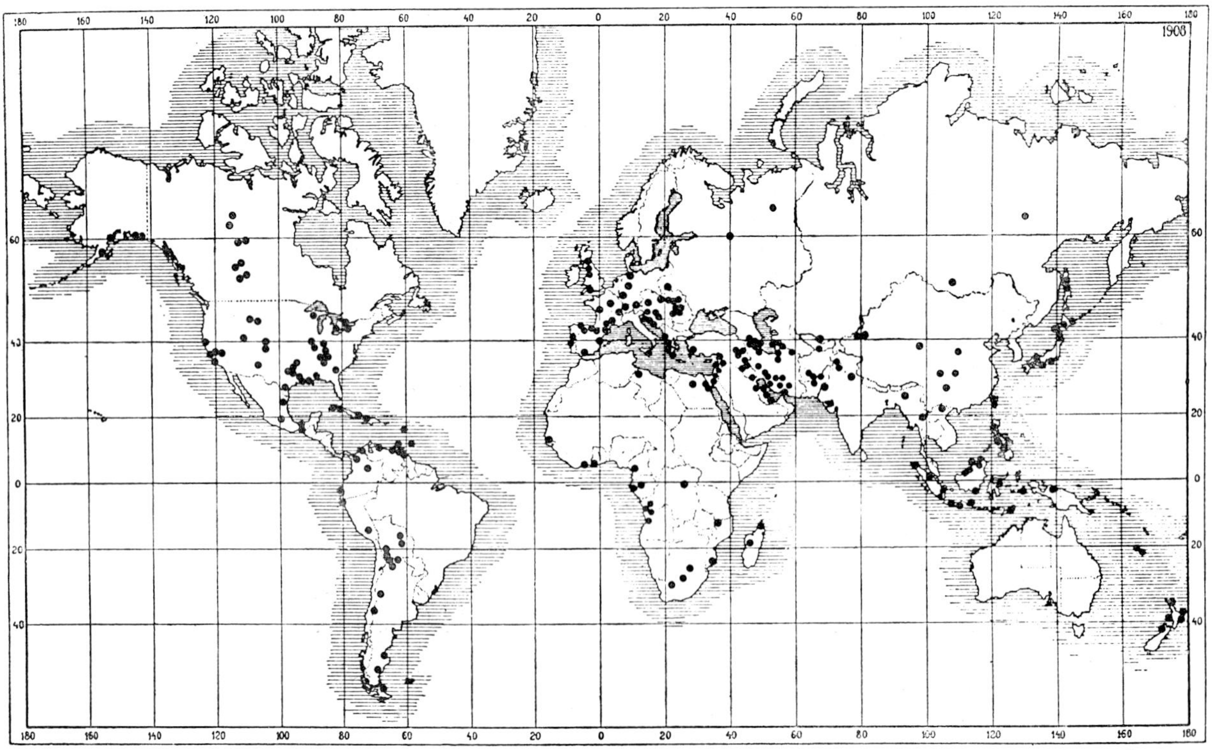

Thus the Diesel engine has doubled the resources of mankind as regards power-production, and has made new and hitherto unutilized products of nature available for motor power. The Diesel engine has thereby exercised a far-reaching influence on the liquid-fuel industry, which is at the present time improving more rapidly than was previously conceivable. This is not the place to discuss the matter in detail, but the author wishes to mention that, owing to the interest which petroleum producers have taken in this important question, new petroleum sources are continually being developed, and new oil districts discovered. Moreover, it has been proved by recent geological researches not only that there is probably on the globe as much, or perhaps even more, liquid fuel than coal, but also that it is more conveniently distributed as regards its geographical position, Fig. 2. These facts, which are indisputable nowadays, have gradually silenced those who objected to too great a development of the Diesel engine for fear of insufficient stores of liquid fuel. Any such anxiety may be relieved by the fact that the world's production of crude oil increases at present $3\frac{1}{2}$ times more quickly than the production of coal, and that the ratio of increase itself is steadily getting higher. Further, that 40 per cent. of the present production of mineral oil is already sufficient to supply the whole of the naval and mercantile fleets of the world with power, if they were worked by Diesel engines; also that with the world's present production the number of Diesel engines now working could be increased about a hundredfold.

It may thus safely be asserted that, with the continual development of new oil-districts, the production of mineral oil will increase much more quickly than the demand for newly-built engines. It is therefore not surprising that, in the last yearly report of the Shell Transport Co., attention was called to the fact that the oil consumption in these engines did not nearly equal the production, and that the company has to look out for other markets to dispose of their superfluous stock. That the auxiliary industries of petroleum production are also considerably influenced is shown by the great increase which the transport industry for liquid fuel has experienced in recent times, especially the great development of tank-vessels which are, or will be, mostly driven by Diesel engines.

But with all this, the influence of the Diesel engine in the world's industry is not exhausted. As early as the year 1899 the author utilized in his engine the by-products of coal distillation and coke plants, such as tar and creosote oils, with the same satisfactory results as with natural liquid fuels, but at that time the quality of these oils was generally too inferior for their use in the Diesel engine, and it was, moreover, subject to continual variations.

These difficulties were then chiefly the following:—

(1) Muddy deposits of solid hydrocarbons, especially naphthalenes, which made the working of the fuel pumps difficult, filled up the pipes and nozzles, and formed a hard crust at the nozzle mouths. These solid hydrocarbons also made higher ignition temperatures necessary.

(2) Continuous change in quality and composition of the crude, uncleaned tar-oils; with each cask fresh variations appeared even in cases where the works guaranteed the use of the same coal, and the carrying out of the same distilling process, so that it was impossible to make scientific observations for drawing any definite conclusions, or making logical experimental arrangements. The characteristics of crude tar-oils were not then exactly known even to the producers; for instance, nobody imagined that differences in the distilling temperature and variations in the nature and position of the retorts gave entirely different tar-products even when the same coal was used. It is only in recent years that the chemical industries interested in the matter have, by improved

methods of fractioning and refining, combined with more careful selection of the material, succeeded in supplying fuel of a constant and regular quality, without the drawbacks of the crude tar-oils used previously. These products—the tar and tar-oils—are thus to-day definitely brought into the sphere of activity of the Diesel engine.

In Appendix I (page 208) is given a Table showing the application of various fuels for Diesel engines, while Appendix II (page 210) contains specifications for tar-oils for Diesel engines, and Appendix III (pages 212–5) gives the properties of tar.

From what has been just stated, it will be seen that the Diesel engine is having an increasing influence on two other industries—the manufacture of gas and coke—the by-products of which have become so important for power production that an enormous business is at present connected with them.

It is especially noteworthy that every town gas-works with a modern installation, and every coke-works, can be completed with an electric power generating plant by using its tars. This will have an excellent effect on many municipal and national works. It would take too much time to enter into the details of this question, but one fact stands out clearly in this connection, namely, that coal, which seemed to be most threatened by the liquid fuels, will, on the contrary, gain a new and wider ground of application through the Diesel engine. As tar and tar-oils are from 3 to 5 times better utilized in the Diesel engine than coal in the steam-engine, a much better and more economical utilization of coal is obtained if, instead of being burned under boilers on grates in a wasteful way, it is first transformed into coke and tar by distillation. Coke is used in metallurgical and other general heating purposes; from a part of the tar the valuable by-products are first extracted, and undergo further processes in the chemical industry, whilst the tar-oils and combustible by-products, and a great part of the tar itself, are burned in the Diesel engine under extraordinarily favourable conditions.

The proper development of the utilization of fuel, which has already been started, and is now making rapid progress, is therefore the following: On the one hand liquid fuel in Diesel engines, and, on the other hand, gas fuel also in the form of gasified coke in the gas-engines; solid fuel as little as possible for steam-power generation, but only in the refined form of coke for all other heating and metallurgical purposes.

The list of fuels applicable to the Diesel engine is not, however, exhausted with these liquid fuels mentioned previously. It is known that brown coal or lignite, the production of which is about 10 per cent. of the pit-coal production, also yields tar by distillation, and these tars when distilled, and worked up on paraffin, produce the so-called paraffin oils as by-products. Not all kinds of brown coal, however, are suitable for this process; but, in any case, these oils have been produced in such quantities that so far they have satisfied a good part of the German demand for liquid fuels for Diesel engines. To these must be added other products, such as shale and similar oils—produced not in great but in sufficient quantities to be of importance for power generation. Some countries—France and Scotland for instance—possess them in considerable quantities, and they are used in a good many Diesel engine plants. It is not generally known that it is also possible to burn fat vegetable oils and animal oils in the Diesel engine without any difficulty.

At the Paris Exhibition in 1900 there was shown by the Otto Company a small Diesel engine which, at the request of the French Government, ran on Arachis-oil,* and worked so smoothly that only very few people were aware of

* Earth-nut or pea-nut—botanical name *Arachis hypogaea*.

it. The engine was constructed for using mineral oil, and was then worked on vegetable oil without any alterations being made. The French Government at the time thought of testing the applicability to power production of the Arachide or earth-nut, which grows in considerable quantities in their African colonies, and which can be easily cultivated there; because in this way the colonies could be supplied with power and industry from their own resources, without being compelled to buy and import coal or liquid fuels.

By courtesy of the Royal Institution

Working Diagrams of Single-acting Diesel Engines.

FIG. 3.—*Four-stroke Cycle.*

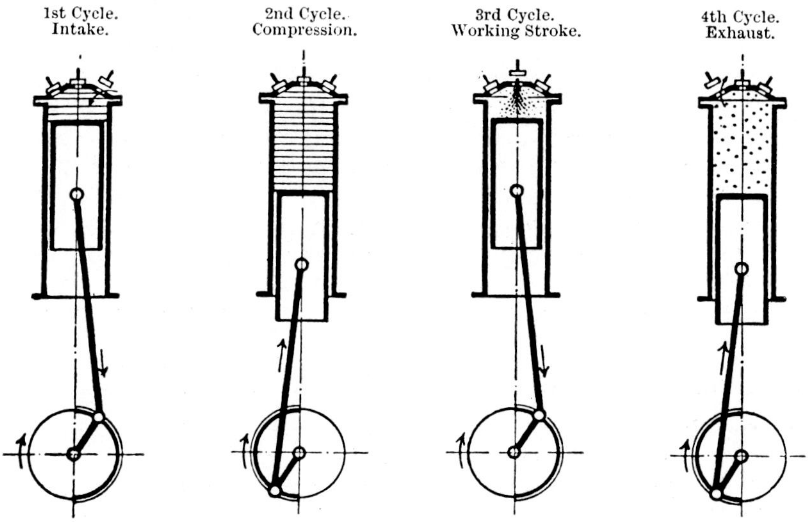

FIG. 4.—*Two-stroke Cycle.*

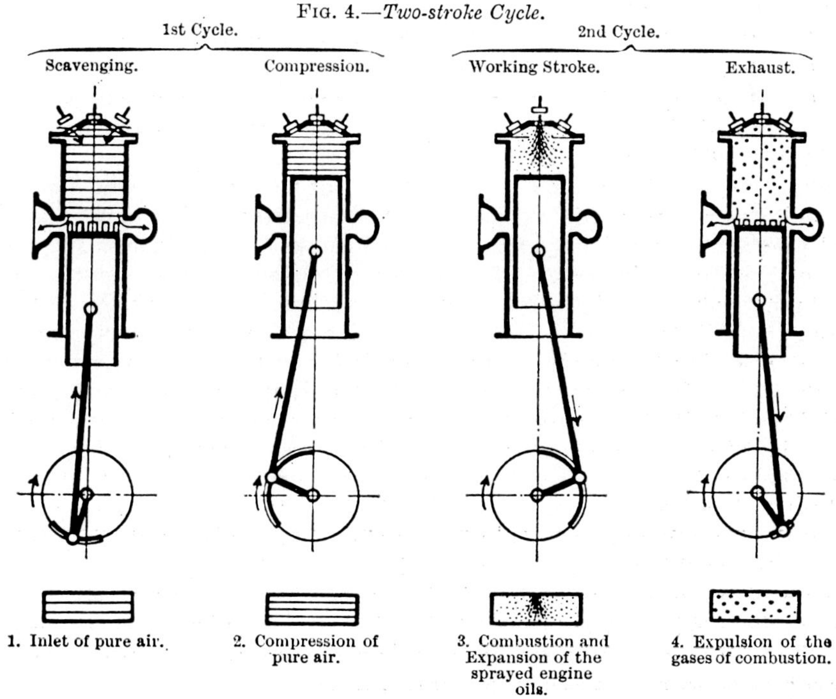

This question has not been further developed in France owing to changes in the Ministry, but the author resumed the trials a few months ago. It has been proved that Diesel engines can be worked on earth-nut oil without any difficulty, and the author is in a position to publish, on this occasion for the first time, reliable figures obtained by tests:—Consumption of earth-nut oil, 240 grammes (0·53 lb.) per b.h.p.-hour; calorific power of the oil, 8,600 calories per kg. (15,480 B.Th.U. per lb.), thus fully equal to tar oils; hydrogen, 11·8 per cent. This oil is almost as effective as the natural mineral oils, and as it can also be used for lubricating oil, the whole work can be carried out with a single kind of oil produced directly on the spot. Thus this engine becomes a really independent engine for the tropics.

Similar successful experiments have also been made in St. Petersburg with castor-oil; and animal oils, such as train-oil, have been used with excellent results. The fact that fat oils from vegetable sources can be used may seem insignificant to-day, but such oils may perhaps become in course of time of the same importance as some natural mineral oils and the tar-products are now. Twelve years ago the latter were not more developed than the fat oils are to-day, and yet how important they have since become.

One cannot at present predict, what part these oils will play in the colonies in the future. In any case, they make it certain that motor power can still be produced from the heat of the sun, which is always available for agricultural purposes, even when all our natural stores of solid and liquid fuels are exhausted. . . .

SPECIAL IMPORTANCE OF THE DIESEL ENGINE FOR GREAT BRITAIN

After this short summary about the importance of the Diesel engine for the world's industry in general and the historical review of Diesel types, the author desires to add a few words on the importance of the Diesel engine, especially for Great Britain. In this connection the three following facts must be borne in mind:—

(1) Great Britain is an exclusively Coal-producing Country;
(2) Great Britain has the largest Colonial Empire in the world; and
(3) Great Britain is the greatest Shipping Nation in the world.

Dealing with (1), Great Britain has had, at least until to-day, no natural liquid fuels of its own; it is an exclusively coal-producing country. Based on these statements, it has often been seriously put forward in recent times that England has no interest in the Diesel engine, and that it is against her most vital interest to participate in the development of this engine, as she would neglect her wealth in coal, and make herself dependent on foreign markets when using liquid fuels. The Exhibition at Turin has shown clearly that England has still very little interest in the Diesel engine as far as stationary plants are concerned owing to her abundance of coal; for in the English Section, with the exception of some very small oil-engines, only generator-gas engines were exhibited; but these were numerous, of first-class workmanship, and were built by the best firms in the country. England has therefore made great progress in power generating with gas, whilst she has up to the present given relatively little attention to power generating with liquid fuels.

But the foregoing arguments in favour of coal are not correct; the contrary is true. Great Britain has the greatest interest in replacing the coal-wasting steam-engine by the more economical Diesel engine, and this, firstly, because she can therewith effect enormous savings in her most valuable treasure—coal, and thus defer the exhaustion of her stock; and, secondly, because she can run her coal industry and the independent chemical industries on more economical lines, when using the coal in the more rational and refined way, as already mentioned. Finally, because she will also make herself free and independent of foreign markets for the supply of liquid fuels by using coal in *this way*, that is, by working the tars and tar-oils in the Diesel engine. It is not intended to imply that the whole demand of fuel for England could be produced in this way in the country itself, but through the inland production, increase of prices for foreign oils and the establishment of trusts and monopoly companies will be prevented. In this sense independence is meant.*

(2) *Great Britain has the largest Colonial Empire in the world.*—It is not possible to foresee to-day what England can obtain for her Colonies from the Diesel engine; even when using only the natural mineral oils, the Diesel engine is the predestined colonial engine, because only about the fourth or sixth part of weight in fuel has to be transported for it to the Colonies and their hinterlands, as compared with the steam-engine. For a colonial engine the cost of freightage for fuel is generally the determining factor. Further, the transport of these liquid fuels is considerably easier and more convenient than the transport of coal, especially when tank-vessels and pipe-lines are used; and finally the difficulty of working a boiler plant will only occasionally come into consideration in the Colonies, especially in the interior, except, of course, for small plants using wood, straw, and the like.

It may be mentioned, in this connection, that a pipe-line of 400 km. (about 250 miles) in length for crude petroleum is being built on the River Congo from Matadi to Leopoldville, by means of which this extensive district will be supplied, in the simplest and cheapest way, with a constant flowing fuel source, from which navigation and the railways, agriculture as well as other industries, will get their element of life, namely, motor power. This magnificent example ought to be followed in the English Colonies; it is not necessary to specify in detail the great importance of such an undertaking for them.

If one considers that the Diesel engine can utilize vegetable oils, entirely new prospects are brought to light for the cultivation and expansion of industry in the Colonies, which are for no other country of such eminent importance as for Great Britain; and this is where she ought to start as soon as possible. The Diesel engine can be worked with the Colonies' own resources, and thus again can influence to a great extent the further expansion of agriculture in districts where it is predominant. This sounds to-day somewhat like a dream, but the author ventures to prophesy, with full conviction, that this way of using the Diesel engine will one day be of the utmost importance.

(3) Finally, *Great Britain is the greatest Shipping Nation in the world.*—When the first successes of the Diesel engine, as a marine engine, were heard of in England recently, and it was published that already numerous small mercantile and war vessels were fitted with Diesel engines, and the possibility of more important plants was mentioned, and when it was realized that larger ocean vessels were destined to be fitted with Diesel engines, and that even a warship with a very large Diesel engine was in construction, it provoked a great movement and excitement in Great Britain which is still fresh in our minds. Moreover, the reports about successful voyages with Diesel vessels under very difficult weather

* In this connection it may be of some interest to state that the tar production of Germany is sufficient for more than 5 milliards of horse-power-hours per year, which means about $1\frac{3}{4}$ millions of horse-power running 300 days of ten hours each all the year. In case of war and cutting off the supply of foreign fuel, this quantity would be entirely sufficient for running the whole fleet, war and mercantile, and for providing in the meantime the power for the inland industry as far as necessary. The author has not the figures for England, but he presumes they are of similar significance.

conditions are already increasing in number. The captains who have had Diesel engines on their vessels report on the great security and comfort in working; shipbuilders publish the figures of their savings. It is unquestionable that one of the greatest evolutions of modern industry will be connected with this development of the Diesel engine, and that Great Britain as the greatest shipping nation of the world will derive the greatest advantage from it.

RUDOLPH DIESEL, 'The Diesel Oil-engine, and its Industrial Importance, particularly for Great Britain', *Proceedings of the Institute of Mechanical Engineers*, 1912, p. 179.

151 Deptford Power Station (1890)

By courtesy of the Science Museum

Photograph at the Science Museum.

16 NATIONALISM IN SCIENCE

16a A Chemist's View of the Great Exhibition

I HAVE shown in my former Lecture, that a rapid transition is taking place in Industry; that the raw material, formerly our capital advantage over other nations, is gradually being equalized in price, and made available to all by the improvements in locomotion; and that industry must in future be supported, not by a competition of local advantages, but by a competition of intellect. All European nations, except England, have recognised this fact; their thinking men have proclaimed it; their governments have adopted it as a principle of state; and every town has now its schools, in which are taught the scientific principles involved in manufactures, while each metropolis rejoices in an Industrial University, teaching how to use the alphabet of Science in reading Manufactures aright. Were there any effects observed in the Exhibition from this intellectual training of their industrial populations? The official reserve, necessarily imposed upon me as the Commissioner appointed to aid the Juries, need exist no longer, and from my personal conviction, I answer without qualification, in the affirmative. The result of the Exhibition was one that England may well be startled at. Wherever—and that implies in almost every manufacture—Science or Art was involved as an element of progress, we saw, as an inevitable law, that the nation which most cultivated them was in the ascendant. Our manufacturers were justly astonished at seeing most of the foreign countries rapidly approaching and sometimes excelling us in manufactures, our own by hereditary and traditional right. Though certainly very superior in our common cutlery, we could not claim decided superiority in that applied to surgical instruments; and were beaten in some kind of edge-tools. Neither our swords nor our guns were left with an unquestioned victory. In our plate-glass, my own opinion—and I am sure that of many others—is, that if we were not beaten by Belgium, we certainly were by France. In flint-glass, our ancient *prestige* was left very doubtful, and the only important discoveries in this manufacture were not those shown on the English side. Belgium, which has deprived us of so much of our American trade in woollen manufactures, found herself approached by competitors hitherto almost unknown; for Russia had risen to eminence in this branch, and the German woollens did not shame their birthplace. In silversmith work we had introduced a large number of foreign workmen as modellers and designers; but, nevertheless, we met with worthy competitors. In calico-printing and paper-staining our designs looked wonderfully French; whilst our colours, though generally as brilliant in themselves, did not appear to nearly so much advantage, from a want of harmony in their arrangement. In earthenware we were masters, as of old; but in china and in porcelain our general excellence was stoutly denied; although individual excellencies were very apparent. In hardware we maintained our superiority, but were manifestly surprised at the rapid advances made by many other nations. Do not let us nourish our national vanity by fondly congratulating ourselves that, as on the whole we were successful, we had little to fear. I believe this is not the opinion of most candid and intelligent observers. It is a grave matter for reflection, whether the Exhibition did not show very clearly and distinctly that the rate of industrial advance of many European nations, even of those who were obviously in our rear, was at a greater rate than our own; and if it were so, as I believe it to have been, it does not require much acumen to perceive that in a long race the fastest-sailing ships will win, even though they are for a time behind. The Exhibition will have produced infinite good, if we are compelled as a nation to acknowledge this truth. The Roman empire fell rapidly, because, nourishing its national vanity, it refused the lessons of defeat, and construed them into victories. All the visitors, both foreign and British, were agreed upon one point, that, whichever might be the first of the exhibiting nations, regarding which there were many opinions, that certainly our great rival, France, was the second. Let us hope that in this there is no historical parallel. After the battle of Salamis the generals, though claiming for each other the first consideration as to generalship, unanimously admitted that Themistocles deserved the second; and the world, ever since, as Smith remarks, has accepted this as a proof that Themistocles was, beyond all question, the first general. Let us acknowledge our defeats when they are real, and our English character and energy will make them victories on another occasion. But our great danger is, that, in our national vanity, we should exult in our conquests, forgetting our defeats; though I have much confidence that the truthfulness of our nation will save us from this peril. A competition in Industry must, in an advanced state of civilization, be a competition of intellect. The influence of capital may purchase you for a time foreign talent. Our Manchester calico-printers may, and do, keep foreign designers in France at liberal salaries. Our glass-works may, and do, buy foreign science to aid them in their management. Our potteries may, and do, use foreign talent both in management and design. Our silversmiths and diamond-setters may, and do, depend much upon foreign talent in art and foreign skill in execution; but is all this not a suicidal policy, which must have a termination, not for the individual manufacturer, who wisely buys the talent wherever he can get it, but for the nation, which, careless of the education of her sons, sends our capital abroad as a premium to that intellectual progress which, in our present apathy, is our greatest danger?

It is well to inquire, in what we are so deficient, and what is the reason of this deficiency. Assuredly it does not consist in the absence of public philanthropy or want of private zeal for education, but chiefly rests in that education being utterly unsuited to the wants of the age. In the thirteenth and fourteenth centuries classical learning was, after its revival, highly esteemed: and its language became the common medium for expression in all nations. A thorough acquaintance with it was an absolute necessity to any one with pretensions to learning. It had a glorious literature, one as fresh as when it grew on the rich soils of Rome and Greece. Its truths were eternal, and were received by us in their traditional mythology, as Bacon beautifully says, like 'the breath and purer spirit of the earliest knowledge floating to us in tones made musical by Grecian flutes.' And why was that bewitching literature made the groundwork of our educational systems? Does it not show that literature, like art, may have a standard excellence; and that we are content to imitate where we cannot surpass. If the main object of life were to fabricate literati, I would not dispute the wisdom of making classics the groundwork of our education. They are not utterly dead, but, like the dry bones of the valley, they may come together, and have breathed into them the breath of life. In the world there is a constant system of regeneration. Theories exist for a time, but like the phœnix, are destroyed, and rise yet more glorious from their ashes. Animals die, and by their decay pass into the atmosphere, whence vegetables derive their nutriment, and thus death becomes the source of life. But in all this there is no incongruity. A phœnix does not from its ashes produce an eagle, but a phœnix as before. The dry bones of dead Literature may vivify into new forms of literary life. Classical Literature and exact Science are, however, wholly antithetic. If classical Literature be sufficient to construct your spinning-jennies and bleach your cottons, your system of instruction is right; but if you are to be braced, and your sinews strengthened, for a hard struggle of industry, is it wise that you should devour poetry, while your competitors eat that which forms the muscles and gives vigour to the sinews? With such different trainings, who in the end will win the race? Science has not, like Literature and Art, a standard of excellence. It is as infinite as the wisdom of God, from whom it emanates. All ordinary powers decrease as you depart from the centre; but the power of knowledge augments the farther it is removed from the human source from which it was transmitted. God has given to man much mental gratification in trying to understand and apply to human uses His laws. The great philosopher of Scripture has said, 'It is to the glory of God to conceal a thing, but the honour of kings to search out a matter.' The poet-prophet of the Bible has also told us, that God 'turneth wise men backward, and maketh their knowledge foolish.' And, therefore, as surely as He is infinite and man finite, until earth passes away, you will have no human standard of scientific knowledge. As this is so, how can we as a nation expect to carry on those manufactures by our sons of Industry, when we do not teach them the nature of the principles involved in their successful prosecution? Solace ourselves as we will with vain thoughts of our gigantic position among nations—Greece was higher than we are, and where is she now? It does not require a lofty stature to see the farthest; for a dwarf on the shoulders of a giant sees farther than the giant,—not that he is less a dwarf, but that he has added the giant's height to his own. The Exhibition showed us many small States which had thus raised themselves on the shoulders of Science within the last few years, while we are merely hovering about its skirts. Let us take care that our excess of pride in the so-termed 'practical' power of our population may not be punished as Arachne was of old. Arachne was wonderfully skilled in needle-work, but presumptuously challenged Minerva to a trial of skill. What chance was there in such an unequal contest? Minerva united Science to her handicraft skill, and this combination insured success. Arachne was justly cast from her proud position among mortals by being changed into a spider, ever spinning the same web in the same way,—the same for wintry blasts as for gentle summer zephyrs.

'You have excelled all other people in the products of Industry. But why? Because you have assisted Industry by Science. Do not regard as indifferent what is your true and greatest glory. Except in these respects, in what are you superior to Athens and Rome? Do you carry away from them the palm in literature and the fine arts? Do you not rather glory, and justly too, in being, in these respects, their imitators? Is it not demonstrated by the nature of your system of public education and by your popular amusements? In what, then, are you their superiors? In everything connected with physical Science; with the experimental arts. These are your characteristics. Do not neglect them. You have a Newton, who is the glory, not only of your own country, but of the human race. You have a Bacon, whose precepts may still be attended to with advantage. Shall Englishmen slumber in that path which these great men have opened, and be overtaken by their neighbours? Say, rather, that all assistance shall be given to their efforts; that they shall be attended to, encouraged, and supported.'—(DAVY.)

All the aspirations of youth are towards Science, especially that depending on observation, but we quench the God-born flame by 'freezing drenches of scholastic lore.' In the language of 'Eōthen,' 'You feel so keenly the delights of early knowledge! You form strange mystic friendships with the mere names of mountains, and seas, and continents, and mighty rivers; you learn the ways of the planets and transcend their narrow limits, and ask for the end of space; you vex the electric cylinder till it yields you, for your toy to play with, that subtle fire in which our earth was forged. You know of the nations that have towered high in the world, and the lives of men who have saved whole empires from oblivion. What more will you ever learn? Yet the dismal change is ordained, and then, thin meagre Latin (the same for everybody) with small shreds and patches of Greek, is thrown, like a pauper's pall, over all your early lore; instead of sweet knowledge, vile, monkish, doggerel grammars and graduses, dictionaries and lexicons, and horrible odds and ends of dead languages, are given you for your portion, and down you fall from Roman story to a three-inch scrap of "*Scriptores Romani*"—from Greek poetry, down, down, to the cold rations of "*Poetæ Græci*," cut up by commentators and served out by schoolmasters.' Is this horrible quenching of all our youthful innate love of God's truth the education for the youth of a nation depending for its country's progress on their developement? How is it possible that dead Literature can be the parent of living Science and of active Industry?

I need not explain myself as meaning that our youthful aspirations point to science as a fit means for developing our intellectual capacities, and that boyhood is scarcely the time rudely to excise all our longings for an acquaintance with the wisdom of creation, or to cramp and torture the mind by the acquisition of dead languages to the exclusion of all other knowledge. In quoting the beautiful language of 'Eōthen,' I intend only to express the violence done to our natural instincts, and not to question the excellence of the means employed in teaching classics. It would ill become me, or any one, to speak disparagingly of the wisdom to be derived from a study of ancient authors, or to deny the immense importance of a knowledge of classical literature to education generally; nor would I like to see that education confined to stern realities, divested of the graces and poetry of polite literature. But I do, at the same time, vehemently protest against the exhaustion of all our youthful years by a mere classical tuition, especially in the case of that large class of the community who, by their exertions in industry, have confided to them, in a great degree, the prosperity of their country. As I do not think the teaching of classical literature as practised in our schools to be worthy of the name of education, neither do I apply that title to the communication of scientific knowledge alone,—and you will observe that I have always spoken of it by the term 'instruction'.—I am propounding no scheme of education, but strongly insisting that instruction in science should form an important part of the education of our youth.

Do not conceal from yourselves that this is the vital difficulty of the question. You may, and I hope you will, soon raise an Industrial University; but this should have its pupils ready trained before it adopts them. Now, it must from itself act downwards, instead of working from the school upwards. Until our schools accept as a living faith that a study of God's works is more fitted to increase the resources of the nation than a study of the amours of Jupiter or of Venus, our Industrial Colleges will make no material headway against those of the Continent. In Paris we find a Central College of Arts and Manufactures, into which the students enter at an average age of nineteen years, already well trained in the elements of Science, and going there to be taught how to use these elements for industrial application. Three hundred of the best youth of France are annually receiving at this College the most elaborate education, and the best proof of its practical value is the great demand among manufacturers for its pupils, a diploma from it being equivalent to assured success in life. Can you wonder at the progress making [*sic*] by France in industry, when she pours every year an hundred and fifty of these highly educated manufacturers into her provinces? A similar education to this is going on in almost all parts of Europe; but in England only one such institution exists. We have our University and King's College, it is true, and they are productive of much good; and similar Colleges exist in Scotland and Ireland; but their instruction in Science terminates just where the Industrial Colleges of the Continent begin. In fact, the latter would be supplementary and a great support to the former. Government, acting on its own perception of right, in its first national recognition of these truths, now happily dawning on England, has established a School of Mines; and the experience of this has shown that it is much appreciated, although it labours under the disadvantage of the want of a preliminary education in its pupils, compelling its professors, in its commencement, to be more elementary in their instruction than is well compatible with the proper objects of such a school. Now, while I urge the impolicy of a mere classical instruction to the youth of this country with all the expression which I can give to a matured conviction, do not suppose that I would wish to put all our youth in one Procrustean bed. I again allege, that it is the present system which follows this singular love of uniformity, and clips or extends the dimension of each youth to one common standard. It is against this very confined system that I protest. I think the glorious wisdom displayed in creation, even in the limited extent to which we are permitted to behold it, forms no unapt means of leading man to a worship of its Creator; and, sympathizing as I do to the utmost in our educational endeavours to unite and not to dissever the acquirement of knowledge from that of religion,—a union which, I think, is at once the glory, the pride, and the peace of England,—I cannot perceive how the mere teaching of profane literature can tend to this end in any degree so much as the reverential teaching of God's wisdom displayed in His works, especially when every step in advance of this knowledge produces a social amelioration of the human race. But, while I would regret to see our Colleges retrograde one step in the teaching of classical Literature, it is truly lamentable that Oxford and Cambridge so little encourage the Sciences; for, until the Colleges throw open their widest portals to these, the schools in the country, deriving their life from them, will do little to reform the present vices of a limited and exclusive education.

In this country we are, in many respects, remarkably unchangeable. Three professions, the Church, the Law, and Medicine, were supposed, some centuries since, to represent learning, and, with a wonderful blindness, they are still accepted as all-sufficient. Industry, to which this country owes her success among nations, has never been raised to the rank of a profession. For her sons there are no honours, no recognised or social position. Her native dignity, if tacitly understood, has never formally been acknowledged. Science, which has raised her to this eminence, is equally unrecognised in position or honours, and, from her very nature, cannot attain the wealth which in Industry solaces the absence of social position. This restriction of learned honours to three recognised professions has a lamentable effect both on the progress of Science and of Industry. Its consequence is, that each profession becomes glutted with ambitious aspirants, who, finding a greater supply than demand, sink into subordinate positions, becoming soured and disappointed, and therefore dangerous to the community. Raise Industry to the rank of a profession,—as it is in other countries,—give to your Industrial Universities the power of granting degrees involving high social recognition to those who attain them, and you will draw off the excess of those talented men, to whom the Church, the Bar, and Medicine, offer only a slender chance of attaining eminence; and by infusing such talent into Industry, depend upon it, the effects will soon become apparent. In foreign countries, professions involving social rank and position arise with their requirements; in our Nation, we are content with a meagre classification, scarcely sufficient for the middle ages, and not even a reflection of our present wants. These considerations are not mean ones, for, as long as ambition exists in the human mind, their good or bad adjustment will exercise a beneficial or pernicious influence on society.

In the establishment of institutions for industrial instruction, you, at the same time, create the wanting means for the advancement of Science in this country. I have alluded in this Lecture, and have shown in another, that the progress of Science and of Industry in countries which have reached a certain stage of civilization ought actually to be synonymous expressions; and hence it follows that it is essentially the policy of a Nation to promote the one which forms the springs for the action of the other. I think it, therefore, no mean advantage to this nation, that the establishment of Industrial Colleges will materially aid the progress of Science by creating positions for its professors and for those who would willingly cultivate Science, but are scared from it by the difficulties they have to encounter in its prosecution. The great Davy says,—'Science, for its progression, requires patronage; but it must be a patronage bestowed, a patronage received with dignity. It must be preserved independent. It can bear no fetters; not even fetters of gold; and, least of all, those fetters in which ignorance or selfishness may attempt to shackle it. And there is no country which ought so much to glory in its progress, which is so much interested in its success, as this happy island.' . . .

L. PLAYFAIR, 'The Chemical Principles involved in the Manufactures of the Exhibition', *Lectures on the Results of the Great Exhibition of 1851*, London, 1852, **1**, 194.

16b Export of English Machinery

Select Committee on Machinery

J. E. Tennent, Esq. 17 May 1841

4435. In what position did you find the Germans generally in those districts which you visited as regards machinery?—I found in every state of Germany, from the north to the south, the greatest possible anxiety to domesticate manufactures amongst themselves, to render themselves independent of the other competing nations of Europe. Those gentlemen with whom I had an opportunity of conversing connected with manufactures, said that that feeling had arisen in a great degree from the vicissitudes which they had experienced during the wars of the French Revolution; that manufactures which had been flourishing previously to the commencement of that era, had been completely trampled down and destroyed by it, that they found their prosperity materially injured, and they ascribed that result to the fact of their not having had those manufactures properly rooted amongst them. I found that, with regard to machinery, they had the greatest possible appetite to obtain English machinery; that all that they there manufactured was of an inferior quality, and cost them much higher than that which they could have from England; and that their skill in using it was inferior to our own.

4436. Did you visit any machine-making establishments at Hamburgh?—I did; and I think the condition of the manufactory which I visited at Hamburgh is a very fair practical illustration of what I have just been saying, of the anxiety which prevails throughout Germany to obtain English machinery. The manufactory to which I refer, is that of Messrs. Gleichman & Busse; this is a manufactory which has been constructed within the last two years, and I found them working with English iron, with English coals, with English models, all their tools English, their director an Englishman, and of the 150 hands who were in their employment, 90 were likewise English; and they informed me, that notwithstanding the expense to which they must be put in all those items of carriage from England, they still were enabled to make machinery so as to under-sell the smuggled machinery which came over from England; and that from using their English materials, and applying their English skill, they found a favourable impression that their work was equal to English.

4463. *Chairman.*] Amongst other places, you visited Berlin?—I did.

4464. Did you examine any of the machine-making establishments there, and in what position did you find them?—I found at Berlin the most enterprising and systematic exertions made on the part of the Government to obtain a command of the manufacture of machinery; I found no expense spared for that purpose; and the exertions quite astonished me. There is one very important institution at Berlin, called the 'Gewerbe Institut,' which is a large establishment for practical education, combining design with almost every branch of manufactures into which science and mechanics enter. The pupils for this institution are selected from similar provincial schools throughout Prussia, and they are sent as a mark of honour to the head institution at Berlin, where they obtain their education, and are likewise even paid a salary during the period of their attendance: in going through the rooms of this institution with the professor, Mr. Wedding, I saw suites of apartments completely filled with models of English machines: the professor informed me that they had in it models of every machine in use in Great Britain for the manufacture of cotton, flax, silk and wool, and likewise a number from America and Germany; that by these means they were not only enabled to have our recent improvements, but, what was a matter of importance which we cannot command, that they were enabled frequently to combine in the same machine two distinct English patents. The system he told me was, that this machinery, as soon as produced in England, was immediately imported at the expense of the Government, and set up at the Gewerbe Institut; that it was proved; that a working model was immediately made from it, to be deposited in the institution, and that the original was presented as an honorary prize by the Government to some manufacturer in Prussia, who had distinguished himself in the peculiar branch to which it was applicable. In the Institut, likewise, the pupils were taught to make the machinery themselves; they were supplied with the tools, and they were permitted to carry away the machines which they themselves had constructed: I cannot but look upon the whole of this system as a most surprising effort made on the part of the Government to obtain a command of the manufacture of machinery, which is at the present moment so important a feature in English manufactures.

Reports from the Select Committee appointed to Inquire into the Operation of Existing Laws affecting the Exportation of Machinery. *B.P.P.*, 1841, (7).

16c Trade and Technical Education

Birmingham Chamber of Commerce. Offices, Exchange Buildings, New Street, 29th January 1868

ANSWERS TO QUESTIONS ON TECHNICAL EDUCATION

First Question.—What trades are now being injured by the want of a technical education?

Answer.—Every trade in Birmingham and the district is being injured by the want of technical education, and those trades the most, in which the cost of the articles produced, consists most of labour, and least of the raw material.

Second Question.—How and in what particulars are they being injured?

Answer.—A large proportion of the multitudinous trades carried on in this district especially suffer because so many of our manufactured articles are composed of a variety of metals and other materials which depend for their successful combination and treatment upon a knowledge of chemistry and other sciences, and the beauty of their form, on a knowledge of art, while our workmen have scarcely any knowledge of either, but are guided in their work by imitation of one another and tradition.

Third Question.—How do other countries, from greater attention to technical instruction, absorb our trade? (Give instances, and, if possible, statistics.)

Answer.—In other countries the workpeople are instructed in science and art; the effect of this is shown in the rapid improvement of their manufactures in beauty of form, excellence of finish, adaptation to the purpose for which they are intended, and cheapness, and these excellencies enable them now to be in the course of largely supplanting us in the markets of the world.

As regards the request that the Chamber would furnish statistics, we beg to point out that the Government have in their power the ready means of ascertaining the great development and present active progress of the manufactures of foreign countries. This Chamber has applied through their late member, Mr. Scholefield, for these statistics, but they have never yet been furnished. We have, however, prepared a list of some articles, for the supply of which the nations of the world but recently depended on Birmingham and the district, and in which we are now in whole, or in part, superseded by the manufacturers of other countries.

Fourth Question.—What plan of technical education would remedy the evil?

The answer to this question was sent on the 13th, being the resolution adopted by this Chamber after two days' debate. In forwarding this resolution, the Council of the Chamber wished it to be understood that they recognise in the term technical education, both artistic and scientific instruction.

Copy of the Resolution:—'That the Council be instructed to request the Association of Chambers of Commerce to inform Lord Robert Montagu, M.P., that, in the opinion of this Chamber, it is of the utmost importance that Government Schools of Science should be established in the great centres of industry for the purpose of giving systematic technical education both to the middle and working classes.'

LIST of some ARTICLES made in *Birmingham* and the Hardware District which are largely replaced in Common Markets of the World by the Productions of other Countries. This List might be immensely extended by farther investigation, which the shortness of time did not permit.

Articles or Class of Articles	Country whose Products are believed to have Replaced those of this District in whole or in part
Carpenters' Tools, as hammers, plyers, pincers, compasses, hand and bench vices	Germany, chiefly
Chains, of light descriptions, where the cost is more in labour than in material, as halter chains and cow ties, and such like	Germany
Fry pans, of fine finish - - - -	France
Wood-handled spades and shovels, an article of very large consumption	United States export them to all our colonies
Hoes for cotton and other purposes, an article of large consumption	United States compete with us for their own use, and to some extent for export
Axes for felling trees, &c., an article of large consumption	United States supply our colonies and the world with the best article
Carpenters' broad axes - - -	Germany and the United States
,, and coopers' axes - -	
Coopers' tools, various sorts - - -	
Shoemakers' hammers and tools - -	
Matchets, for cutting sugar canes, an important article	Believed to be now Germany
Nails:—Cut - - - - - -	United States export to South America and our colonies
Wrought - - - - -	Belgium
Point de Paris (wire nails) - -	French and Belgian largely supersede English
Horse nails - - - - -	Beautifully made by machinery in the United States
Pumps, of various sorts - - - -	Largely exported by United States *Note.*—An American pump finding water for the Abyssinian expedition
Agricultural Implements:—Ploughs, cotton gins, cultivators, kibbling machines, corn crushers, churns, rice hullers, mowing machines, hay rakes	Many articles similar to these are exported by United States to common markets
Sewing Machines - - - - -	United States
Lamps for use with petroleum, now an article of very large consumption; lamps for the table	The United States petroleum lamps supplant the English in India and China; French even imported to England
Tin Ware:—Tinned spoons and cooks' ladles; various culinary articles of fine manufacture and finish	France
Locks:—Door locks, chest locks, draw locks, cupboard locks, in great variety	United States, France, and Germany
Door latches, in great variety - - -	United States export to Canada
Curry Combs - - - - - -	United States and France
Traps, rat, beaver, and fox - - -	United States export to Canada
Hinges, in wrought iron, for doors, grates, &c., in great variety	
Gimlets and augers (twisted) - - -	United States export to Canada and probably elsewhere

Articles or Class of Articles	Country whose Products are believed to have Replaced those of this District in whole or in part
Brass-foundry (cast), as hinges, brass hooks and castors, great variety; door buttons, sash fasteners, and a great variety of other articles	These articles, in great variety, are now extensively exported from Germany and France
Brass-foundry (stamped), as curtain pins and bands, cornices, gilt beading, and a great variety of other brass-foundry	- - ditto - - ditto
Needles:—An article of large consumption -	Mostly for Germany (Rhenish Prussia); even imported to England
Fish Hooks - - - - - - -	Believed Germany
Guns:—A great variety of sporting guns, articles of large consumption, formerly entirely from Birmingham	Now exported largely from Liege, Belgium, and St. Etienne, France
Breech-loading muskets, revolver pistols -	Some United States are exported
Watches - - - - - - - -	Switzerland and France, even imported into England
Clocks - - - - - - - -	United States and France *Note.*—Watches made in United States interchangeable by machinery
Iron - - - - - - - - - -	Belgium
Glass, for windows, an article of large consumption; spectacle and all other glasses	Belgium supplants ours in our own colonies
Table Glass - - - - - - -	Believed to be Belgium and France
Swords - - - - - - - -	Prussia and Belgium
Jewellery:—Gold and gilt; fancy steel; these in very great variety	France and Germany; these articles are even imported into England
Small Steel Trinkets, as bag and purse clasps, steel buttons and chains, key rings, and other fastenings, and many others in great variety	France and Germany; many of these even imported into England
Leather Bags, with clasps, purses, &c., courier bags, &c.	Austria, France, and Prussia. We believe about all these articles sold in England are imported
Buttons:—Mother o' pearl - - - -	Vienna, imported to England
Horn - - - - - -	France - - ditto
Porcelain (formerly Minton's, of Stoke)	France entirely superseded English, and imported to England largely
Steel buttons (formerly Bolton & Watts')	France
Florentine or lasting coat buttons -	Germany
Steel Pens: Penholders - - - - - -	France
Brass Scales and Weights - - - -	- ditto
Cutlery, in great variety; scissors; light edge tools, such as chisels, &c.	Germany
Pins for Piano Strings, and other small fittings for pianos	France
Silver Wire for binding the bass strings of pianos, &c.	- ditto
Iron Gas Tubing - - - - - - -	Germany
Elastic Belts, with metal fastenings - -	- ditto
Brass Chandeliers and gas fittings - - -	Prussia and France
Harness Buckles and furniture - - -	- - ditto
German Silver spoons, forks, &c. - - -	France, Austria, and Prussia
Locks:—Best trunk locks, best door and cabinet locks	Prussia and France

Articles or Class of Articles	Countries whose Products are believed to have Replaced those of this District in whole or in part
Umbrella furniture - - - - -	France and Prussia
Horn Combs - - - - - - -	Prussia
Pearl and Tortoiseshell articles - - -	France and Austria
Iron Wire - - - - - - - -	Prussia and Belgium
Iron and Brass hooks and eyes - - -	Prussia and France
Bronzed Articles - - - - - -	Prussia and France
Hollowares, enamelled - - - - - -	- - ditto
Optical instruments - - - - -	France, Austria, and Bavaria
Mathematical instruments - - - -	- - ditto - ditto
Japanned wares - - - - - -	Germany and France
Bits and Stirrups - - - - - -	Belgium and France
Coach Springs and axletrees - - - -	France
Electro-plated wares, customers preferring French goods	- ditto
Gas Fittings - - - - - -	United States
Weighing Machines - - - - -	- ditto
Plumbers' Brass foundry - - - -	- ditto
Table Glass ware - - - - -	- ditto
Doors, Locks - - - - - -	- ditto
Machines for domestic purposes, as sausage-machines, coffee-mills, and washing-machines	- ditto
Nuts and Bolts - - - - - -	- ditto
Penknives and Scissors - - - - -	- ditto
Stamped Brass Ware (certain kinds) - -	- ditto
American 'Notions,' as buckets, clothes-pegs, washing-machines, agricultural machines	- ditto

Answers from Chambers of Commerce to Queries of the Vice-President of the Council as to Technical Education. *B.P.P.*, 1867–8, (54).

16d 'Education and War' (1870)

THE dogs of war are again let loose, and in the two most highly civilised countries of Europe, where, a little time ago, science, education and commerce were in full sway, all the arts of peace are already neglected, and in prospect have gone back a quarter of a century. We can hardly yet realise that at the present moment railways are being torn up, lighthouses dismantled, lightships towed into harbour, and monuments of engineering skill, such as the bridge over the Rhine at Kiel, undermined, so that they may be destroyed at a moment's notice. But these, after all, are calamities of the second order; education is stopped; science schools are broken up; both professors and pupils are forsaking the laboratory and the class-room, and the whole machinery of progress has come to a standstill.

Science has little to do with politics: the function of science is to unite the whole human family, whereas the function of politics seems to be, both in the case of men and nations, to create parties and to emphasise them as much as possible, the object in each case being place for the partisans—whether that place be an income of a few thousands a year in one case or increased territory in the other.

As science advances such policies will be overridden—when science and education have taken their proper position—when the sword has given place to brain—when more of the best men of each nation take part in each nation's counsels, the dreadful thirst after blood will give way to something better; monarchs will see the folly of being surrounded merely with empty helmets, or at all events if they do not, others will; and much will have been done when the pampering of armed men shall cease.

There is one point, however, in connection with the coming war which cannot be pointed out too strongly—one duty which England owes to herself, and which, if it be well done, may make her after all a gainer from the dreadful strife. It has been already stated, and the statement is not an exaggeration, that the war will throw the countries engaged in it back a quarter of a century. Now, England at the present moment, be the cause what it may, is in many things a quarter of a century behind France and Prussia, notably in education of all kinds and especially in scientific education.

The following extract from the Report of Mr. Samuelson's Committee on Scientific Education—a report which, we believe, has not even yet been taken into consideration by our Legislature—is so much to the point that we give it here:—

> Nearly every witness speaks of the extraordinarily rapid progress of Continental nations in manufactures, and attributes that rapidity, not to the model workshops which are met with in some foreign countries, and are but an indifferent substitute for our own great factories, and for those which are rising up in every part of the Continent; but, besides other causes, to the scientific training of the proprietors and managers in France, Switzerland, Belgium and Germany, and to the elementary instruction which is universal amongst the working population of Germany and Switzerland. There can be no doubt, from the evidence of Mr. Mundella, of Professor Fleeming Jenkin, of Mr. Kitson and others, and from the numerous reports of competent observers, that the facilities for acquiring a knowledge of theoretical and applied science are incomparably greater on the Continent than in this country, and that such knowledge is based on an advanced state of secondary education.
>
> All the witnesses concur in desiring similar advantages of education for this country, and are satisfied that nothing more is required, and that nothing less will suffice, in order that we may retain the position which we now hold in the van of all industrial nations. All are of opinion that it is of incalculable importance economically that our manufacturers and managers should be thoroughly instructed in the principles of their arts.
>
> They are convinced that a knowledge of the principles of science on the part of those who occupy the higher industrial ranks, and the possession of elementary instruction by those who hold subordinate positions, would tend to promote industrial progress by stimulating improvement, preventing costly and unphilosophical attempts at impossible inventions, diminishing waste and obviating in a great measure ignorant opposition to salutary changes.
>
> Whilst all the witnesses concurred in believing that the economical necessity for general and scientific education is not yet fully realised by the country, some of them consider it essential that the Government should interfere much more actively than it has done hitherto, to promote the establishment of scientific schools and colleges in our great industrial centres.

It is impossible that we can say anything stronger than this in favour of taking the fullest advantage of the opportunity of regaining our intellectual and therefore our commercial prestige.

If England is to prepare for war, the abnormal condition, so let it be; but surely, *á fortiori* she should *prepare for peace*, the normal one, as well. This has never struck her ministers, and the reason is not far to seek.

But this is not all; the same disregard for science, arising from the ignorance of science among our rulers, has probably placed us in another position of disadvantage. While France and Prussia have been organising elaborate systems of scientific training for their armies, a recent Commission has destroyed what little chance there was of our officers being scientifically educated at all. As there is little doubt that a scientific training for the young officer means large capabilities for combination and administration when that officer comes to command, we must not be surprised if the organisation of our army, if it is to do its work with the minimum of science, will, at some future time, again break down as effectually as it did in the Crimea, or that our troops will find themselves over-matched should the time ever come when they will be matched with a foe who knows how to profit to the utmost from scientific aids.

While, therefore, the Continent is being deluged with blood, let us *prepare for peace* as well as for war; let us prepare ourselves for victories in the arts, conquests over nature; let us, by means of a greater educational effort, more science schools, a truer idea of the mode in which a nation can really progress, fit ourselves to take our place among the nations when peace returns. Surely if there be statesmen among us, such a clear line of policy will not be overlooked.

Education and Science at the present moment are England's greatest needs.

N. LOCKYER, *Education and National Progress*, London, 1906, p. 1.

16e A Victorian Retrospect (*see following page*)

'Science of the Reign', *Illustrated London News*, Jubilee edition 1897, p. 33.

Science of the Reign.

When the Queen ascended the throne a considerable capital of scientific knowledge had been collected. In her time it has grown with the pleasant and ever-increasing rapidity of a sum at compound interest. Science in England has become less insular and more and more cosmopolitan. The Victorian Era has been a time of giants in science; but that is not the peculiarity of the period. Darwins, Daltons, Faradays, Stephensons, Kelvins, and Listers there have been in every century, but never before has the world seen the growth and organisation of a vast army of scientific workers that add their quota to the general sum and disappear nameless into the unknown. They are as necessary to the giants as workers to the queen bee. Scientific literature is now so vast, and expanding so astoundingly, that it threatens, like an avalanche, to descend and bury its votaries.

The sum and substance of all the efforts of Victorian science has been to put the body and mind of man more at ease. Thanks to the Stephensons, the Brunels, and their disciples, to Faraday, Siemens, Kelvin, Rowland Hill, and their followers, space has been almost annihilated, and the ends of the earth brought together. Simpson, Lister, the Continental descendants of the great Jenner, Parkes, and the pioneer band of English sanitarians have alleviated the sting of disease, and put our bodies more at ease. More than any nationality, Englishmen have been busy to put our minds at rest by teaching us the nature of the world we live in, and making us less the creatures of chance. The seas have been charted, the depths of space fathomed, the heavens mapped, the earth and its plants and animals examined. Darwin, Lyell, Spencer, and Huxley have replaced empirical theories by a history written in rocks, bones, and living tissue.

By courtesy of the London Library

17 SOCIAL RESPONSIBILITY

17a 'Science in War'

Science in War

CHAPTER I: INTRODUCTION

THE tragic events which have led to the dangerous pass in which we now stand have run their course not so much for lack of will or action, as for lack of thought, and particularly for lack of scientific thought and foresight. Science, it cannot be urged strongly enough, is something which does not stand outside the ordinary ways of acting and thinking. It is simply the most orderly expression of those ways. There is therefore a greater need for science in the present situation than there ever has been in the quiet days of the past. It is not sufficient merely to use science to solve problems that are seen to exist. What is much more important is to use science to discover what the problems are and what is their order of importance.

This does not involve any deep or mysterious application of scientific method. The Nazis, whose mental processes are simple and easily apprehended, have shown for the last ten years that they were treating the subject of warfare in a scientific way. They openly stated their intentions and their methods in books they published, and they illustrated their military ideas in the campaigns of Spain and Poland. Merely to counter these methods, let alone discover ways of our own, requires the application of at least as much science as they put into the problem.

In the development of science and its application we usually distinguish between fundamental science and applied science. This distinction is mainly based on the time-interval that separates the acquirement of knowledge from its effects in practice. A fundamental scientific discovery may take anything from twenty to one hundred years before it influences the lives of the majority of people. A particular scientific study of some aspect of a manufacturing process may, on the other hand, be effective in production in from twenty to one hundred days. In war, and particularly in the acute phase of war in which we are now placed, it would seem, at first sight, that only short-range work matters, and many would go further, and say that scientific research in any sense can be of little use now. According to this opinion it is sufficient to follow established and traditional methods of thought and action. Actually, however, it is just in acute warfare that traditional methods break down most completely. They presuppose conditions in which materials, men, and institutions are working in a regular way. But such a pre-supposition has no possible justification in war. The usual materials are not to be found. Men, particularly skilled men, are few. The ordinary institutions of business are working in an unfamiliar war economy. Not to use science under these conditions is not just obstinate stupidity, which at worst could lead to stagnation; it is a sure way to breakdown, to confusion, and defeat. Only scientific methods can deal with the problems brought about by new conditions. And we have available to help in the solution of these problems a large force of scientists normally occupied in the long-range problems of civil life.

Even in the most acute crisis, therefore, there is an urgent need for making the full use of science. It may be impossible to develop in a short time entirely new weapons of an elaborate type, such as the tank in the last war. But scientific research based on existing fundamental knowledge can have immediate effects on current production. Every productive process is a complex of a large number of operations. One of these may limit the rate of production by its slowness, or by the scarcity of some necessary material. Scientific research concentrated at such points could help to remove bottlenecks by indicating a change in operation, and in many cases it could shorten the whole process of production by simplifying the product without detracting from its usefulness. Science is also required in the general task of surveying the whole field of needs in war, and for indicating the best directions in which effort should be expended. In a war-economy the only limiting factor, in principle, is the supply of materials and men. Theoretically the commercial and financial restrictions of peace no longer exist. Whether or not they may in fact still operate, it is one of the advantages of the scientific approach that a survey of needs and means can be carried out without regarding them as a primary consideration. Later they stand revealed as the irrational obstructions that they are.

We have to maintain the life and health of the civil and military population, to secure them and their productive machinery and agriculture from physical destruction, and to enable them to continue their work. And we must do this in spite of drastic limitations in available men and materials. At different stages of the war one or more of these aspects may appear the most urgent. At one time the chief fear may be invasion, and the first consideration purely military action; at another aerial bombardment may be the main problem; in the longer run, the urgent issue may be the possibility of starvation. But every one of these eventualities must be treated as part of a general scheme. To consume all available materials to repel one attack so that nothing is left for the next stage of the war is as sure a way to defeat as would be failure to repel the attack itself.

A survey of needs, to be of any practical use, involves measurement. We now possess in newly developed statistical methods means that are far more accurate and rapid than mere commonsense for sizing-up any situation, however complicated, and for measuring the actual effect of any steps taken to deal with it. So far these methods have by no means been fully used. We do not yet possess a statistical bureau where information is collected on every war problem on which statistics exist, and where new inquiries can be set on foot as and when needed. As a result, once available information is required for official purposes outside its own particular department, and sometimes even inside it, elaborate research has often to be undertaken, and research which is often fruitless, as there is no mechanism for making available later the relevant information, even though it may be securely filed in some safe place.

It is only on a basis of quantitative knowledge that problems can be formulated and the need for new research properly determined. In practice major problems of war seem to arise only after some disaster has warned people that things are amiss, or when the enemy discloses some new weapon. If the problem is definite it can be tackled, and often tackled with success, as notably in the case of the magnetic mine. An acute shortage of a particular material will also prompt research for substitutes, and many of these researches have been fruitful in the past. What is lacking, however, is not the ability to deal with these *ad hoc* problems as they arise, but the ability to foresee them and to have solutions ready, so as to obviate as far as possible the inevitable delays of research and development.

At least as important is the necessity for recognizing less striking problems of a general nature, to analyse their causes, and to provide the remedies for intangible but vital failings, such, for example, as disorganizing delays in communications, or a lowering of a nutritional standard. These problems require far more organized vigilance than they enjoy at present. It is not sufficient to depend on the various councils and advisory committees which deal with separate problems. In relatively few cases are these committees doing more than examining specific issues put up to them by Government departments. The task of general scientific and technical survey and prevision seems to be nobody's business. Even when a problem is seen, it is often difficult to get it effectively dealt with, although there is in the country a great wealth of expert knowledge outside existing service departments. The knowledge is largely unused and almost wholly unorganized. What has been done in many fields is the result of voluntary initiative of teams of research workers themselves. But all such work suffers from the disadvantages that those concerned have difficulties in obtaining necessary information as to the actual state of affairs, and even greater difficulties in seeing that the results of their researches are put into practice on the scale they merit. How far, for example, the knowledge won during the past twenty years through research on nutrition has failed to get across in a situation which must demand its application is shown in a later chapter.

The needs of the situation and the capacities for science to deal with them seem so obvious that it is difficult for people to believe that what is being done is far less than what could be done. One might think that a mere statement of the value of the *approach* we have outlined would secure its adoption; but it is not as simple as that. It is not merely the novelty of science and the unfamiliarity of the scientific approach that prevent its adoption. Very real external and internal obstacles stand in its way, and it would be mere wishful thinking to pretend that these obstacles can easily be removed.

The essential difficulty is that a general and organized scientific approach to the problems of national existence is foreign to the tradition of both the economic and political management of this country. Furthermore, a great number of scientists themselves, brought up in this tradition, have neither the desire nor the capacity to see that effective use is being made of their science. These obstacles lie very deep in our social system. In commerce special knowledge is something from which profit may be derived. The idea of a general survey of the needs of the population and of the planning of methods of satisfying them, irrespective of particular interests, meets opposition at every point. In the first place, commercial secrecy still prevents us from knowing what industrial processes, and even what scientific knowledge, there are available in the country. And although the powers of the Government are in theory complete, in practice every firm has divided loyalty; part national, and the rest the firm's interests. There is an enormous power of scientific research locked up in industrial laboratories and largely wasted through lack of common knowledge and coordinated action.

As the war develops, more and more control passes out of the hands of private industry into that of Government departments. There another obstacle (which is considered again in a later section) becomes manifest. Government departments are still, to a very large extent, run by civil servants and by executives of large industry who have acquired many of the permanent civil servant's characteristics. The tradition of civil servants belongs to the age of Victorian Liberalism and is one of *laissez faire* and of Government non-interference. The whole tradition tends to prevent things being done. An even more serious objection is that high administrative officers have a classical training and are almost completely ignorant of technical matters. Having little conception of what scientific research is, they fail to see how it can be practically used. In technical matters they rely on their permanent technical advisers who, for various reasons, are not infrequently cut off from the more active streams of scientific thought. There is also usually a complete lack of initiative within the civil service itself, due mainly to the cautious desire not to make mistakes. And over the whole system lies, first, the blight of treasury control—the fear of allowing expenditure that may be difficult to justify—and, second, the knowledge that failure to act rapidly and decisively is not the responsibility of the civil servant, but of his minister.

In the present period of transition from private to Government enterprise these difficulties are accentuated because of the long-standing mutual suspicions of civil servants and business men—civil servants fearing that the Treasury will be defrauded, and therefore checking everything by an elaborate system of specification and inspection—and the business men fearing ever-increasing interference with the conduct of their affairs. In such an atmosphere it is extremely difficult to build an effective and operative scientific service. At best the attitude of both business men and civil servants towards scientists is still that they must be treated as consultants. They can only be asked particular questions, and their answers must be collated with each other and judged in the light of established custom before they can be used as a basis for the decisions taken by administrators.

On the whole scientists are excluded from discussions on policy, and even more so from decisions. As a result, even the most authoritative scientific committees often flounder in the dark, and their work, however excellent and to the point, may lie unused for vital months. When it is used it is not infrequently distorted by the technical ignorance of those who put their recommendations into practice. Part of the blame for all this rests with the scientists themselves. Many scientists have grown to accept the position, and find it difficult to conceive of any other. They are perfectly willing to answer questions if asked, and feel their responsibility begins and ends there. They are not called upon to survey the situation, to make decisions, or to see that they are carried out. The position of scientists in regard to the war is, indeed, a very complex and fluid one. At the present moment scientists fall into five groups: (i) the scientists who were employed before the war in permanent Government service—for example, in the Department of Scientific and Industrial Research; (ii) those who have joined those departments from the outside since the war started; (iii) those who sit on Government Advisory Bodies and Committees, or conduct extra-mural war work in their laboratories; (iv) those who in industrial laboratories are responsible for the production of war materials; and finally (v) those scientists who still stand outside all war work. Broadly it can be said that the first group is in a position to know most about the situation, but, from its position and tradition, is also less able to affect it; whereas the last group, which contains some of the most powerful minds in science, still has no knowledge, except in the most general terms, of the problems by which the country is faced.

The scientific societies, which might have integrated the scientific effort of the country, have almost gone into hibernation, and preserve a precarious existence carrying on their pre-war functions and assisting, somewhat ineffectively, in the operation of the National Register. It might be thought that the central position occupied by the advisory scientists would overcome these difficulties. Unfortunately, this is not so. They seem to have been much more concerned with agreeing with Government decisions than with challenging them in the name of science. This attitude can be summed up in the naïve Press report of the formation of one such committee: 'the Ministry of —— has appointed a committee of eminent authorities to prove scientifically the validity of the policy that it has adopted'. In the long run the 'yes men' of science are likely to be at least as dangerous as the 'yes men' of politics.

These difficulties, though serious, are not insuperable, and as conditions worsen they are more and more appreciated by people inside and outside the scientific world. This book is itself an attempt to focus attention on some deficiencies of scientific action, in the hope that they will be rapidly remedied. There is great power of individual and co-operative activity in the scientific world, and if it is helped by the pressure of intelligent public opinion, we may obtain, perhaps even in time, the effective utilization of science to save the situation. But the need is very urgent indeed. There is a shortage in applied intelligence in this country, perhaps greater than the shortage of anything else. This shortage will not be met until we make use of the large quantity of available, but at present unemployed, scientific intelligence we have at the moment.

ANON., *Science in War*, London, 1940, p. 7.

17b 'Bernalism'

COUNTERBLAST TO BERNALISM

[*In this article Dr. John Baker, the eminent Oxford biologist, attacks the doctrine expounded in Professor Bernal's 'Social Function of Science.' Next week we shall publish a reply by Professor Bernal himself.*]

'. . . *a higher degree of Reputation is due to* Discoverers, *than to the* Teachers *of* Speculative Doctrines . . .' (Sprat, 1667.)

BERNALISM is the doctrine of those who profess that the only proper objects of scientific research are to feed people and protect them from the elements, that research workers should be organised in gangs and told what to discover, and that the pursuit of knowledge for its own sake has the same value as the solution of crossword puzzles. Professor Bernal will no doubt permit the immortalisation of his name by the introduction of this new word into the English language.

Opponents of Bernalism, well aware of the way in which great discoveries have been made in the past, are appalled at the thought that if once Bernalists get power the fount of discovery will be dried up. If science is left free, incalculable benefits for human welfare, material and mental, will accrue as they have accrued in the past; but it would be as undesirable to try to force organisation on scientists as on composers, artists or choreographers. It is true that music can be composed under duress. One knows that Mozart, whose antipathy to the harp was only equalled by his positive dislike for the flute, was persuaded to write a very tolerable concerto for these two instruments. Yet even de Guisnes would not have been able to get him to submit to the organisation which Bernalists seek to impose, far less to confine himself to the composition of music of the greatest economic value (*i.e.*, presumably, that which would give employment to the greatest number of musicians and musical instrument makers).

Let the reader try to imagine Charles Darwin told off by the Bernalist-in-charge to organise a gang of Ph.D. students. The mind reels before the thought. Among other famous workers in the biological sciences who showed by their lives a particularly obvious antipathy to the principles of Bernalism one may mention T. H. Huxley, Russel Wallace (co-founder with Darwin of the theory of evolution by natural selection), Mendel (founder of the Science of Heredity), Fabre (student of insect behaviour), Jenner (student of the cuckoo and discoverer of vaccination), Schwann (founder of the cell-theory), Linnaeus (inventor of the scientific method of naming plants and animals), van Leeuwenhoek (discoverer of blood-capillaries, spermatozoa, etc.), and Malpighi (discoverer of the air-cells in the lungs, etc., etc.). These are men whose names will be revered when even the derivation of the word Bernalism has become obscure.

Many famous workers in other sciences besides biology could never have been forced into the Bernalist rut. Supreme as an example of these was Cavendish, discoverer of the fact that water is formed by the combination of the two gases which were later named hydrogen and oxygen. Could Cavendish have worked in a gang under orders and confined himself to economic research? So far was he from being capable of collaboration, that he could not bear the proximity of others and had his library built four miles from his house, so that those whom he permitted to borrow books should not come near him when they did so. That he worked mainly for his own enlightenment is shown by his never bothering to publish his monumental researches in electricity, which were only brought to light after his death. Newton, it will be recollected, was almost equally unconcerned as to whether others got to know of certain of his most important discoveries. This was indeed regrettable, but it shows that geniuses have pursued knowledge for its own sake; and in doing so they have thrown off much that has been of tremendous significance for the practical affairs of mankind.

Science has two functions—to serve human welfare in material ways, and to increase man's capacity to comprehend the universe. To many who are contemplative there is nothing more worth-while in life than the increase of knowledge for its own sake. To pretend that this is 'escapism' and comparable to interest in crossword puzzles is nonsense. No amount of solving puzzles increases anything worth increasing; but every discovery in science separates falsehood from truth and makes an accretion to that vast body of demonstrable knowledge whose possession is the most valid criterion of distinction between cultured and savage communities. There are those who are unrewarded by the contemplation of nature just as there are those who find nothing in music or art. Nevertheless, knowledge and music and art are among the ultimate things in life for many people who regard just keeping alive and healthy as merely means to an end.

One of the keenest and most satisfying pleasures to many people is the possession of some understanding of the landscape which lies before them when they travel, whether in this country or abroad. This may be taken as an example of how science can serve as an end in itself. To walk in an unfamiliar part of England with Trueman's 'Scenery of England and Wales' and Tansley's 'Types of British Vegetation' in one's haversack is to experience something which is denied to those whose appreciation of landscape lacks understanding and is entirely æsthetic. When the walker opens his haversack to eat, he has the Bernalists' approval, because eating is useful. When he opens it to fill his mind and not his stomach, does he necessarily sink to the level of the solver of the crossword puzzle? Is he really to fall down and worship the grocer who sold him his lunch, and pretend to ignore his debt to Professors Trueman and Tansley? Whether subsequently he does or does not make 'use' of the added knowledge which his study of scenery has given him, the study is worth while as an end in itself. In the American Ambassador's memorable words, spoken at Manchester a few days ago, 'that desire to find out, which is unencumbered by any desire for a practical cash return, has given us many of civilisation's great blessings.'

Perversely enough, the keenest Bernalists are often those whose own research has been the most meagre in supplying the material wants of man. Let them devote themselves for a few years to the scientific investigation of some really practical matter, such as making two blades of grass grow where one grew before, or one baby grow where two grew before. They might then be more useful members of society and waste less of their time in yapping at those who are of a more contemplative nature than themselves. What a scientist ought to do is an ethical concern for the judgment of his own conscience. Those to whom one listens with respect when they speak of verifiable matters (*e.g.*, in crystallography) compel attention much less inevitably when they try to lay down the law on moral issues.

There is room for all in science. Let the gangsters work always in gangs and order one another about and improve whatever it may be that they tell one another to improve. Everyone will be thankful for their services. Let there be freedom, nevertheless, for those who lack the gang instinct but possess that insatiable curiosity, sneered at by Bernalists as puerile, which is the source of all real advance in science, whether pure or applied. If these can be left free and unhampered, science and its applications will flourish. Let Bernalists remember the words with which Faraday answered one of their number who inquired the use of his experiments: 'Madam, will you tell me the use of a new-born child?' Let them ponder also the magnificent lines which Prebendary Sprat wrote of their forerunners in 1667: '. . . they [the Bernalists] are to know, that in so large, and so various an *Art* as this of *Experiments*, there are many degrees of usefulness: some may serve for real, and plain *benefit*, without much *delight*: some for *teaching* without apparent *profit*: some for *light* now, and for *use* hereafter; some only for *ornament*, and *curiosity*. If they will persist in contemning all *Experiments*, except those which bring them immediate *gain*, and a present *harvest*: they may as well cavil at the Providence of God, that he has not made all the seasons of the year to be times of *mowing*, *reaping*, and *vintage*.'

JOHN R. BAKER

PROFESSOR BERNAL REPLIES

[*Professor Baker's 'Counterblast to Bernalism' appeared in our issue of July 29th.*]

According to Dr. Baker, 'Bernalism is the doctrine of those who profess that the only proper object of scientific research is to feed people and protect them from the elements, that research workers should be organised in gangs and told what to discover, and that the pursuit of knowledge for its own sake has the same value as the solution of crossword puzzles.' Now I am not concerned in defending this doctrine: it is not mine and never has been. Moreover, I protest against the addition of this barbarous perversion of the language. I do not know where Dr. Baker has met the Bernalists; I have never seen one myself and I don't believe they exist.

It would be sheer waste of time and ink to attempt to continue a controversy on Dr. Baker's terms. But the issue itself is real enough, though it cannot be seriously discussed till the caricature of it is cleared away. What science is for and how it can best be carried on, are vital questions to discuss and to act on. The first question has really two distinct parts, according to whether we consider science as a social activity, part of the general complex of such activities, or as the personal interests of an individual. The failure to separate these two aspects is the ground for most of Dr. Baker's animus against the monster of Bernalism. This type of confusion is not confined to science. There are many who are just as unclear in their minds as to whether agriculture is a technique for raising animal and plant products or a means of providing a simple and ennobling life close to the soil. But though the questions are closely related, failure to separate them prevents any clear understanding of either. There is, in effect, no contradiction between considering the practical use of science to a community as a justification for encouraging it and considering that the scientist carries on his work for what he thinks is a sheer love of truth and beauty. The one is a question of social economics, the other of psychology. That science has a social function is no new doctrine but palpable and admitted fact. That that function is largely economic under present conditions and is likely to become even more so, is clear to any student of the history of science. It is written plainly in the present organisation of science, with its governmental and industrial laboratories, and in the relative importance, even in academic science, of those branches such as chemistry which have the greatest and most immediate economic importance. What is new is a growing conscious recognition of the nature of that function, brought about by the rapid development of science and by the glaring contradiction between the actual and potential effects of scientific activity on our civilisation. No question arises as to whether or not science should be used for social ends. It is being so used, but badly and largely for bad ends. We have not to decide whether science is or is not to be a directed activity, but only what its direction should be.

Now what ultimate ends do we set for science? They are the old indissoluble human trinity, goodness, truth and beauty; or commoditie, firmeness and delighte. In stressing the utility of science, we imply the accuracy of its truths and the enjoyment of its discoveries as certainly as if they were taken as the prime considerations. Nor can a man fully appreciate the truths of science unless he can relate them to their fruit in the satisfaction of human needs. But this is far from demanding, as Dr. Baker seems to think, that all scientific work should deal with immediate research for practical ends. On the contrary, those of my way of thinking have always been consistent in our claims for the full co-ordination of theory with practice and for the greatest possible development of fundamental and apparently abstract research.

Scientific research cannot, of its very nature, be restricted to any set of problems, whether they be practical or academic, without ultimate futility and pedantry. Knowledge increases and discoveries are made from an interplay between intrinsic development of disciplines, such as mathematics or genetics, and the problems of practical life, such as the production of power or the cure of disease. Without the former, discoveries cannot be made; without the latter, they will not be. If the quarrel about ends is an illusory one, that about means seems real enough. Should science advance by the casual co-ordination of the work of gifted individuals, each following his inner light, or by groups or gangs of workers mutually assisting each other and integrating their work according to some preconceived though flexible plan? Now, if Dr. Baker could descend from the ideal to the actual, he would see that this is also an illusory distinction. For modern science is already in a state of transition from one mode of working to the other. It is simply a question of how, under given conditions of knowledge and social organisation, to advance our understanding and ability in the most effective way. Up to the present century, where the workers were few and the means of work simple, the individual method was clearly the best in many fields. The authorities Dr. Baker quotes bear ample witness to that. Even then, however, there were advantages to be reaped from working in schools and in organised laboratories. Pythagoras, Aristotle, Tycho Brahe, Galileo, Boerhaave, Linnaeus, Liebig, and Pasteur are names enough to show that the monks of science were as significant as its hermits. But none of this is relevant to-day. People are working together and they have got to work together in most fields of science if they are to get results. Techniques have become too multiplex and costly for any other means.

The problem is no longer to decide whether science should be organised or not, but how to organise it, so as to avoid confusion and obscurity without damaging the freedom of thought and activity of the individual worker, on which ultimately everything depends. Dr. Baker thinks this is undesirable and impossible. I think it must and can be done. For a scientist of to-day to confine himself to purely individual work and not to attempt to integrate his work with that of his fellows is deliberately to restrict his capacity for discovery and to treat the overcoming of unnecessary difficulty as something to be pursued for its own sake. Such occupation may justly be compared to crossword-puzzle solving or rock-climbing—good and enjoyable diversion, but not life work. Of course, there will always be a small number of highly original or cross-grained minds that cannot fit in with others, and because of the possibility that their work may be of the utmost importance, a place must be kept for them. But this cannot be the way for the great majority. Team work will tend increasingly to be the mode of scientific research. This does not, however, mean the forceful disciplining of gangs by gang leaders. Scientists will work together and are working together, because for each of them this means the greatest scope for original discovery. The keys to the successful organisation of science are freedom and democracy. Nowadays freedom for individual research means more than the absence of prohibition or even of compulsion; it implies the provision of adequate material and help. Thus the limitation of subsidies and organisation for science is probably the most effective way to stifle discovery, for discovery and fruitful application can no longer be counted on to come of themselves; they need intelligent provision. The consciousness of a common aim binds people together far better than any autocratic ordering. But the present organisation of science is not free and democratic, it is still in form authoritarian and hierarchical, and any new organisation must base itself far more on voluntary association and on the executive power of councils of all grades of workers.

To work for such organisation as part of the great struggle of our time for order and liberty is not to forget the more individual and subjective values of science, but accepting them, to break down the barriers that make those values an exclusive and sterile possession of a few, so that they form part of a far wider and more conscious civilisation.

J. D. BERNAL

J. BAKER, 'Counterblast to Bernalism', *New Statesman*, 29 July 1939; J. D. BERNAL, 'Professor Bernal Replies', *New Statesman*, 5 August 1939.

17c Letter from Einstein to Roosevelt on the Building of an Atomic Bomb, 1939

August 2nd, 1939

F. D. Roosevelt
President of the United States
White House
Washington, D.C.

Sir:

Some recent work by E. Fermi and L. Szilard, which has been communicated to me in manuscript, leads me to expect that the element uranium may be turned into a new and important source of energy in the immediate future. Certain aspects of the situation which has arisen seem to call for watchfulness and, if necessary, quick action on the part of the Administration. I believe therefore that it is my duty to bring to your attention the following facts and recommendations:

In the course of the last four months it has been made probable—through the work of Joliot in France as well as Fermi and Szilard in America—that it may become possible to set up a nuclear chain reaction in a large mass of uranium, by which vast amounts of power and large quantities of new radium-like elements would be generated. Now it appears almost certain that this could be achieved in the immediate future.

This new phenomenon would also lead to the construction of bombs, and it is conceivable—though much less certain—that extremely powerful bombs of a new type may thus be constructed. A single bomb of this type, carried by boat or exploded in a port, might very well destroy the whole port together with some of the surrounding territory. However, such bombs might very well prove to be too heavy for transportation by air.

The United States has only very poor ores of uranium in moderate quantities. There is some good ore in Canada and the former Czechoslovakia, while the most important source of uranium is Belgian Congo.

In view of this situation you may think it desirable to have some permanent contact maintained between the Administration and the group of physicists working on chain reactions in America. One possible way of achieving this might be for you to entrust with this task a person who has your confidence and who could perhaps serve in an inofficial [*sic*] capacity. His task might comprise the following:

(a) to approach Government Departments, keep them informed of the further development, and put forward recommendations for Government action, giving particular attention to the problem of securing a supply of uranium ore for the United States;

(b) to speed up the experimental work, which is at present being carried on within the limits of the budgets of University laboratories, by providing funds if such funds be required, through his contacts with private persons who are willing to make contributions for this cause, and perhaps also by obtaining the co-operation of industrial laboratories which have the necessary equipment.

I understand that Germany has actually stopped the sale of uranium from the Czechoslovakian mines which she has taken over. That she should have taken such early action might perhaps be understood on the ground that the son of the German Under-Secretary of State, von Weizsäcker, is attached to the Kaiser Wilhelm-Institut in Berlin where some of the American work on uranium is now being repeated.

Yours very truly,
A. EINSTEIN

A. EINSTEIN, *Tongues of Conscience*, 1939, p. 72.

17d Extract from a Framework for Government Research and Development

The Organisation and Management of Government R. & D. By Lord Rothschild, *Head of the Central Policy Review Staff*

CHAPTER 1: INTRODUCTION

1. One might expect that a review of Government research and development (R. & D.) would be greatly concerned with answers to such questions as:

(*a*) Are we doing too much, too little, or about the right amount of R. & D.?
(*b*) Is the balance between pure and applied research about right?
(*c*) Should we do more intramural and less extramural R. & D., or *vice versa*?
(*d*) What R. & D. are we doing which should not be done?
(*e*) What R. & D. should we do which is not being done?
(*f*) Is there adequate machinery, at the centre, critically to evaluate the overall R. & D. scene?

2. These questions, and possible answers to them, are not systematically discussed in this report because to try and do so would not help to provide a framework in which the efficiency of Government R. & D. can be maximised. The questions either relate to an out-of-date concept of R. & D. management, or they are unanswerable. The resolution of the dilemma, which is more apparent than real, is to ensure that the organisation and management of R. & D. is logical, flexible, humane and decentralised, the prerequisites of an efficient system. Furthermore, that each person in the system has clearly defined responsibilities. This report is, therefore, concerned with the establishment of the principles on which such an efficient system must rest and the consequences of accepting and implementing them.

3. Some comments may, however, be made about the six questions enumerated above. (*a*) is discussed later; (*b*) is not a meaningful question because there cannot be a correct 'balance' between the resources devoted to pure and applied research. The amount of pure research done at Universities and elsewhere depends on the talent and money available; but one can make certain assertions about the proportion of pure research permissible or necessary within an applied research programme. Questions (*c*), (*d*) and (*e*) are automatically answered when a satisfactory organisation for the management and prosecution of R. & D. exists and it is one of the purposes of this report to recommend one. Question (*f*) is difficult to answer, because it is doubtful whether any central body can or should try critically to evaluate the Government's R. & D. as a whole. Nevertheless, some focal point, i.e. the Chief Scientific Adviser, is necessary at the centre, not only to satisfy the general desire for such focal points, but also to co-ordinate trans-departmental and trans-national subjects involving R. & D. The phrase 'subjects involving R. & D.' is deliberate and intended to emphasise that applied R. & D., with which this report is mainly concerned, is not an activity in its own right; it is part of a wider process, for reasons that are discussed at length in the following pages.

TOO MUCH R. & D., TOO LITTLE, OR JUST ABOUT RIGHT?

4. In spite of many books and papers, some of a mathematical nature, no-one in industry or Government, has found a method of determining the 'right' amount of R. & D. to do. In fact, most research directors and R. & D. administrators admit that decisions on this matter are an art rather than a science. Nevertheless, the person who has to pay for the end-product of applied R. & D. is, for obvious reasons, in the best position to decide how much he can afford. Depending on his qualifications, he may need to take the advice of scientists, mathematicians or engineers skilled in the subject under consideration.

5. Some people believe that eminent scientists, mathematicians and engineers should have an opportunity to express an overall view on the nation's R. & D. to those ultimately in charge. But such an overall view, whether greatly desired or even ordered, is of questionable value. Some comfort—or discomfort—may, for example, be gained from comparisons with other countries, see Table 1.

The countries in this Table almost certainly define R. & D. in different ways; they have widely different populations; and they are at different stages of industrial development. The reader may, therefore, feel that the figures should be divided by some factor or factors to take account of these differences; but the question is, by what? If this exercise is not undertaken, if only because it is unclear how to do it, the reader may derive some satisfaction from the United Kingdom being 'about in the middle'. Why, on the other hand, does France spend more than us? Should this be a matter for satisfaction or discomfort? Leaving aside this difficult or, perhaps, unanswerable question, one must remember that the R. & D. effort of a country will have no effect on that country's welfare for at least seven years after the date of the effort in question; so that no inferences can be made from the figures in Table 1 until, at the earliest, 1977. Even then the inferences will be of questionable value because it is not, for example, known whether an increase in G.N.P. is in part caused by an increase in R. & D. expenditure, or whether the latter is a consequence of increased prosperity. Writers sometimes seem confused about this point.

TABLE 1: PUBLIC EXPENDITURE ON RESEARCH AND DEVELOPMENT IN VARIOUS COUNTRIES (1970 figures except where stated)

	Total (£m.)	Civil (£m.)
United States (1969)	6,652	3,393
Germany	725	600
France	735	510
United Kingdom	581	353
Japan (1969)	351	344
Italy	191	185
Netherlands	128	122
Belgium	52	50

LORD ROTHSCHILD, *Cmnd.* 4814 *H.M.S.O. The Rothschild Report*, p. 1.

17e Statement of Aims of the British Society for Social Responsibility in Science

Aims of the Society

THIS organisation believes that the development of science is not predetermined but should depend upon the social choices of the community and the individual choices of the scientist. In furtherance of this belief the organisation has the following aims:

(a) To stimulate amongst scientists an awareness of the social significance of science and of their corresponding social responsibilities both individually and collectively.

(b) To draw the attention of all to the political, social and economic pressures affecting the development of science.

(c) To draw public attention to the implications and consequences of scientific development and thus to create an informed public which can exercise choice in these matters.

(d) To seek international exchange on these matters with similar groups in other countries.

Activities of the Society

(a) Identify those developments in science and technology which are likely to have major effects on human life and its environment.

(b) Carry out research into the nature of these effects and make its conclusions known to the community of scientists, to the public and to government.

(c) Urge changes in public policy relating to science and technology when the actual or potential consequences of such policy appear undesirable.
(d) Similarly urge the reform of science education at all levels so as to encourage an understanding of the social role of science and of the responsibilities of scientists.
(e) Encourage the formation of local groups to relate the aims of the society to specific local issues.
(f) Circulate a regular news bulletin.

Leaflet issued by the British Society for Social Responsibility in Science, 1969.